A. Liemant · K. Matthes · A. Wakolbinger

Equilibrium Distributions of Branching Processes

Mathematical Research · Mathematische Forschung

Wissenschaftliche Beiträge
herausgegeben von der
Akademie der Wissenschaften der DDR
Karl-Weierstraß-Institut für Mathematik

Band 42

Equilibrium Distributions of Branching Processes

by

A. Liemant
K. Matthes
A. Wakolbinger

Equilibrium Distributions of Branching Processes

by

Alfred Liemant
Klaus Matthes
Anton Wakolbinger

Akademie-Verlag Berlin 1988

Autoren:
Dr. sc. Alfred Liemant
Prof. Dr. rer.nat.habil. Klaus Matthes
Karl-Weierstraß-Institut für Mathematik
der Akademie der Wissenschaften der DDR, Berlin

Doz. Dr. Anton Wakolbinger
Institut für Mathematik der
Johannes-Kepler-Universität, Linz

Die Titel dieser Schriftenreihe werden vom Originalmanuskript der Autoren reproduziert

ISBN 3-05-500453-1
ISSN 0138-3019

Erschienen im Akademie-Verlag Berlin, DDR-1086 Berlin, Leipziger Str.3-4

Lizenznummer: 202 · 100/412/88
Printed in the German Democratic Republic
Gesamtherstellung: VEB Kongreß- und Werbedruck, 9273 Oberlungwitz
Lektor: Dr. Reinhard Höppner
LSV 1075
Bestellnummer: 763 827 1 (2182/42)
03400

Preface

The present monograph deals mainly with one aspect of stochastic branching models, namely their equilibrium distributions. Any model of this kind has the trivial equilibrium distribution, which is concentrated on the "void population". Unless, however, the branching model corresponds to an "independent substochastic translation" or may be reduced in a certain sense to this case (cf. section 2.9.), *infinite* populations occur with positive probability in any non-trivial equilibrium situation.The results presented here are thus in the intersection of two theories: stochastic infinite particle systems and branching models.

From the point of view of stochastic infinite particle systems our model is elementary, because the independence hypotheses, which are characteristic of branching models, are rather restrictive. These restrictive hypotheses, however, admit a number of conclusions, which, at least partially, can serve as a "case study" for more complicated stochastic evolutions.

On the other hand, from the viewpoint of classical branching models, the introduction of infinite particle systems leads to significant complications. Nevertheless the subject treated here should be regarded as a natural one in the theory of branching models.

The theory of random point fields (and in particular that of infinitely divisible distributions of random point fields) is an important *tool* in the present investigations.

The whole monograph is restricted to models in discrete time, in order to avoid questions connected with the existence and the description of stochastic branching dynamics in continuous time (cf. Ikeda, Nagasawa, Watanabe (1968,1969), Nagasawa (1977), Asmussen, Hering (1983)).

Also, we do not touch upon the theory of measure valued branching processes. Parts of that theory are parallel to our setting (cf. Hermann (1981)), to which it is also connected through asymptotic considerations.

From the historic viewpoint, the present investigations have two sources. The first is the theory of spatially homogeneous branching models, which has been presented in detail in Matthes, Kerstan, Mecke (1978) (quoted in the sequel by [MKM]). The second is the general theory of spatially inhomogeneous substochastic translations, especially the papers of Kerstan et al. (starting with Kerstan, Debes (1969)), Shiga and Takahashi (1974) and Liggett (1978). The spatially homogeneous branching models with different "types" of individuals (cf. eg. Prehn, Röder (1977)) can be considered as an intermediate stage between the spatially homogeneous case and the general case. The present text contains no specific chapter on the spatially homogeneous theory. This is for the sake of conciseness and also because a number of important questions in the spatially homogeneous case still remain open. This case has yet to be studied from the point of view of the advanced general theory. A first impression of how the general theory developped here affects the spatially homogeneous case is provided by Fleischmann, Hermann, Matthes (1982).

The results presented in this text considerably improve upon the hitherto existing publications on the subject. As a systematic presentation, it clarifies the theory, eliminates superfluous assumptions of past treatments, and creates some useful concepts such as that of a "family equilibrium distribution." Terminology and notation are modified in order to facilitate reading. Examples play an important role in the whole presentation, some of which are unbroken threads running throughout the text. In the first chapter basic facts from the theory of random point fields are reviewed, for the most part without proof, but with reference to [MKM] and to Kallenberg (1983) (cited by [K]).

References and attributions of results are collected in an appendix "Comments and References", unless these results are "already standard" or are due to some of the present authors (whose relevant publications, however, are included in the list of references).

We cordially thank Miss Irene Steininger for her patient and diligent work in preparing the camera-ready manuscript, and R. Siegmund-Schultze, whose contributions were essential for the development of the present theory.

Berlin and Linz, March 1987

The authors

Contents

1. AN INTRODUCTION TO RANDOM POINT FIELDS AND BRANCHING DYNAMICS

Let us consider a system of particles or rather a population of individuals which are indistinguishable up to their location, and assume that the population is subject to a stochastic branching dynamics in discrete time in the following way: each individual in the population produces, independently of the other individuals, a random offspring, whose distribution depends on the individual's location, and the population in the subsequent generation is the superposition of all these offsprings. This monograph deals with the time evolution of such systems, in particular investigating the structure of equilibrium distributions and establishing convergence-to-equilibrium results for certain initial distributions.

As a "phase space" (i.e. the space which the particles or individuals live in) we admit an arbitrary complete, separable metric space $[A, \rho_A]$. A population of individuals in a phase space $[A, \rho_A]$ might be described by the set S of the individuals' locations. Thus, as a mathematical model of populations all those subsets S of A could be considered which obey the finiteness condition $\#(S \cap X) < +\infty$ for all bounded subsets X of A. Such a modelling, however, does not allow to describe populations with more than one individual at a site. Even if such multiplicities do not play much role in the modelling of "real" populations, they do occur in certain mathematical idealizations, e. g. in stochastic branching models with countable phase space. It is therefore appropriate to associate, with an individual at the site $a \in A$, not the set $\{a\}$ but the corresponding Dirac measure δ_a. The whole population is then described by the sum of those δ_a's, i.e. by a measure Φ on the σ-algebra $\mathcal{A}$ of Borel subsets of the phase space. The multiplicity of individuals at some site $a \in A$ is now given by $\Phi(\{a\})$; this is how often the summand δ_a occurs in Φ.

Populations are thus modelled by measures Φ which take values in $\mathbf{Z}_+$ on bounded Borel subsets of the phase space; such measures are called counting measures. The superposition of populations is now described no more by the union of sets, but by the addition of the corresponding counting measures. The set of counting measures will be equipped with a natural topological and measurable structure.

The metric ρ_A comes into the above concepts only via the topology which it generates and the ρ_A-boundedness of subsets of A. If two complete metrics coincide in these respects, then they are equivalent for the whole theory.

According to the above statements, random counting measures - also called random point fields - may be understood as special random measures. Many concepts and results from the theory of random point fields may be extended in a natural way to random measures. This is more than a formal generalization: it turns out - also in the present monograph - that the study of random counting measures leads to random measures in its own right, e. g. via the Cox transformation.

A stochastic branching dynamics is constructed from a so called clustering field $(\kappa_{(a)})_{a\in A}$, where $\kappa_{(a)}$ is the distribution of the random offspring of one individual at site $a\in A$. The basic assumption for a stochastic branching dynamics is that the offsprings of the single individuals in the system are generated in a stochastically independent way according to the given clustering field. This independence assumption is the principal reason why favorite candidates for equilibrium distributions of stochastic branching dynamics are the so called infinitely divisible distributions. These are probability distributions of such random populations which may be represented as an independent superposition of an arbitrary number of identically distributed random populations. Via the canonical representation of infinitely divisible distributions, also infinite measures on the space of counting measures come into the theory.

Chapter 1 is an introduction into those parts of the theory of random point fields which are relevant for our purposes. Concerning the proofs, we usually refer to the comprehensive monographs [MKM] or [K]. There, one finds also detailed historical remarks.

In the subsequent chapters, basic concepts and standard results will be used sometimes without mentioning this introduction explicitely. It is also the purpose of the introduction to explain certain procedures in an exemplaric way, in particular the definition of new phase spaces (e. g. $[\mathbb{A}, \rho_{\mathbb{A}}]$ in 1.6., $[A^\wedge, \rho_{A^\wedge}]$ in 1.7.) in order to draw new consequences from already known facts.

The reader's attention should be drawn in particular to the presentation of Kallenberg's "backward technique" in 1.9., where special emphasis is put on the intuitive content of the formulae.

1.1. Random Measures

Let $[A, \rho_A]$ be a complete separable metric space, called **phase space**, $\mathcal{A}$ the corresponding σ-field of all Borel subsets of A, and $\mathcal{B}$ the ring of all bounded sets X in $\mathcal{A}$. Further, let N denote the set of all measures ν on $\mathcal{A}$ taking only finite values on $\mathcal{B}$. Any $\nu\in N$ may be viewed as a "mass allocation" on A. In the sequel, the set N will be furnished with order, topological and measurable structure.

With the natural semi-ordering $\leq$ of measures, N forms a conditionally complete distributive lattice with zero measure o on $\mathcal{A}$ as smallest element. The greatest lower bound resp. least upper bound of $\{\sigma, \rho\}$ in $[N, \leq]$ is denoted by $\sigma\wedge\rho$ resp. $\sigma\vee\rho$. Any decreasing sequence (ν_n) of measures in N has a greatest lower bound ν. In this case there holds $\nu_n(X) \downarrow \nu(X)$, $X\in\mathcal{B}$, and we write $\nu_n \downarrow \nu$. The least upper bound of any increasing sequence (γ_n) exists in $[N, \leq]$ iff $\sup_{n=1,2..} \gamma_n(X) < +\infty$, $X\in\mathcal{B}$. In this case there holds $\gamma_n(X) \uparrow \gamma(X)$, $X\in\mathcal{A}$, and we write $\gamma_n \uparrow \gamma$.

Let F denote the set of all $\mathcal{A}$-measurable mappings from A into $[0,+\infty]$, and F_c be the subset of all bounded, ρ_A-continuous mappings $f\in F$ with bounded support supp f. For all $f\in F$ and all measures ν on $\mathcal{A}$ we put

$$\langle \nu, f\rangle := \int f(a)\,\nu(da) .$$

Let $\mathcal{N}$ be the σ-field on N generated by the mappings $\nu \rightarrow \nu(X)$, $X\in\mathcal{B}$, and let τ_N denote the **vague topology** on N, i. e. the topology generated by all mappings $\nu\rightarrow\langle\nu, f\rangle$, $f\in F_c$; $\xrightarrow{N}$ will stand for the corresponding convergence.

1.1.1. [MKM, Prop. 3.2.1., 3.2.3.] $[N,\tau_N]$ is a Polish space (i. e. τ_N is generated by a complete, separable metric). The corresponding σ-algebra of Borel subsets coincides with $\mathcal{N}$.

For all measures ν on $\mathcal{A}$, the ring of all $X\in\mathcal{B}$ with the property $\nu(\partial X) = 0$ is denoted by $\mathcal{B}_\nu$.

1.1.2. [MKM, Prop. 3.2.2.] $\nu_n \xrightarrow{N} \nu$ iff $\nu_n(X) \rightarrow \nu(X)$, $X\in\mathcal{B}_\nu$.

1.1.3. [MKM, Prop. 3.2.5.] A subset Y of N is τ_N-relatively compact iff it obeys, for all ρ_A- bounded and closed subsets X of A, the following conditions:

a) $\sup_{\nu\in Y} \nu(X) < +\infty$.

b) For any $\varepsilon>0$ there exists a ρ_A-compact subset B of X such that $\sup_{\nu\in Y} \nu(X\setminus B) < \varepsilon$.

$\nu_n \uparrow \nu$ as well as $\nu_n \downarrow \nu$ imply the convergence $\nu_n \xrightarrow{N} \nu$. Moreover, $\nu_n \xrightarrow{N} \nu$, $\gamma_n \xrightarrow{N} \gamma$, $\nu_n \leq \gamma_n$, n=1,2,..., leads to $\nu \leq \gamma$.

In virtue of 1.1.3. this yields

1.1.4. For all $\sigma\in N$ the set $\{\nu: \nu\in N, \nu \leq \sigma\}$ is τ_N - compact.

By a **random measure with phase space** $[A, \rho_A]$ we mean a random element of the measurable space $[N, \mathcal{N}]$; this can be interpreted as a random mass allocation on the phase space. Any random measure induces a probability distribution on $\mathcal{N}$; the probability distributions and - more generally - the measures on $\mathcal{N}$ will be our primal object of interest for the rest of this section.

For all natural numbers k, $\nu \to \nu^{\otimes k}$ is a mapping from N into the set of measures on $\mathcal{A}^{\otimes k}$. As the mapping $\nu \to \nu^{\otimes k}(S)$ is $\mathcal{N}$-measurable for all $S \in \mathcal{A}^{\otimes k}$, we may form, for each measure H on $\mathcal{N}$, the measure

$$\Lambda^{(k)}{}_H(.) := \int \nu^{\otimes k}(.)\, H(d\nu)$$

on $\mathcal{A}^{\otimes k}$, which is called k^{th} **moment measure** of H. We say that H is **of kth order** if $\Lambda^{(k)}{}_H(X^k)$ is finite for all $X \in \mathcal{B}$. For k=1, we write Λ_H instead of $\Lambda^{(1)}{}_H$ and speak of the **intensity measure** of H. For all $f \in F$ there holds

$$\langle \Lambda_H , f \rangle = \iint f(a)\, \nu(da)\, H(d\nu) \ .$$

If H is interpreted as the law of a random mass allocation, then $\Lambda_H(X)$ is the expectation of the random mass in X. For simplicity, we further put $\mathcal{B}_H := \mathcal{B}_{\Lambda_H}$.

Let **G** denote the linear space of all finite signed measures on $\mathcal{N}$. Together with the **convolution operation**

$$G_1 * G_2 := (G_1 \otimes G_2)((\nu_1+\nu_2) \in (.)) \qquad (G_1, G_2 \in \mathbf{G}) \ ,$$

G becomes a commutative real algebra with unit element δ_o. For all $G \in \mathbf{G}$ and $k \in \mathbf{Z}_+$, we denote the k-th convolution power of G by G^k. In particular, there holds $G^0 = \delta_o$.

For abbreviation, we put, for all $X \in \mathcal{A}$ and all measures ν on $\mathcal{A}$,

$$_X\nu := \nu((.) \cap X) \ .$$

For all $X \in \mathcal{B}$, the mapping

$$G \to {}_XG := G({}_X\nu \in (.))$$

is a homomorphism of the algebra **G** into itself.

We will always write $\|q\|$ for the **total variation** of a finite signed measure q on an arbitrary measurable space.

With the norm $\| \ \|$, **G** is a complete normed algebra. Instead of $\|_XG\|$ we also write $_X\|G\|$ $(G \in \mathbf{G}, X \in \mathcal{A})$. $\mathbf{G}_+$ denotes the set of all finite measures on $\mathcal{N}$, $\tau_{\mathbf{G}_+}$ stands for the weak topology on $\mathbf{G}_+$, and $\Rightarrow$ for the corresponding convergence. $[\mathbf{G}_+, \tau_{\mathbf{G}_+}]$ is again a Polish space.

For all finite sequences $X_1,...,X_m$ of sets in $\mathcal{B}$ and all $G \in \mathbf{G}$ we put

$$G_{X_1,...,X_m} := G([\nu(X_1),...,\nu(X_m)] \in (.)) \ .$$

Obviously there holds

$$(G_1 * G_2)_{X_1,...,X_m} = (G_1)_{X_1,...,X_m} * (G_2)_{X_1,...,X_m} \qquad (G_1, G_2 \in \mathbf{G}_+) \ ,$$

where $*$ on the right hand side means the usual convolution of finite measures on the σ-algebra $(\mathcal{B}_+)^{\otimes m}$ of Borel sets on $\mathbf{R}_+{}^m$.

1.1.5. [MKM, Prop. 7.2.4.] In G_+ there holds $G_n \Rightarrow G$ iff for all finite sequences $X_1,...,X_m$ of sets in $\mathcal{B}_G$ the weak convergence

$$(G_n)_{X_1,...,X_m} \Rightarrow G_{X_1,...,X_m}$$

in the space of all finite measures on $(B_+)^{\otimes m}$ takes place.

Due to $\mathcal{B}_{G_1 * G_2} = \mathcal{B}_{G_1} \cap \mathcal{B}_{G_2}$ there follows

1.1.6. In G_+, $G_n \Rightarrow G$ and $H_n \Rightarrow H$ imply that $G_n * H_n \Rightarrow G*H$.

Moreover, from 1.1.5. it is easy to conclude

1.1.7. $G_n \Rightarrow G$ and $X \in \mathcal{B}_G$ imply that $\Lambda_G(X) \leq \liminf_{n\to\infty} \Lambda_{G_n}(X)$.

Another consequence of 1.1.5. is

1.1.8. For all measures G, G_1, G_2,... of first order in G_+ such that $G_n \Rightarrow G$, the vague convergence $\Lambda_{G_n} \xrightarrow{N} \Lambda_G$ takes place iff for all $X \in \mathcal{B}$ the condition

$$\sup_{n=1,2,...} \int_{\{v:v(X)>c\}} v(X)\, G_n(dv) \xrightarrow[c\to\infty]{} 0$$

is fulfilled.

Let V denote the set of all probability distributions on $\mathcal{N}$ and V be the smallest σ-algebra of subsets of V with respect to which all mappings $Q \to Q(Y)$, $Y \in \mathcal{N}$, are measurable. V is a weakly closed subset of G_+, and the σ-algebra V coincides with the σ-algebra of Borel subsets of V with respect to the weak topology τ_V, i.e. the restriction of τ_{G_+} to V. Obviously, $v \to \delta_v$ is a homeomorphic imbedding of $[N, \tau_N]$ into $[V, \tau_V]$. The following simple estimate will sometimes be useful:

1.1.9. For all $Q_1, Q_2 \in V$ and all $X \in \mathcal{A}$ there holds

$$\begin{aligned} {}_X\|Q_1 - Q_2\| &\leq 2\,(Q_1(v(X)>0) + Q_2(v(X)>0)) \\ &\leq 2\,(\Lambda_{Q_1}(X) + \Lambda_{Q_2}(X)) . \end{aligned}$$

Together with Q, obviously all of the $Q_{X_1,...,X_m}$ are probability distributions. In this case, these are called **finite dimensional distributions** of Q. On the other hand, with any probability distribution p on $(B_+)^{\otimes m}$ we may associate via

$$Q := p\left(\sum_{1\leq i\leq m} x_i \delta_i \in (.)\right)$$

the distribution of a random measure with phase space {1,...,m}, for which then holds

$$p = Q_{\{1\},...,\{m\}} .$$

Hence one may identify random elements of $\mathbf{R}_+^m$ in a natural way with random measures on the phase space {1,...,m}.

In V, a semi-order $\preceq$ will be introduced as follows: we say that $Q_1 \preceq Q_2$ if there exists a distribution H on $\mathcal{N}^{\otimes 2}$ with the properties

$$H(v_i \in (.)) = Q_i \ \text{ for } i = 1,2 \quad ; \ H(v_1 \leq v_2) = 1 \quad .$$

This is equivalent to the existence of an $\mathcal{N}, \boldsymbol{V}$ - measurable mapping $v \rightarrow L(v)$ from N into V with the property $Q_1 = \int L(v)\,(.)\, Q_2\,(dv)$; $(L(v))\,(\gamma \leq v) = 1$ for all $v \in N$.

1.1.10. $V_n \Rightarrow V$, $Q_n \Rightarrow Q$ and $V_n \preceq Q_n$ for $n = 1,2,...$ imply that $V \preceq Q$.

If A coincides with $\{1,...,m\}$ and **V** is identified in the above mentioned sense with the set of all distributions on $(\boldsymbol{B}_+)^{\otimes m}$, then $\preceq$ is just the well-known relation "stochastically smaller" (cf. [Stoyan (1983)]), which we continue to denote by $\preceq$:

$p_1 \preceq p_2$ iff there exists a distribution q on $(\boldsymbol{B}_+)^{\otimes 2m}$ such that

$$q([x_1,...,x_m] \in (.)) = p_1 \ ; \ q([x_{m+1},...,x_{2m}] \in (.)) = p_2 \ ; \ q(x_i \leq x_{i+m} \text{ for } 1 \leq i \leq m) = 1.$$

By means of 1.1.10. there results

1.1.11. In **V** there holds $V \preceq Q$ iff for all finite sequences $X_1,...,X_m$ of sets in $\mathcal{B}$ the inequality

$$V_{X_1,...,X_m} \preceq Q_{X_1,...,X_m}$$

holds true.

Now let $(Q_i)_{i \in I}$ be some at most countable family of distributions on $\mathcal{N}$. For $I=\emptyset$ we define $\underset{i \in I}{*} Q_i$ as the unit element δ_o of **G**. Otherwise we form $H := \bigotimes_{i \in I} Q_i$ and put

$$\underset{i \in I}{*} Q_i \ := H\,((\sum_{i \in I} v_i) \in (.)) \ ,$$

provided that $H\,((\sum_{i \in I} v_i) \in N) = 1$. (With this definition, $\underset{i \in I}{*} Q_i$ is the distribution of the sum of independent random measures with distribution Q_i, $i \in I$.) The intensity measure of $Q = \underset{i \in I}{*} Q_i$ is $\Lambda_Q = \sum_{i \in I} \Lambda_{Q_i}$. On the other hand, the convolution $\underset{i \in I}{*} Q_i$ certainly exists if $\sum_{i \in I} \Lambda_{Q_i}$ belongs to N.

If V is a distribution on $\mathcal{N}$ and k is a natural number, then there exists at most one convolution root of V in **V**. If there exists a k-th convolution root for any $k \in \mathbf{Z}_+$, then V is called **infinitely divisible.** Let **I** denote the set of all infinitely divisible V in **V**.

With any measure G in $\mathbf{G}_+$, we associate in form of the "Poissonian mixture"

$$\mathcal{E}_G := e^{-G(N)} \sum_{k \geq 0} (k!)^{-1}\, G^k$$

a distribution in **I**, satisfying the relations

$$\mathcal{E}_{G_1} * \mathcal{E}_{G_2} = \mathcal{E}_{G_1+G_2} \quad \text{for all } G_1, G_2 \in \mathbf{G}_+ \ ; \ \Lambda_{\mathcal{E}_G} = \Lambda_G \quad \text{for all } G \in \mathbf{G}_+ \ .$$

In view of $\mathcal{E}_G = \mathcal{E}_{G((.)\setminus\{o\})}$ we may restrict ourselves here to measures G in $\mathbf{G}_+$ which have no mass in o .

Let U denote the set of all possibly infinite measures on $\mathcal{N}$ satisfying the conditions

$$U(\{o\}) = 0 \ ; \ \int (1-e^{-v(X)})\, U(dv) < \infty \ \text{ for all } X \in \mathcal{B} \ .$$

It can be shown that for an arbitrary at most countable family $(G_i)_{i \in I}$ of measures in G_+ with the property $G_i(\{o\}) = 0$, $i \in I$, the convolution $V := \ast_{i \in I} \mathcal{E}_{G_i}$ exists iff $U := \sum_{i \in I} G_i$ belongs to **U**. In this case, V depends only on U and not on the special representation of U. Taking into consideration that any $U \in \mathbf{U}$ has such a representation $U = \sum_{i \in I} G_i$, we may define, for all $U \in \mathbf{U}$, the distribution $\mathcal{E}_U$ by $\mathcal{E}_U := V$. Again there holds

1.1.12. $\mathcal{E}_{U_1} * \mathcal{E}_{U_2} = \mathcal{E}_{U_1+U_2}$ $(U_1, U_2 \in \mathbf{U})$; $\Lambda_{\mathcal{E}_U} = \Lambda_U$ $(U \in \mathbf{U})$.

Hence all distributions $\mathcal{E}_U$, $U \in \mathbf{U}$, are infinitely divisible.

In view of $\delta_v = (\delta_{k^{-1}v})^k$ for $v \in N$, $k = 1,2,\ldots$, also δ_v, $v \in N$, is an element of **I**.

From the two above mentioned special types of infinitely divisible distributions on $\mathcal{N}$, all distributions in **I** may be obtained in the following way:

1.1.13. [K, Prop.6.1.] The mapping $[v,U] \to \delta_v * \mathcal{E}_U$ establishes a 1-1 correspondence between N×**U** and **I**, the (unique) factorization

$$V = \delta_v * \mathcal{E}_U \qquad (v \in N, U \in \mathbf{U})$$

is called **canonical representation** of V, and the measure U figuring there is called **canonical measure** of V, written as $U = \tilde{V}$. Obviously there holds $\Lambda_V = v + \Lambda_{\tilde{V}}$.

1.2. Random Point Fields

A measure Φ in N is called **counting measure** if it takes only nonnegative integer values on $\mathcal{B}$. The set of all counting measures in N will be denoted by M.

1.2.1. [MKM, Prop.1.1.2.] Any $\Phi \in M$ is of the form $\Phi = \sum_{a \in \mathrm{supp}\Phi} \Phi(\{a\})\delta_a$.

Hence any counting measure Φ may be interpreted as an (at most countable) system of identical particles or individuals in the phase space $[A, \rho_A]$. For all $X \in \mathcal{A}$, $\Phi(X)$ "counts" the number of particles in the system which are situated in X. Specifically, $\Phi(\{a\})$ is the number of particles at the site $a \in A$. If, for all $a \in A$, $\Phi(\{a\})$ is either zero or one, then Φ is called **simple**. Obviously, the mapping $\Phi \to \mathrm{supp}\Phi$ establishes a 1-1 correspondence between the set of all simple counting measures and the system of all subsets X of A with the property $\#(X \cap S) < +\infty$, $S \in \mathcal{B}$.

Being a subset of N, the set M inherits the order, topological and measurable structure of N defined in 1.1. Let $\mathcal{M}$ be the σ-algebra on M generated by the mappings $\Phi \to \Phi(X)$, $X \in \mathcal{B}$, and let τ_M denote the **vague topology** on M, i. e. the topology generated by all the mappings $\Phi \to \langle \Phi, f \rangle$, $f \in F_c$; $\xrightarrow{M}$ will stand for the corresponding convergence. Obviously, τ_M is the restriction of τ_N to M, and $\mathcal{M}$ is the restriction of $\mathcal{N}$ to M.

1.2.2. [MKM, Prop. 3.2.4.] The set M is τ_N-closed.

This together with 1.1.1. shows:

1.2.3. $[M, \tau_M]$ is a Polish space. The corresponding σ-algebra of Borel subsets coincides with $\mathcal{M}$.

The mapping $a \to \delta_a$ provides a homeomorphic imbedding of the phase space into $[M, \tau_M]$. One may conclude easily from 1.1.2. that vague convergence $\nu_n \xrightarrow{N} \nu$ in $[N, \tau_N]$ can be characterized as follows : Let (X_m) be some ascending sequence in $\mathcal{B}_\nu$ such that any $X \in \mathcal{B}$ is covered by some X_m. Then $\nu_n \xrightarrow{N} \nu$ is equivalent to the weak convergence of $(\chi_{X_m} \nu_n)$ towards $(\chi_{X_m} \nu)$ in the space of finite measures on $\mathcal{A}$ for all $m \in \mathbf{Z}_+$.

On the other hand, weak convergence of finite counting measures can be characterized in an intuitive way:

1.2.4. [Kerstan, Matthes, Mecke(1982), Prop. 1.9.10.] In the set of finite counting measures on $\mathcal{A}$, weak convergence $\Phi_n \Rightarrow \Phi$ takes place iff, for all suitably large natural numbers n, $\Phi_n(A) = \Phi(A)$, and there exist representations $\Phi_n = \sum_{1 \le i \le k} \delta_{a_{n,i}}$ such that

$a_{n,i} \xrightarrow[n \to \infty]{} a_i$, $1 \le i \le k$, and $\Phi = \sum_{1 \le i \le k} \delta_{a_i}$.

By a **random point field** (or random counting measure) **with phase space** $[A, \rho_A]$ we mean a random element of the measurable space $[M, \mathcal{M}]$. Because of $M \in \mathcal{N}$ and $\mathcal{M} = M \cap \mathcal{N}$ we may interpret the random point fields as random measures whose realizations are almost surely counting measures. Any random point field induces a distribution on $\mathcal{M}$; for the rest of this section we will deal with distributions and, more generally, with measures on $\mathcal{M}$.

Let **E** denote the linear space of all finite signed measures on $\mathcal{M}$. Together with the **convolution operation**

$$E_1 * E_2 := (E_1 \otimes E_2)((\Phi_1 + \Phi_2) \in (.)) \qquad (E_1, E_2 \in \mathbf{E})$$

and the variation norm $\| \ \|$, **E** becomes a commutative, complete normed real algebra, which may be identified with the subalgebra of **G** consisting of all G which are concentrated on M. In the same way, the set $\mathbf{E}_+$ of all finite measures on $\mathcal{M}$ will be viewed as a subset of $\mathbf{G}_+$.

Let τ_{E_+} denote the weak topology on E_+ and $\Rightarrow$ the corresponding convergence. The set E_+ is τ_{G_+}-closed, and τ_{E_+} coincides with the restriction of τ_{G_+} to E_+.

1.2.5. [MKM, Prop. 3.2.7.] A subset **L** of E_+ is τ_{E_+}-relatively compact iff it has the following two properties:

a) $\sup_{L\in \mathbf{L}} L(M)<+\infty$,

b) for each closed set X in $\mathcal{B}$ and each $\eta>0$ there exists a natural number $n_{X,\eta}$ as well as a compact subset $B_{X,\eta}$ of X such that

$$\sup_{L\in \mathbf{L}} L(\Phi(X) \geq n_{X,\eta}) < \eta \ ; \ \sup_{L\in \mathbf{L}} L(\Phi(X\setminus B_{X,\eta})>0) < \eta \ .$$

Let **P** denote the set of all distributions on $\mathcal{M}$ and $\boldsymbol{P}$ denote the σ-algebra generated by the mappings $P \to P(Y)$, $Y\in \mathcal{M}$. In the above mentioned sense **P** is viewed as a subset of **V**, which leads to the equality $\boldsymbol{P} = \mathbf{P}\cap \boldsymbol{V}$.

τ_P denotes the weak topology on **P**, i. e. the restriction of τ_{E_+} to the τ_{E_+}-closed set **P**. The σ-algebra of Borel subsets of the Polish space $[\mathbf{P}, \tau_P]$ coincides with $\boldsymbol{P}$.

1.2.6. [MKM, Prop. 3.2.8.] If **L** is a subset of **P** such that $\{\Lambda_L\}_{L\in \mathbf{L}}$ is a τ_N-relatively compact subset of N, then **L** itself is τ_P-relatively compact.

In view of 1.1.4. this leads to

1.2.7. For all $\sigma\in N$, $\{P:P\in \mathbf{P}, \Lambda_P\leq\sigma\}$ is τ_P-compact.

Note that, for $P\in \mathbf{P}$, the **finite dimensional distributions** $P_{X_1,\dots,X_m}$ may be viewed as distributions on $\mathcal{P}(Z_+^m)$; on the other hand the distributions on $\mathcal{P}(Z_+^m)$ can be identified with the distributions of random point fields with phase space $\{1,\dots,m\}$.

If a distribution P on $\mathcal{M}$ has, for a natural number k, a k-th convolution root L in **P**, then we write $L=:\sqrt[k]{P}$. It may happen that a k-th convolution root of P exists in **V** but not in **P**. For any counting measure Φ, δ_Φ is an element of **I**, but $\sqrt[k]{\delta_\Phi}$ exists iff for all points $a\in \operatorname{supp}\Phi$ the number $\Phi(\{a\})$ is an integer multiple of k. **T** will denote the set of all **distributions** on $\mathcal{M}$ which are **infinitely divisible** in **P**, i. e. the set of all $P\in \mathbf{P}$ for which all convolution roots $\sqrt[k]{P}$, k=1,2,..., exist. The set **T** is contained in **I**, but, as mentioned above, is different from $\mathbf{I}\cap\mathbf{P}$.

1.2.8. [MKM, Prop. 3.2.11.] The set **T** is τ_P-closed.

Let **W** denote the set of all possibly infinite measures W on $\mathcal{M}$ obeying the two conditions

$$W(\{o\}) = 0 \quad ; \quad W(\psi(X)>0) < +\infty \text{ for all } X\in \mathcal{B} \quad .$$

As a counterpart to 1.1.13. there holds:

1.2.9. [MKM, Prop. 2.1.10.] The mapping $W \rightarrow \mathcal{E}_W$ provides a 1-1 correspondence between the sets **W** and **T**.

Obviously **W** may be identified with the set of those U∈**U** which are concentrated on M. Instead of $P=\mathcal{E}_W$ we also write $W=:\tilde{P}$, calling $\tilde{P}$ the **canonical measure** of P.

For all P∈**T** and r∈$\mathbf{R}_+$, one puts $P^r := \mathcal{E}_{r\tilde{P}}$, thus obtaining a one-parameter convolution semigroup $(P^r)_{r\in \mathbf{R}_+}$ (cf. section 2.4. in [MKM]).

Because of $\Lambda_{\tilde{P}} = \Lambda_P$, a distribution P in **T** is of first order iff its canonical measure is of first order. Further note that a measure H on $\mathcal{M}$ of first order is contained in **W** iff it has no mass in o .

1.2.10. [MKM, Prop. 2.2.4.] For all P∈**T** and all X∈$\mathcal{A}$, also ${}_XP$ is an element of **T** and there holds

$$\widetilde{({}_XP)} = \tilde{P}({}_X\psi\in(.),\ \psi(X)>0) \quad .$$

In virtue of 1.2.10. and the definition of $W \rightarrow \mathcal{E}_W$, for any given P∈**T** and X∈$\mathcal{A}$ such that $P(\Phi(X)=0) = \exp(-\tilde{P}(\psi(X)>0)) > 0$, the distribution ${}_XP$ may be represented as Poissonian mixture

$${}_XP = e^{-Q(M)} \sum_{k\geq 0} (k!)^{-1}Q^k \ , \quad Q := \tilde{P}({}_X\Psi\in(.),\ \Psi(X)>0)$$

1.2.11. [MKM, Prop. 2.2.6.] For all distributions P∈**T** of first order there holds

$$\Lambda^{(2)}{}_P = \Lambda^{(2)}{}_{\tilde{P}} + \Lambda_P\otimes\Lambda_P \ .$$

With any measure ν∈N we associate via

$$\mathcal{Q}_\nu := \nu(\delta_a\in(.))$$

a measure $\mathcal{Q}_\nu$ ∈**W** with intensity measure ν, and call

$$\Pi_\nu := \mathcal{E}_{\mathcal{Q}_\nu}$$

the **Poisson distribution with intensity measure** ν.

1.2.12. [MKM, Prop. 2.2.14.] For all ν∈N, a distribution P on $\mathcal{M}$ coincides with Π_ν iff for any finite sequence $X_1,\dots,X_m$ of pairwise disjoint sets in $\mathcal{B}$ there holds

$$P_{X_1,\dots,X_2} = \pi_{\nu(X_1)} \otimes\dots\otimes \pi_{\nu(X_m)} \quad .$$

1.2.13. [MKM, Prop. 3.3.7.] The mapping $\nu \rightarrow \Pi_\nu$ provides a homeomorphic imbedding from $[N, \tau_N]$ into $[\mathbf{P}, \tau_\mathbf{P}]$.

With any $G \in \mathbf{G}$, we associate via

$$\mathfrak{I}(G) := \int \Pi_v(.)\, G(dv)$$

a signed measure $\mathfrak{I}(G)$ on $\mathcal{M}$, the so called **Cox transform** of G.

1.2.14. [MKM, Prop. 7.1.2.] The Cox transform $\mathfrak{I}$ is an isomorphism from the algebra **G** onto a subalgebra of **E**.

Obviously, $\mathfrak{I}(G)$ is in $\mathbf{E}_+$ resp. in **P**, if G is in $\mathbf{G}_+$ resp. in **V**. In this case there holds $\Lambda_{\mathfrak{I}(G)} = \Lambda_G$. Distributions of the form $\mathfrak{I}(G)$, $G \in \mathbf{V}$, are called **Cox distributions.**

1.2.15. [MKM, Prop. 7.2.1. , 7.2.2.] The Cox transform $\mathfrak{I}$ provides a homeomorphism from $[\mathbf{G}_+, \tau_{\mathbf{G}_+}]$ into a closed subset of $[\mathbf{E}_+, \tau_{\mathbf{E}_+}]$.

Any counting measure Φ may be written in the form $\Phi = \sum_{i \in I} \delta_{a_i}$.
For all natural numbers k we put

$$\Phi^{(k)} := \sum_{i_1,\dots,i_k \in I \text{ pairwise distinct}} \delta_{[a_{i_1},\dots,a_{i_k}]} \ .$$

Obviously there holds $\Phi^{(k)} \leq \Phi^{\otimes k}$.

1.2.16. For all simple $\Phi \in M$ we have

$$\Phi^{(k)} = \Phi^{\otimes k}\left((.) \cap \{[x_1,\dots,x_k] \in A^k : x_i \neq x_j \text{ for } 1 \leq i < j \leq k\}\right) \ .$$

Let $\rho_{\mathbf{Z}_+}$ be the metric of $\mathbf{Z}_+$ defined by $\rho_{\mathbf{Z}_+}(i,j) = 0$ if $i=j$ and 1 if $i \neq j$, and denote, for the moment, the direct product of $[A, \rho_A]$ and $[\mathbf{Z}_+, \rho_{\mathbf{Z}_+}]$ by $[A', \rho_{A'}]$. (All objects belonging to the new phase space A' will be distinguished by a prime.)

1.2.17. The mapping

$$\Phi \to \Phi' := \sum_{a \in \operatorname{supp}\Phi} \ \sum_{0 \leq j < \Phi(\{a\})} \delta_{[a,j]}$$

is a bijective and in both directions continuous mapping from M onto the set of simple counting measures in M'.

Noting that $\Phi^{(k)}$ may be reconstructed from $(\Phi')^{(k)}$ in an easy way, we conclude from 1.2.16. and 1.2.17. that the mapping $\Phi \to \Phi^{(k)}(S)$ is $\mathcal{M}$-measurable for all $S \in \mathcal{A}^{\otimes k}$. Hence we may form, for any measure H on $\mathcal{M}$ and all natural numbers k, by

$$\Gamma^{(k)}{}_H := \int \Phi^{(k)}(.) H(d\Phi)$$

the k-th **factorial moment measure** $\Gamma^{(k)}{}_H$ of H. Obviously there holds

1.2.18. $\Lambda_H = \Gamma^{(1)}{}_H$, $\Lambda^{(2)}{}_H = \Gamma^{(2)}{}_H + \int \delta_{[a,a]}(.) \Lambda_H(da)$.

Together with 1.2.11. this implies

1.2.19. For all distributions P of fist order in **T** there holds

$$\Gamma^{(2)}{}_P = \Gamma^{(2)}{}_{\tilde{P}} + \Lambda_P \otimes \Lambda_P \ .$$

1.3. The Laplace Transformation

For all $G \in \mathbf{G}_+$ we define via

$$\mathcal{L}_G(f) := \int \exp(-\langle \nu, f\rangle) G(d\nu) \qquad (f \in F)$$

its **Laplace transform** $\mathcal{L}_G$ (where $\exp(-\infty) := 0$). It is immediate from the definition that

1.3.1. $\mathcal{L}_{c_1G_1+c_2G_2}(f) = c_1\mathcal{L}_{G_1}(f) + c_2\mathcal{L}_{G_2}(f) \qquad (f \in F\,;\, c_1, c_2 \in \mathbf{R}_+\,;\, G_1, G_2 \in \mathbf{G}_+)$.

1.3.2. $\mathcal{L}_{G_1 * G_2}(f) = \mathcal{L}_{G_1}(f) \cdot \mathcal{L}_{G_2}(f) \qquad (f \in F\,;\, G_1, G_2 \in \mathbf{G}_+)$.

The mapping $G \to \mathcal{L}_G$ is 1-1; there even holds

1.3.3. For all $G_1, G_2 \in \mathbf{G}_+$,

$$\mathcal{L}_{G_1}(f) = \mathcal{L}_{G_2}(f) \qquad (f \in F_c)$$

implies that $G_1 = G_2$.

By the Laplace transformation, weak convergence in $\mathbf{G}_+$ is carried into pointwise convergence for all $f \in F_c$. Sharpening 1.3.3., there holds

1.3.4. [K, Prop. 4.2.] For all $G, G_1, G_2, \ldots$ in $\mathbf{G}_+$, weak convergence $G_n \Rightarrow G$ is equivalent to the validity of

$$\mathcal{L}_{G_n}(f) \to \mathcal{L}_G(f) \qquad (f \in F_c) \ .$$

For all measures ν on $\mathcal{A}$, let F_ν denote the set of all $f \in F$ with the property $\langle \nu, f\rangle < +\infty$. For all $G \in \mathbf{G}_+$ and all $f, g \in F_{\Lambda_G}$ there holds

$$\left| \int \exp(-\langle \nu, f\rangle)\, G(d\nu) - \int \exp(-\langle \nu, g\rangle) G(d\nu) \right|$$
$$\leq \int \left| \langle \nu, f\rangle - \langle \nu, g\rangle \right| G(d\nu) \quad \leq \quad \int \langle \nu, |f-g|\rangle \cdot G(d\nu) \ ,$$

which leads to the elementary inequality

1.3.5. For all $G \in \mathbf{G}_+$ and all $f, g \in F_{\Lambda_G}$ there holds

$$\left| \mathcal{L}_G(f) - \mathcal{L}_G(g) \right| \leq \langle \Lambda_G, |f-g| \rangle \ .$$

For all $G \in \mathbf{G}_+$ and all $X_1, \ldots, X_m \in \mathcal{B}$ we obtain for all $s_1, \ldots, s_m \in \mathbf{R}_+$:

$$\mathcal{L}_{G_{X_1,\ldots,X_m}}(s_1, \ldots, s_m)$$
$$= \int (\exp(-\sum_{1 \leq i \leq m} s_i x_i) G_{X_1,\ldots,X_m}(d[x_1, \ldots, x_m]) = \mathcal{L}_G(\sum_{1 \leq i \leq m} s_i 1_{X_i}) \ .$$

Hence follows

1.3.6. For all $G, G_1, G_2, \ldots \in \mathbf{G}_+$ and all finite sequences $X_1, \ldots, X_m$ of sets in $\mathcal{B}$,

$$(G_n)_{X_1,\ldots,X_m} \Rightarrow G_{X_1,\ldots,X_m}$$

is equivalent to

$$\mathcal{L}_{G_n}(\sum_{1 \leq i \leq m} s_i 1_{X_i}) \to \mathcal{L}_{G_n}(\sum_{1 \leq i \leq m} s_i 1_{X_i}) \qquad (s_1, \ldots, s_m \in \mathbf{R}_+) \ .$$

Assume now that in G_+ the convergence relation $G_n \Rightarrow G$ holds true, and that, for some $\nu \in N$, the inequalities $\Lambda_{G_n} \leq \nu$, $n = 1, 2, \ldots$, are fulfilled. Subject to these assumptions, also $\Lambda_G \leq \nu$ holds true. Furthermore, for any $f \in F_\nu$ and any $\varepsilon > 0$ there exists some $g \in F_c$ such that $\langle \nu, |f-g| \rangle < \varepsilon$, and we obtain by 1.3.4. and 1.3.5.

$$\limsup_{n\to\infty} |\mathcal{L}_{G_n}(f) - \mathcal{L}_G(f)|$$

$$\leq \limsup_{n\to\infty} |\mathcal{L}_{G_n}(f) - \mathcal{L}_{G_n}(g)| + \lim_{n\to\infty} |\mathcal{L}_{G_n}(g) - \mathcal{L}_G(g)| + |\mathcal{L}_G(f) - \mathcal{L}_G(g)|$$

$$\leq \limsup_{n\to\infty} \langle \Lambda_{G_n}, |f-g| \rangle + 0 + \langle \Lambda_G, |f-g| \rangle \leq 2\langle \nu, |f-g| \rangle| \leq 2\varepsilon .$$

Hence we recognise the validity of

1.3.7. If in G_+ the weak convergence $G_n \Rightarrow G$ takes place, and if for some $\nu \in N$ the inequalities $\Lambda_{G_n} \leq \nu$, $n=1,2,\ldots$, are fulfilled, then there holds

$$\mathcal{L}_{G_n}(f) \to \mathcal{L}_G(f) \quad (f \in F_\nu) .$$

In view of 1.3.6. one further gets:

1.3.8. Subject to the assumptions of 1.3.7. there holds $(G_n)_{X_1,\ldots,X_m} \Rightarrow G_{X_1,\ldots,X_m}$ for any finite sequence $X_1, \ldots, X_m$ of sets in $\mathcal{B}$.

The canonical representation of infinitely divisible distributions on $\mathcal{N}$ now takes the following form:

1.3.9. [K, Prop. 6.1.] For all $\nu \in N$ and all $U \in \mathbf{U}$ there holds

$$\mathcal{L}_{\delta_\nu * \mathcal{E}_U}(f) = \exp(-\langle \nu, f\rangle - \int (1 - \exp(-\langle \gamma, f\rangle)) U(d\gamma)) .$$

Hence results

1.3.10. For all $P \in \mathbf{T}$ one has

$$\mathcal{L}_P(f) = \exp(-\int (1 - \exp(-\langle \psi, f\rangle)) \tilde{P}(d\psi)) \qquad (f \in F) .$$

Particulary there hold

1.3.11. $\mathcal{L}_{\Pi_\nu}(f) = \exp(-\int (1 - \exp(-f(a))) \nu(da)) \qquad (\nu \in N;\ f \in F)$.

1.3.12. $\mathcal{L}_{\mathfrak{I}(Q)}(f) = \mathcal{L}_Q(1 - \exp(-f)) \qquad (Q \in V;\ f \in F)$.

In V, 1.3.2. generalizes to infinite convolution products:

1.3.13. In V, the relation $Q = \underset{i\in I}{*} Q_i$ implies

$$\mathcal{L}_Q(f) = \prod_{i \in I} \mathcal{L}_{Q_i}(f) \quad (f \in F) .$$

For some computations, the set F will be too large and will be replaced by the set F_b of all $\mathcal{A}$-measurable bounded nonnegative real-valued functions on A with bounded support. Obviously there holds

$$0 < \mathcal{L}_Q(f) \leq 1 \qquad (Q \in V;\ f \in F_b) .$$

1.4. Clustering Fields and Branching Dynamics

Let $[A', \rho_{A'}]$ be some other complete, separable metric space. All objects which refer to the phase space $[A', \rho_{A'}]$ will be indicated by a prime (e.g. $\mathcal{A}'$, $\mathcal{B}'$,...)

A **clustering field on** $[A, \rho_A]$ **with phase space** $[A', \rho_{A'}]$ is an $\mathcal{A}$, P'-measurable mapping $\kappa_{(.)}$ from A into $\mathbf{P}'$: with each site in A one associates a distribution of a random element (a "cluster") χ_a in $[M', \mathcal{M}']$. χ_a may be interpreted as random offspring of an individual at site a. A clustering field κ will be called **stochastic** resp. **substochastic** if $\kappa_{(a)}(\chi(A)=1) = 1$ resp. $\kappa_{(a)}(\chi(A)\leq 1) = 1$ for all $a \in A$.

Having in mind the interpretation of $\kappa_{(a)}$ as an "individual's random offspring distribution", we put for $\Phi = \sum_{i\in I} \delta_{a_i}$

$$\kappa_{(\Phi)} := \mathop{*}_{i\in I} \kappa_{(a_i)} ,$$

if the convolution on the right hand side exists. Of course one may avoid a particular indexing of points in Φ, putting

$$\kappa_{(\Phi)} := \mathop{*}_{a\in \mathrm{supp}\Phi} (\kappa_{(a)})^{\Phi(\{a\})} .$$

Our definition of the "population's random offspring distribution" $\kappa_{(\Phi)}$ is based on an **assumption** which is typical for stochastic branching models, namely, that the random offspring χ of the population Φ is an independent superposition $\sum_{i\in I} \chi_{a_i}$ of the random offsprings χ_{a_i} of the individuals in Φ, the family $(\chi_{a_i})_{i\in I}$ being distributed according to $\bigotimes_{i\in I} \kappa_{(a_i)}$, i. e. the offsprings of the different individuals in Φ are generated in a stochastically independent way according to the given clustering field κ. If κ is stochastic rep. substochastic, then the random offspring χ arises from Φ via a so called **stochastic** resp. **substochastic translation.**

If, for some $\Phi \in M$, $\kappa_{(\Phi)}$ does not exist, then this means that the superposition $\sum_{i\in I} \chi_{a_i}$ violates, with positive probability, the finiteness condition which is characteristic for measures in N, i. e. in this case there exists an $X \in \mathcal{B}$ such that

$$(\bigotimes_{i\in I} \kappa_{(a_i)})(\sum_{i\in I} \chi_{a_i}(X) = +\infty) > 0 .$$

Let ${}_{\kappa}M$ denote the set of all $\Phi \in M$ for which $\kappa_{(\Phi)}$ is defined (which are, so speak, "sufficiently thin").

1.4.1. [MKM, Prop. 4.1.2.] Let κ be a clustering field on $[A, \rho_A]$ with phase space $[A', \rho_{A'}]$. Then $\Phi \in M$ belongs to ${}_{\kappa}M$ iff $\langle \Phi, \kappa_{(.)}(\chi(X)>0)\rangle < +\infty$ for all $X \in \mathcal{B}$.

Obviously, ${}_{\kappa}M$ is a descending set relative to the natural semi-order $\leq$. 1.4.1. shows that ${}_{\kappa}M$ belongs to $\mathcal{M}$; there even holds

1.4.2. [MKM, Prop. 4.1.3.] Subject to the assumptions of 1.4.1., $\Phi \to \kappa_{(\Phi)}$ is an $\mathcal{M}, \boldsymbol{P}'$-measurable mapping from ${}_{\kappa}M$ into $\mathbf{P}'$.

The mapping $\Phi \to \kappa_{(\Phi)}$ is called **stochastic branching dynamics** on $[M, \mathcal{M}]$ induced by the clustering field κ. If H is some measure on $\mathcal{M}$ with the property $H(\Phi \notin {}_{\kappa}M) = 0$, then we form via

$$H_{\kappa} := \int \kappa_{(\Phi)}(.)\, H(d\Phi)$$

the "clustered" measure H_{κ} on $\mathcal{M}'$. If a distribution $P \in \mathbf{P}$ is interpreted as a distribution of a random population Φ_0 of individuals in the phase space $[A, \rho_A]$, then P_{κ}, if it exists, may be interpreted as distribution of the succeeding generation Φ_1.

The set $\{P{:}P \in \mathbf{P},\ P({}_{\kappa}M) = 1\}$ of those distributions P on $\mathcal{M}$, for which P_{κ} exists, belongs to $\boldsymbol{P}$. There even holds

1.4.3. [MKM, Prop. 4.2.4.] Subject to the assumptions of 1.4.1. , $P \to P_{\kappa}$ is a $\boldsymbol{P}, \boldsymbol{P}'$-measurable mapping from $\{P{:}\ P \in \mathbf{P},\ P({}_{\kappa}M = 1)\}$ into $\mathbf{P}'$.

The clustering operation is monotone in the following sense:

1.4.4. Let P_1, P_2 be distributions in $\mathbf{P}$ such that $P_1 \lesssim P_2$ and $(P_2)_{\kappa}$ exists. Then also $(P_1)_{\kappa}$ exists, and $(P_1)_{\kappa} \lesssim (P_2)_{\kappa}$.

Clustering commutes with convolution:

1.4.5. [MKM, Prop. 4.3.1.] If κ is a clustering field on $[A, \rho_A]$ with phase space $[A', \rho_{A'}]$, and $(P_i)_{i \in I}$ is an at most countable family of distribution on $\mathcal{M}$ whose convolution P exists, then there holds

$$P_{\kappa} = \mathop{*}_{i \in I} (P_i)_{\kappa} \quad ,$$

where the existence of one side implies the existence of the other.

A simple but frequently used tool in dealing with clustered infinitely divisible distributions is the so called "clustering theorem":

1.4.6. [MKM, Prop. 4.3.3.] Let P be a distribution in $\mathbf{T}$ and κ be a clustering field on $[A, \rho_A]$ with phase space $[A', \rho_{A'}]$. Then P_{κ} exists iff $\tilde{P}_{\kappa}$ exists and $\tilde{P}_{\kappa}((.)\backslash\{o\})$ belongs to $\mathbf{W}'$. In this case P_{κ} belongs to $\mathbf{T}'$,and there holds

$$\widetilde{P_{\kappa}} = \tilde{P}_{\kappa}((.)\backslash\{o\}) \ .$$

By 1.3.13. there follows

1.4.7. If P is a distribution on $\mathcal{M}$ and κ is a clustering field on $[A, \rho_A]$ with phase space $[A', \rho_{A'}]$ such that P_{κ} exists, then there holds

$$\mathcal{L}_{P_{\kappa}}(f) = \mathcal{L}_P(-\ln \mathcal{L}_{\kappa_{(.)}}(f)) \quad , \quad f \in F\,, \quad \text{where } \ln 0 := -\infty \ .$$

In view of 1.3.11. one gets

1.4.8. If, under the assumptions of 1.4.7., P is a Poisson distribution Π_v, then one has

$$\mathcal{L}_{(\Pi_v)_\kappa}(f) = \exp(-\langle v, 1-\mathcal{L}_{\kappa_{(.)}}(f)\rangle) \qquad (f \in F) \ .$$

Let now, in addition to $[A, \rho_A]$ and $[A', \rho_{A'}]$, a third complete separable metric space $[A'', \rho_{A''}]$ be given. If κ is a clustering field on $[A, \rho_A]$ with phase space $[A', \rho_{A'}]$ and ω is a clustering field on $[A', \rho_{A'}]$ with phase space $[A'', \rho_{A''}]$ such that $(\kappa_{(a)})_\omega$ exists for all $a \in A$, then we put

$$(\omega \circ \kappa)_{(a)} := (\kappa_{(a)})_\omega \qquad (a \in A) \ ,$$

thus obtaining, in virtue of 1.4.2. and 1.4.3., a clustering field $\omega \circ \kappa$ on $[A, \rho_A]$ with phase space $[A'', \rho_{A''}]$. We may interprete $(\omega \circ \kappa)_{(\Phi)}$ as distribution of the random "grandchildren's generation" of a population Φ, i. e. $\omega \circ \kappa$ may be conceived as a composition of the stochastic branching dynamics κ and ω.

1.4.9. [MKM, Prop. 4.3.4.] Let H be a measure on $\mathcal{M}$, κ be a clustering field on $[A,\rho_A]$ with phase space $[A', \rho_{A'}]$ and ω be a clustering field on $[A', \rho_{A'}]$ with phase space $[A'', \rho_{A''}]$. If both the clustering field $\omega \circ \kappa$ and the measure H_κ exist, then there holds

$$(H_\kappa)_\omega = H_{\omega \circ \kappa} \ ,$$

where the existence of one side entails the existence of the other.

In the special case $[A', \rho_{A'}] = [A, \rho_A]$ we simply speak of **clustering fields** κ **on** $[A, \rho_A]$.

For a clustering field κ on $[A, \rho_A]$, we introduce step by step its **clustering powers** $\kappa^{[n]}$, $n \in \mathbf{Z}_+$:

$$\kappa^{[0]}{}_{(a)} := \delta_{\delta_a} \quad (a \in A) \ ; \quad \kappa^{[n+1]} := \kappa \circ \kappa^{[n]} \quad (n \in \mathbf{Z}_+) \ .$$

It can easily be shown by 1.4.9. that

$$\kappa^{[m]} \circ \kappa^{[j]} = \kappa^{[m+j]} \qquad (j,m \in \mathbf{Z}_+) \ ,$$

provided that all clustering powers of κ exist.

1.5. Intensity Kernels

A **kernel** from a measurable space $[E_1, \mathcal{E}_1]$ to a measurable space $[E_2, \mathcal{E}_2]$ is a mapping K from $E_1 \times \mathcal{E}_2$ into $[0, +\infty]$ with the two properties

a) For all $e \in E_1$, $K(e,(.))$ is a measure on $\mathcal{E}_2$.

b) For all $X \in \mathcal{E}_2$, $K((.),X)$ is an $\mathcal{E}_1$-measurable mapping from E_1 into $[0, +\infty]$.

For any measure γ on $\mathcal{E}_1$ and any $\mathcal{E}_2$-measurable mapping f from E_2 into $[0, +\infty]$, one puts

$$(\gamma * K)(.) := \int K(e,(.))\gamma(de) \quad ; \quad (K*f)(.) := \int f(z)K((.),dz) \quad ,$$

thus obtaining a measure $\gamma * K$ on $\mathcal{E}_2$ and an $\mathcal{E}_1$-measurable mapping $K*f$ from E_1 into $[0,+\infty]$. There always holds

1.5.1. $\int f(z)\,(\gamma * K)(dz) \;=\; \int (K*f)(e)\,\gamma(de)$.

If J is a kernel from $[E_2, \mathcal{E}_2]$ to a measurable space $[E_3, \mathcal{E}_3]$, then we obtain via

$$(K*J)(e,(.)) := K(e,(.))*J \qquad (e \in E_1)$$

a kernel $K*J$ from $[E_1, \mathcal{E}_1]$ to $[E_3, \mathcal{E}_3]$.

A kernel K is called **stochastic** resp. **substochastic** if for all $e \in E_1$ there hold $K(e,E_2) = 1$ resp. $K(e,E_2) \leq 1$.

If K is a kernel from $[E, \mathcal{E}]$ to $[E, \mathcal{E}]$, then the kernels $K^{[n]}$, $n \in \mathbb{Z}_+$, are defined by

$$K^{[0]}(a,(.)) := \delta_a \qquad (a \in A) \quad ; \quad K^{[n+1]} := K*K^{[n]} \qquad (n \in \mathbb{Z}_+) \ .$$

Obviously there holds

$$K^{[m]}*K^{[k]} = K^{[m+k]} \qquad (m,k \in \mathbb{Z}_+) \ .$$

For any clustering field κ on $[A, \rho_A]$ with phase space $[A', \rho_{A'}]$, we define a kernel J_κ from $[A, \mathcal{A}]$ to $[A', \mathcal{A}']$ by

$$J_\kappa(a,(.)) := \Lambda_{\kappa(a)} \quad ,$$

calling J_κ the **intensity kernel** of κ.

If a measure L on $\mathcal{M}$ is "clustered" by κ, then this corresponds, on the level of intensity measures, to a transformation of Λ_L by J_κ:

1.5.2. [MKM, Prop. 4.2.2.] If κ is a clustering field on $[A, \rho_A]$ with phase space $[A', \rho_{A'}]$ and L is a measure on $\mathcal{M}$ such that L_κ exists, then

$$\Lambda_{L_\kappa} = \Lambda_L * J_\kappa \ .$$

Conversely, we can conclude the existence of L_κ by inspection of J_κ and Λ_L:

1.5.3. [MKM, Prop. 4.2.3.] If κ is a clustering field on $[A, \rho_A]$ with phase space $[A', \rho_{A'}]$ and L is a measure on $\mathcal{M}$ such that $\Lambda_L * J_\kappa \in N'$, then L_κ exists.

1.5.2. and 1.5.3. yield

1.5.4. If κ is a clustering field on $[A, \rho_A]$ with phase space $[A', \rho_{A'}]$ and ω is a clustering field on $[A', \rho_{A'}]$ with phase space $[A'', \rho_{A''}]$, then

a) the existence of $\omega \circ \kappa$ implies the equality $J_{\omega \circ \kappa} = J_\kappa * J_\omega$.

b) $(J_\kappa * J_\omega)(a, (.)) \in N''$ $\quad (a \in A)$ $\quad$ implies the existence of $\omega \circ \kappa$.

This leads immediately to

1.5.5. If a clustering field τ on $[A, \rho_A]$ obeys the condition

$$(J_\tau)^{[n]}(a,(.)) \in N \qquad (a \in A;\ n = 1,2,\ldots) \quad ,$$

then there exist all clustering powers $\tau^{[n]}$, $n \in \mathbb{Z}_+$, and there holds

$$J_{\tau^{[n]}} = (J_\tau)^{[n]} \qquad (n \in \mathbb{Z}_+) \quad .$$

Henceforth, we will write $J_\tau^{[n]}$ instead of $(J_\tau)^{[n]}$.

As can be seen easily, the assumptions

$$P(\Phi \text{ is simple}) = 1;\ P_\kappa \text{ exists} \ ;\ \kappa_{(a)}(\psi \text{ is simple}) = 1 \text{ for all } a \in A$$

do not yet imply that $P_\kappa(\psi$ is simple$) = 1$. But in virtue of [MKM; Prop. 1.5.5.] one obtains the sufficient criterion:

1.5.6. Let κ be a clustering field on $[A, \rho_A]$ with phase space $[A', \rho_{A'}]$ and P be a distribution on $\mathcal{M}$ such that P_κ exists. Assume that, for all $a \in A$

$$\kappa_{(a)}(\psi \text{ is simple}) = 1 \quad ; \quad J_\kappa(a, \{b\}) = 0 \quad (b \in A) \quad .$$

Then $P_\kappa(\psi$ is simple$) = 1$.

Obviously J_κ is a stochastic resp. substochastic kernel if κ is a stochastic resp. substochastic clustering field (but not conversely!) A substochastic clustering field κ may be recovered from J_κ as follows:

1.5.7. [MKM; Prop. 4.4.1.] With any substochastic kernel K from $[A, \mathcal{A}]$ to $[A', \mathcal{A}']$, there may be associated a substochastic clustering field $\tau(K)$ on $[A, A]$ with phase space $[A', \mathcal{A}']$ via

$$(\tau(K))_{(a)} := (1 - K(a,A))\delta_o + \varrho_{K(a,(.))} \quad (a \in A) \quad .$$

The mapping $K \to \tau(K)$ provides a 1-1 correspondence between the set of all substochastic clustering fields on $[A, \rho_A]$ with phase space $[A', \rho_{A'}]$ and the set of all substochastic kernels from $[A, \mathcal{A}]$ to $[A', \mathcal{A}']$; its inverse mapping is given by $\kappa \to J_\kappa$.

(Note that "dying" of a particle in $a \in A$ is described by the transition into the distribution δ_o, which is concentrated on the measure o corresponding to the void population.)

The "clustering theorem" 1.4.6. yields, together with 1.5.2. and 1.5.3.:

1.5.8. For a substochastic kernel K from $[A, \mathcal{A}]$ to $[A', \mathcal{A}']$ and a measure $\lambda \in N$, the distribution $(\Pi_\lambda)_{\tau(K)}$ exists iff $\lambda * K$ belongs to N'; in this case there holds

$$(\Pi_\lambda)_{\tau(K)} = \Pi_{\lambda * K} \ .$$

In the following we need the lemma

1.5.9. a) Let κ be a clustering field on $[A, \rho_A]$ with phase space $[A', \rho_{A'}]$. Then there holds for all $h \in F$

$$-\ln \mathcal{L}_{\kappa_{(.)}}(h) \le J_\kappa * h \ .$$

b) If, in addition, λ is a measure on $\mathcal{A}$, one has

$$(-\ln \mathcal{L}_{\kappa_{(.)}}(f)) \in F_\lambda \quad \text{for all } f \in F'_{\lambda * J_\kappa} \ .$$

Proof. By Jensen's inequality there follows

$$\begin{aligned} -\ln \mathcal{L}_{\kappa_{(.)}}(h) &= -\ln \int \exp(-\langle \Phi, h\rangle)\kappa_{(.)}(d\Phi) \\ &\le \int \langle \Phi, h\rangle \kappa_{(.)}(d\Phi) = J_\kappa * h \ , \end{aligned}$$

which proves a). Concerning b), one observes that, due to a),

$$\langle \lambda, -\ln \mathcal{L}_{\kappa_{(.)}}(f)\rangle \le \langle \lambda, J_\kappa * f\rangle = \langle \lambda * J_\kappa, f\rangle < +\infty \quad . \blacksquare$$

The following monotonicity relation will be of considerable importance:

1.5.10. Let γ be a measure in N and κ be a clustering field on $[A, \rho_A]$ with phase space $[A', \rho_{A'}]$ such that $\gamma * J_\kappa$ belongs to N'. Then there holds for all $f \in F$

$$\mathcal{L}_{\Pi_{\gamma * J_\kappa}}(f) \le \mathcal{L}_{(\Pi_\gamma)_\kappa}(f) \ .$$

Proof. Taking into consideration that for each at most countable family $(c_i)_{i \in I}$ of numbers in $[0,1]$ we have

$$1 - \prod_{i \in I}(1 - c_i) \le \sum_{i \in I} c_i \ ,$$

we get, using 1.4.8. in the last equality:

$$\begin{aligned} \mathcal{L}_{\Pi_{\gamma * J_\kappa}}(f) &= \exp\left(-\int (1 - \exp(-f(a)))(\gamma * J_\kappa)(da)\right) \\ &= \exp\left(-\iiint (1 - \exp(-f(y)))\Phi(dy)\kappa_{(x)}(d\Phi)\gamma(dx)\right) \\ &\le \exp\left(-\iint 1 - \prod_{y \in \operatorname{supp}\Phi} (1 - (1 - \exp(-f(y))))^{\Phi(\{y\})}\kappa_{(x)}(d\Phi)\gamma(dx)\right) \\ &= \exp\left(-\int (1 - \mathcal{L}_{\kappa_{(a)}}(f))\gamma(da)\right) = \mathcal{L}_{(\Pi_\gamma)_\kappa}(f) \quad . \blacksquare \end{aligned}$$

The next proposition, which results immediately from 1.3.4., 1.3.7., 1.4.7., 1.5.2., 1.5.3. and 1.5.9., states that clustering is weakly continuous on a set of distributions whose intensity measures are dominated in a certain sense:

1.5.11. Let κ be a clustering field on $[A, \rho_A]$ with phase space $[A', \rho_{A'}]$ and λ be a measure on $\mathcal{N}$. If in **P** the weak convergence $P_n \Rightarrow P$ takes place, and if there holds

$$\Lambda_{P_n} \leq \lambda \quad \text{for } n=1,2,\ldots; \quad \lambda * J_\kappa \in N' ,$$

then there exist P_κ, $(P_1)_\kappa$, $(P_2)_\kappa$, ..., and there holds

$$\mathcal{L}_{(P_n)_\kappa}(f) \to \mathcal{L}_{P_\kappa}(f) \quad (f \in F'_{\lambda * J_\kappa})$$

and, a fortiori,

$$(P_n)_\kappa \Rightarrow P_\kappa .$$

Towards the end of this section we turn, once again, to general statements on kernels.

Let $[E_i, \tau_{E_i}]$, $i = 1,2,3$, be some Polish spaces, and E_i be the corresponding σ-algebras of Borel sets.

1.5.12. Let ν be a σ-finite measure on E_1 and K be a kernel from $[E_1, E_1]$ to $[E_2, E_2]$ such that also $\gamma := \nu * K$ is a σ-finite measure. Then there exists a stochastic kernel D from $[E_2, E_2]$ to $[E_1, E_1]$ such that for all $E_1 \otimes E_2$-measurable mappings f from $E_1 \times E_2$ into $[0, +\infty]$ there holds

$$\iint f(e_1, e_2) K(e_1, de_2) \nu(de_1) = \iint f(e_1, e_2) D(e_2, de_1) \gamma(de_2) .$$

For two stochastic kernels D_1, D_2 with these properties there holds

$$D_1(e_2, (.)) = D_2(e_2, (.)) \quad \gamma\text{-almost all } e_2 .$$

A kernel D with the properties of 1.5.12. is called a **dual stochastic kernel of K with respect to** ν. Note that D is a "regular desintegration" of the σ-finite measure C on $E_1 \otimes E_2$ defined by

$$C(B_1 \times B_2) = \int 1_{B_1}(e_1) K(e_1, B_2) \nu(de_1) \qquad (B_1 \in E_1 ; B_2 \in E_2)$$

with respect to its second marginal measure γ, i. e. $D(e_2, B_1)$ $(e_2 \in E_2 ; B_1 \in E_1)$ is a "stochastic kernel version" of the Radon-Nikodym derivatives $(dC(B_1 \times (.))/d\gamma)(e_2)$, and it may be constructed like in the standard constructions of regular conditional probabilities (see, e. g. [Parthasarathy (1967), page 146].)

Now let g be some E_2-measurable mapping from E_2 into $[0, +\infty]$. Putting

$$\sigma := \int_X g(e_2) \gamma(de_2) ,$$

we obtain for all $X \in E_1$:

$$(\sigma * D)(X) = \int 1_X(e_1) (\sigma * D)(de_1) = \int (D * 1_X)(e_2) \, g(e_2) \, \gamma(de_2)$$

$$= \iint 1_X(e_1) \, g(e_2) \, K(e_1, de_2) \nu(de_1) = \int_X (K * g)(e_1) \nu(de_1) .$$

This yields:

1.5.13. Under the assumptions of 1.5.12., for any E_2-measurable mapping g from E_2 into $[0, +\infty]$ the measure $\int_{(.)} g(e_2)\gamma(de_2)) * D$ has density $K*g$ with respect to ν.

A direct computation leads to

1.5.14. Let ν be a σ-finite measure on E_1. Further let K_1 resp. K_2 be a kernel from $[E_1, E_1]$ to $[E_2, E_2]$ resp. from $[E_2, E_2]$ to $[E_3, E_3]$ such that $\gamma := \nu * K_1$ as well as $\gamma * K_2$ are σ-finite. If D_1 is a dual stochastic kernel of K_1 with respect to ν and D_2 is a dual stochastic kernel of K_2 with respect to γ, then $D_2 * D_1$ is a dual stochastic kernel of $K_1 * K_2$ with respect to ν.

A fortiori, this implies

1.5.15. Let ν be a σ-finite measure on E_1 and K be a kernel from $[E_1, E_1]$ to $[E_1, E_1]$ such that $\nu = \nu * K$. If D is a dual stochastic kernel of K with respect to ν, then, for all $n \in \mathbb{Z}_+$, $D^{[n]}$ is a dual stochastic kernel of $K^{[n]}$ with respect to ν.

1.6. Family Structures

Assume that a (possibly void) population $\Phi \in M$ is partitioned into non-void subpopulations or "families" $\Phi_i \in M$ via $\Phi = \sum_{i \in I} \Phi_i$. Then, for each $X \in \mathcal{B}$, only finitely many of the Φ_i's have individuals in X. On the other hand, any collection $(\Phi_i)_{i \in I}$ with this property builds up a population Φ via $\Phi = \sum_{i \in I} \Phi_i$. This indicates that the "family structures" in the non-void populations on A may be described as counting measures on $M \setminus \{o\}$ equipped with a suitable metric.

We put $\mathbb{A} := M \setminus \{o\}$ and $\mathbb{X} := \{\Phi \in \mathbb{A} : \Phi(X) > 0\}$ $(X \in \mathcal{A})$.

1.6.1. [MKM, Prop. 3.3.2.] There exists a complete separable metric $\rho_{\mathbb{A}}$ in $\mathbb{A}$ which generates the restriction of the topology τ_M to $\mathbb{A}$ and has the property that a subset of $\mathbb{A}$ is $\rho_{\mathbb{A}}$-bounded iff it is contained in $\mathbb{X}$ for a suitable $X \in \mathcal{B}$.

In the following, such a metric will be kept fixed. All objects referring to the new phase space $[\mathbb{A}, \rho_{\mathbb{A}}]$ will be indicated by contoured letters.

From the definition of $\mathbb{N}$ and W there follows immediately:

1.6.2. The mapping $W \to W((.) \cap \mathbb{A})$ establishes a 1-1correspondence between W and $\mathbb{N}$.

In the sequel, any measure $W \in \mathbf{W}$ will be identified with the corresponding measure in $\mathbf{N}$. Thus, the vague topology τ_N is carried into a topology τ_W, and the corresponding convergence is denoted by $\xrightarrow{W}$.

1.6.3. [MKM, Prop. 3.3.6.] The bijective mapping $P \rightarrow \tilde{P}$ from $\mathbf{T}$ onto $\mathbf{W}$ is continuous in both directions with respect to τ_P and τ_W.

As a counterpart to 1.1.5. and 1.1.8., respectively, one obtains

1.6.4. [MKM, Prop. 3.3.5.] In $\mathbf{W}$ there holds $W_n \xrightarrow{W} W$ iff for all finite sequences $X_1, \ldots, X_m$ of sets in $\mathcal{B}_W$ the convergence

$$W_n([\Psi(X_1), \ldots, \Psi(X_m)] \in (.), \Psi(X_1 \cup \ldots \cup X_m) > 0)$$
$$\Rightarrow W([\Psi(X_1), \ldots, \Psi(X_m)] \in (.), \Psi(X_1 \cup \ldots \cup X_m) > 0)$$

takes place.

1.6.5. [MKM, Prop. 10.1.1.] Let $W, W_1, W_2, \ldots$ be measures in $\mathbf{W}$ such that $W_n \xrightarrow{W} W$. Then vague convergence $\Lambda_{W_n} \xrightarrow{N} \Lambda_W$ holds iff for all $X \in \mathcal{B}$ the condition

$$\sup_{n=1,2,\ldots} \int_{\{\Psi: \Psi(X) > k\}} \Psi(X) W_n(d\Psi) \underset{k \to \infty}{\longrightarrow} 0$$

is fulfilled.

By definition, any counting measure $\Phi \in \mathbf{M}$ may be written as

(*) $\qquad \Phi = \sum_{i \in I} \delta_{\Psi_i}$, where I is at most countable and $\Psi_i \in A$ for all $i \in I$.

Thus, the counting measures $\Phi \in \mathbf{M}$ may be interpreted as (possibly void) systems of non-void families of individuals in $[A, \rho_A]$. The finiteness condition

(**) $\qquad \Omega(X) < +\infty$,

which is characteristic for measures $\Omega \in \mathbf{N}$, means in the special case of measures Φ of the form (*) that in any $X \in \mathcal{B}$ there may be situated only individuals of finitely many of the families Ψ contained in Φ.

For all measures Φ of the form (*) we put

$$sp\Phi := \sum_{i \in I} \Psi_i$$

The finiteness condition (**) then takes the simple form

$$sp\Phi \in \mathbf{M} .$$

Passing from Φ to $sp\Phi$ means abandonning the family structure. Obviously $sp\Phi$ is zero measure $o \in \mathbf{M}$ iff the system Φ is "void", i. e. coincides with zero measure $o \in \mathbf{M}$.

1.6.6. sp is a τ_M, τ_M-continuous mapping from M onto M.

Proof. Let (Φ_n) be a sequence in M converging vaguely towards $\Phi \in M$. We have to show that $sp\Phi_n \to sp\Phi$, i. e.

$$\int f(a)(sp\Phi_n)(da) \xrightarrow[n\to\infty]{} \int f(a)(sp\Phi)(da) \quad (f \in F_c) \; .$$

Now take $f \in F_c$ and choose $X \in \mathcal{B}_{sp\Phi}$ such that supp $f \subseteq X$. Noting that $\partial(X) \subseteq \{\Phi \in A : \Phi(\partial X) > 0\}$, it is easily checked that $\Phi(\partial(X)) = 0$ in virtue of $sp\Phi(\partial X) = 0$. Hence, the assumed convergence $\Phi_n \xrightarrow{M} \Phi$ leads, by 1.1.2., to ${}_X\Phi_n \Rightarrow {}_X\Phi$. This, by 1.3.4., means that there exist, for all suitable large natural numbers n, representations ${}_X\Phi_n = \sum_{1\le i\le k} \delta_{\Phi_{n,i}}$ such that $\Phi_{n,i} \xrightarrow{M} \Phi_i$ for $1\le i\le k$, and ${}_X\Phi = \sum_{1\le i\le k} \delta_{\Phi_i}$.

Thus we obtain, for suitably large n,

$$\int f(a)\,(sp\Phi_n)(dx) \;=\; \sum_{1\le i\le k} \int f(a)\Phi_{n,i}(da)$$

which for $n\to+\infty$ tends towards

$$\sum_{1\le i\le k} \int f(a)\Phi_i(da) \;=\; \int f(a)(sp\Phi)(da) \; . \blacksquare$$

In particular, sp is an $\mathcal{M}$, $\mathcal{M}$-measurable mapping, and we may form, for an arbitrary measure H on $\mathcal{M}$, the measure

$$spH := H(sp\Phi \in (.))$$

on $\mathcal{M}$. 1.6.6. implies immediately:

1.6.7. The mapping $G \to spG$ from $\mathbb{P}$ onto P is $\tau_{\mathbb{P}}$, τ_P-continuous.

There further holds:

1.6.8. A subset Y of M is τ_M-relatively compact iff sp(Y) is τ_M-relatively compact.

Proof: The "only if" direction is clear from 1.6.6. Concerning the "if" direction, we note that, in view of 1.1.3., we have to show that, for any $\rho_{\mathbf{A}}$-bounded and closed subset Z of A, there holds

$$(*) \qquad \sup_{\Phi\in Y} \Phi(Z) < +\infty$$

and

(**) there exists a τ_A-compact subset C of Z such that $\Phi(Z\setminus C) = 0 \quad (\Phi \in Y)$.

Indeed, the assumed τ_M-relative compactness of sp(Y) implies, again by 1.1.3., that for any ρ_A-bounded, closed subset X of A there holds

$$(+) \qquad \sup_{\Phi\in Y} sp\Phi(X) < +\infty$$

and

(++) there exists a τ_A-compact subset B of X such that $sp\Phi(X\setminus B) = 0 \quad (\Phi \in Y)$.

Now take a ρ_A-bounded, closed subset X of A such that $Z \subseteq X$. Then, in view of

$$\Phi(Z) \leq \Phi(X) \leq sp\Phi(X) \quad (\Phi \in Y),$$

(+) implies (*).

Next, take an ascending sequence (X_m) in $\mathcal{B}$ such that any bounded subset of A is covered by some X_m. Put. for each m, $r_m := \sup_{\Phi \in Y} sp\Phi(X_m)$ (which is finite by (+)); and choose, in virtue of (++), a compact subset B_m of A such that $sp\Phi(X_m \setminus B_m) = 0$ for all $\Phi \in Y$. Then, again by 1.1.3.,

$$D := \{\Phi \in M : \Phi(X_m \setminus B_m) = 0 \,;\, \Phi(X_m) \leq r_m \,;\, m=1,2,...\}$$

is a τ_M-relatively compact subset of M, hence

$$C := Z \cap \tau_M\text{-closure of } D$$

is a τ_A-compact subset of A.
For all $\Phi \in Y$ the counting measure $sp\Phi$ belongs to the descending set D, which implies that $\Phi(A \setminus D) = 0$ and finally leads to $\Phi(Z \setminus C) = 0$, thus yielding (**). ■

1.6.7., 1.6.8. and a straightforward application of Prokhorov's theorem [Billingsley (1968), Thm. 6.1.] show:

1.6.9. A subset L of $\mathbb{P}$ is $\tau_{\mathbb{P}}$-relatively compact iff sp(L) is τ_P-relatively compact.

Putting

$$\alpha_{(\Phi)} := \delta_\Phi \qquad (\Phi \in A),$$

we obtain a "deterministic" clustering field on $[A, \rho_A]$ with phase space $[A, \rho_A]$, for which there holds

1.6.10. $\alpha_{(\Phi)} = \delta_{sp\Phi} \quad (\Phi \in \mathbb{M})$.

1.6.11. For all measures H on $\mathcal{M}$, one has

$$spH = H_\alpha .$$

The infinitely divisible distributions on $\mathcal{M}$ may be derived in a canonical way from the Poisson distributions on $\mathcal{M}$ by the **standard clustering representation** described below:

1.6.12. The mapping $P \to \Pi_{\tilde{P}}$ is a bijective and with respect to $\tau_P, \tau_{\mathbb{P}}$ in both directions continuous mapping from T onto the set of all Poisson distributions on $\mathcal{M}$. Its inverse mapping is given by $G \to spG$.

Proof. 1. In view of the identification of W and N and in virtue of 1.2.9. and 1.2.13., the mapping $P \to \Pi_{\tilde{P}}$ is bijective from T onto the set of all Poisson distributions on $\mathcal{M}$. 1.6.3. and 1.2.13. together imply that this mapping is in both directions continuous with respect to $\tau_P, \tau_{\mathbb{P}}$.

2. For all $W \in \mathbf{W}$ we obtain from 1.6.11. that

$$P := \mathrm{sp}\Pi_W = (\Pi_W)_\alpha \quad .$$

Due to the "clustering theorem" 1.4.6., P belongs to **T**, and there holds

$$\tilde{P} = (Q_W)_\alpha((.)\backslash\{o\}) = \int \delta_\Psi(.)W(d\Psi) = W \quad . \blacksquare$$

Identifying quite generally the measures on $\mathcal{A}$ with the measures on $\mathcal{M}$ having no mass in o, one obtains by a simple computation:

1.6.13. For all measures H on $\mathcal{M}$, there holds $\Lambda_{\Lambda_H} = \Lambda_{\mathrm{sp}H}$.

For all $X \in \mathcal{B}$ there holds

$$\Lambda_H(X) \leq \Lambda_{\mathrm{sp}H}(X) \quad .$$

Hence H is certainly of first order if $\mathrm{sp}H$ has this property.

For any distribution G on $\mathcal{M}$ and any $X \in \mathcal{B}$, G_X has a simple interpretation: it is the distribution of the random number of families Φ in the system Φ distributed according to G, which have points in X. The corresponding expectation value is $\Lambda_G(X)$.

With any clustering field κ on $[A, \rho_A]$, we associate a substochastic kernel K_κ from $[\mathbb{A}, \mathcal{A}]$ to $[\mathbb{A}, \mathcal{A}]$ by putting

$$K_\kappa(\Phi,(.)) = \begin{cases} \kappa_{(\Phi)}((.)\backslash\{o\}) & \text{if } \Phi \in {}_\kappa M\backslash\{o\} \\ o & \text{if } \Phi \in \mathbb{A}\backslash{}_\kappa M \end{cases} \quad .$$

The kernel K_κ describes the stochastic development of families Φ according to the branching dynamics induced by κ. The possible "dying out" of the family Φ is modelled by a genuine substochastic behavior of K_κ.

We now put

$$\boldsymbol{\kappa} := \tau(K_\kappa) \quad ,$$

thus obtaining a substochastic clustering field on $[\mathbb{A}, \rho_{\mathbb{A}}]$ which induces a stochastic dynamics $\Phi \to \boldsymbol{\kappa}_{(\Phi)}$ (called **family dynamics**) on the space $\mathbb{M}$ of family systems Φ.

For a measure G on $\mathcal{M}$, the "catastrophic case" $\Phi \in \mathbb{A}\backslash{}_\kappa M$ is of no relevance in forming G_κ if G has the property $\Lambda_G(\Phi \notin {}_\kappa M) = 0$. A stronger but sometimes more tractable condition is $(\mathrm{sp}G)(\Phi \notin {}_\kappa M)=0.$. Particulary, there follows

1.6.14. Let G be a distribution on $\mathcal{M}$ such that $(\mathrm{sp}G)_\kappa$ exists. Then there hold

$$\mathrm{sp}(G_\kappa) = (\mathrm{sp}G)_\kappa \;\; ; \;\; \Lambda_{G_\kappa} = (\Lambda_G)_\kappa((.)\backslash\{o\}) \, .$$

1.7. Campbell Measures

Let ρ_M be some bounded, separable, complete metric on M which generates the vague topology τ_M. $[A^\wedge, \rho_{A^\wedge}]$ will denote the direct product of $[A, \rho_A]$ and $[M, \rho_M]$. All objects referring to the new phase space $[A^\wedge, \rho_{A^\wedge}]$ will be indicated by the upper index ^.

1.7.1. [MKM, Prop. 10.1.2.] The mapping

$$\Phi \rightarrow \Phi^\wedge := \Phi \otimes \delta_\Phi \quad (\Phi \in M)$$

from M into M^ is $\tau_M, \tau_{M^\wedge}$-continuous.

The passage from Φ to $\Phi^\wedge = \sum_{a \in \mathrm{supp}\Phi} \Phi(\{a\})\delta_{[a,\Phi]}$ may be interpreted in attaching to each individual in the family Φ the "mark" Φ. In this way, obviously only ordered pairs $[a, \Phi] \in A^\wedge$ with the property $a \in \mathrm{supp}\Phi$ are obtained.

The mapping $\Phi \rightarrow \Phi^\wedge$ is $\mathcal{M}, \mathcal{M}^\wedge$-measurable. Thus, for any measure H on $\mathcal{M}$, via

$$H^\wedge := H(\Phi^\wedge \in (.))$$

a measure H^ on $\mathcal{M}^\wedge$ may be defined. Obviously, H can be reconstructed from H^.

For all measures H on $\mathcal{M}$ we put

$$C_H := \Lambda_{H^\wedge} \; ,$$

calling the measure C_H on $\mathcal{A}^\wedge = \mathcal{A} \otimes \mathcal{M}$ the **Campbell measure** of H. Immediately from this definition one gets

1.7.2. For any at most countable family $(H_i)_{i \in I}$ of measures on $\mathcal{M}$ and all $r_i \in \mathbf{R}_+$, $i \in I$, there holds

$$C_{\sum_{i \in I} r_i H_i} = \sum_{i \in I} r_i C_{H_i} .$$

1.7.3. For all $\mathcal{A} \otimes \mathcal{M}$-measurable mappings h from A×M into $[0, +\infty]$ there holds

$$\int h(a,\Phi) C_H(d[a,\Phi]) = \iint h(a,\Phi)\Phi(da)H(d\Phi) \; .$$

The Campbell measure C_H may be conceived as a refinement of the intensity measure Λ_H, due to the redundancies coming into the definition of H^.

1.7.4. [MKM; Prop. 5.1.10.] For two measures H_1, H_2 there holds $C_{H_1} = C_{H_2}$ iff

$$H_1((.)\backslash\{o\}) = H_2((.)\backslash\{o\}) \; .$$

1.7.5. [MKM, Prop. 5.1.11.] The Campbell measure of a measure H on $\mathcal{M}$ is σ-finite iff $H((.)\backslash\{o\})$ has this property.

If (X_n) is some ascending sequence of sets in $\mathcal{B}$ whose union is A, then (X_n) is an ascending sequence of sets in $\mathcal{M}$ with union $M\setminus\{o\}$. Bearing in mind that for all $W \in \mathbf{W}$ there holds

$$W(X_n) < +\infty \text{ for } n=1,2,\ldots; \quad W(\{o\}) = 0 \ ,$$

one obtains from 1.7.5.

1.7.6. For all $W \in \mathbf{W}$, C_W is a σ-finite measure.

For any σ-finite measure S on $\mathcal{A}^\wedge$ and any σ-finite measure H on $\mathcal{M}$ we put

$$S \circ H := (S \otimes H)([a, \chi_1 + \chi_2] \in (.)) \ ,$$

thus obtaining a measure $S \circ H$ on $\mathcal{A}^\wedge$.

1.7.7. [MKM, Prop. 5.2.3.] If the convolution P of an at most countable family $(P_i)_{i \in I}$ of distributions on $\mathcal{M}$ exists, then

$$C_P = \sum_{i \in I} C_{P_i} \circ \mathop{*}_{j \in I - \{i\}} P_j \ .$$

1.7.8. [MKM, Prop. 5.2.5., 5.2.6.] If S is some σ-finite measure on $\mathcal{A}^\wedge$, then the functional equation

$$C_P = S \circ P$$

has a solution in **P** iff there exists a distribution $T \in \mathbf{T}$ such that $S = C_{\tilde{T}}$. In this case, the above functional equation has exactly one solution, that is, T itself.

In particular there holds, for all $T \in \mathbf{T}$,

$$C_T = C_{\tilde{T}} \circ T \ .$$

Obviously, C_H belongs to $N^\wedge$ iff the measure H is of first order, i. e. $H((.)\setminus\{o\})$ is a measure of first order in **W**.

Among the measures $S \in N^\wedge$, those of the form C_H are a very special subclass which may be characterized by certain symmetry properties (or, equivalently, by the solvability of the equation $C_P = S \circ P$ in **P**).

The **generalized Palm-Khinchin-theorem** says:

1.7.9. [MKM, Prop. 10.1.5., 10.1.6.] Let $W, W_1, W_2, \ldots$ be measures of first order in **W**. Then the convergence $C_{W_n} \xrightarrow{N^\wedge} C_W$ holds iff both convergences $W_n \xrightarrow{\mathbf{W}} W$ and $\Lambda_{W_n} \xrightarrow{N} \Lambda_W$ take place.

Further there holds

1.7.10. [MKM, Prop. 10.1.7.] The set $\{C_W\}_{W \in \mathbf{W}, \Lambda_W \in N}$ is $\tau_{N^\wedge}$-closed.

For all $\nu \in N$ we obtain

$$C_{\widetilde{\Pi_\nu}} = C_{\mathcal{Q}_\nu} = \int \delta_{[a,\delta_a]}(.)\nu(da)$$

and hence according to 1.7.8.

$$C_{\Pi_\nu} = (\int \delta_{[a,\delta_a]}(.)\nu(da)) \circ \Pi_\nu \ .$$

It turns out that here as well as in certain other topics, e. g. in the general theory of Gibbs distributions (cf.[K, ch. 13]) a slight modification of the Campbell measure C_H leads to significant simplifications. Like in the definition of the factorial moment measures, this construction relies on the assumption that H is a measure on $\mathcal{M}$ and not, more generally, on $\mathcal{N}$.

We put

$$\Phi^{\bullet} := \sum_{a \in \operatorname{supp}\Phi} \Phi(\{a\})\delta_{[a,\Phi-\delta_a]} \ ,$$

thus suppressing some redundant information which appears in the definition of $\Phi^\wedge$, namely that each $[a, \Psi] \in \operatorname{supp}\Phi^\wedge$ obeys the condition $\Psi(\{a\})>0$. On the analogy of 1.7.1. there follows the τ_M, τ_M-continuity of the mapping $\Phi \to \Phi^{\bullet}$, and we may, for any measure H on $\mathcal{M}$, introduce the measure

$$H^{\bullet} := H(\Phi^{\bullet} \in (.))$$

on $\mathcal{M}^\wedge$. We call

$$C^!_H := \Lambda_{H^{\bullet}}$$

the **reduced Campbell measure** of H.

For all $\nu \in N$ there follows

$$C^!_{\mathcal{Q}_\nu} = \int \delta_{[a,o]}(.)\nu(da) = \nu \otimes \delta_o \ .$$

Hence we obtain by means of 1.7.8.:

1.7.11. For all $\nu \in N$, a distribution P on $\mathcal{M}$ coincides with Π_ν iff it obeys the functional equation

$$C^!_P = \nu \otimes P \ .$$

This latter equation admits several modifications characterizing other important classes of distributions; for one of these see [MKM, Theorem 5.4.2.].

In terms of the reduced Campbell measure, 1.7.2. is now replaced by

1.7.12. For all $\mathcal{A}\otimes\mathcal{M}$-measurable mappings h from $A \times M$ into $[0, +\infty]$ there holds

$$\int h(a,\Phi+\delta_a) C^!_H(d[a,\Phi]) = \iint h(a,\Phi)\Phi(da)H(d\Phi) \ .$$

Assertions 1.7.9. and 1.7.10., too, admit reformulations in terms of the reduced Campbell measure.

1.8. Palm Distributions

If H is a measure on $\mathcal{M}$, then, for all $X\in\mathcal{B}$ and $Y\in\mathcal{M}$, the value $C_H(X\times Y)$ may be interpreted as mean value of the number of pairs $[a,\Phi]$ with $a\in X$ and $\Phi\in Y$, where the population Φ is distributed according to H, and a is an individual in Φ. Assuming that H is of first order, we may form a regular desintegration of the measure C_H with respect to its first marginal measure Λ_H, which, in the terminology of 1.5., is a dual stochastic kernel $[a,Y]\to H_a(Y)$ of the kernel $[\Phi,X]\to\Phi(X)$ with respect to the measure H. Again let us note that $H_a(Y)$ is a "stochastic kernel version" of the Radon-Nikodym derivatives $(dC_H((.)\times Y)/d\Lambda_H)(a)$ $(a\in A,\ Y\in\mathcal{M})$. Noting that for all $B\in\mathcal{B}$ there holds $(dC_H((.)\times\{\Phi(B)>0\})/\ d\Lambda_H)(a) = 1$ for Λ_H-almost all $a\in B$, the mapping $a\to H_a(.)$ may still be modified in such a way that

$$H_a(\Phi(\{a\})>0)=1 \quad (a\in A)\ .$$

In summary, we get

1.8.1. For any measure H on $\mathcal{M}$ of first order there exists an $\mathcal{A}$, $\boldsymbol{P}$-measurable mapping $a\to H_a$ from A into **P** with the property

$$H_a(\Phi(\{a\})>0)=1 \quad (a\in A)$$

and such that for all $\mathcal{A}\times\mathcal{M}$-measurable mappings h from $A\times M$ into $[0, +\infty]$ there holds

$$\int h(a,\Phi)C_H\, d[a,\Phi] = \iint h(a,\Phi)\, H_a(d\Phi)\Lambda_H(da)\ .$$

Two mappings of this kind coincide Λ_H-almost everywhere.

A family $(H_a)_{a\in A}$ with the properties of 1.8.1. is called a **family of Palm distributions** of H. In the special case of a distribution P of first order with the property

$$P(\Phi \text{ is simple}) = 1\ ,$$

an interpretation of $P_a(d\Phi)\,\Lambda_P(da)$ as

$$P(\, d\Phi \mid \Phi(\{a\})=1)\cdot P(\Phi(da)=1)$$

may be justified (cf. [K, section 1.2.4.]).

A well known "differentation theorem" yields that for any sequence $(\mathcal{B}_n)$ of nested partitions of A, where each $\mathcal{B}_n$ consists of countably many bounded sets and $\bigcup_{n\in Z_+}\mathcal{B}_n$ generates $\mathcal{A}$, and for any $Y\in\mathcal{M}$ there holds

$$(dC_P((.)\times Y)/\ d\Lambda_P(.))(a) = \lim_{n\to\infty} C_P(B_n(a)\times Y)/\Lambda_P(B_n(a)) \qquad \Lambda_P\text{-almost all } a\in A\ ,$$

where $B_n(a)$ is the set in $\mathcal{B}_n$ containing the point a.

Using this"constructive approach" in the definition of the mapping $[P,a]\to P_a(.)$, one arrives at

1.8.2. The exists a $\boldsymbol{P}\otimes\mathcal{A}$, $\boldsymbol{P}$-measurable mapping $[P,a]\to P_a$ from $\{P\in\mathbf{P} : \Lambda_P\in N\}\times A$ into **P** such that, for any fixed $P\in\mathbf{P}$ of first order, $(P_a)_{a\in A}$ is a family of Palm distributions of P.

If furtheron there occur Palm distributions of variable distributions P, we will tacitly use a version of $[P,a]\to P_a$ with the above mentioned measurability property.

As a consequence of 1.7.8. one gets

1.8.3. For all distributions P of first order in **T** there holds Λ_P-almost everywhere

$$P_a = \widetilde{P}_a * P \ .$$

Next we show the validity of

1.8.4. For all measures H on $\mathcal{M}$ of first order and for any $\mathcal{A}\otimes\mathcal{A}$-measurable mapping h from A×A into $[0, +\infty]$ there holds

$$\int h(a,b)\Lambda^{(2)}{}_H(d[a,b]) = \iint h(a,b)\Lambda_{H_a}(db)\,\Lambda_H(da)$$

Proof. Obviously $\Lambda_{H_{(.)}}(X)$ is $\mathcal{A}$-measurable for all $X\in\mathcal{A}$, i. e. $[a, X]\to\Lambda_{H_a}(X)$ is a kernel from $[A, \mathcal{A}]$ to $[A, \mathcal{A}]$. We obtain

$$\int h(a,b)\Lambda^{(2)}{}_H(d[a,b]) \ = \ \iiint h(a,b)\Phi(db)\Phi(da)H(d\Phi)$$

$$= \ \iiint h(a,b)\Phi(db)H_a(d\Phi)\,\Lambda_H(da) \ = \ \iint h(a,b)\Lambda_{H_a}(db)\Lambda_H(da) \ . \blacksquare$$

For all measures W of first order in **W** we put for abbreviation

$$W_{(f)} := \int f(a)W_a(.)\Lambda_W(da) \quad (f\in F_c) \ .$$

Note that $W_{(f)}$ does not depend on the special choice of the version $(W_a)_{a\in A}$ and has finite total mass $\langle\Lambda_W,f\rangle$.

The generalized Palm-Khinchin theorem 1.7.9. may now be reformulated as follows:

1.8.5. Let W, W_1, W_2,... be measures of first order in **W**. Then the convergences

$$(W_n)_{(f)} \Rightarrow W_{(f)} \qquad (f\in F_c)$$

hold iff both convergences

$$W_n \overset{\mathbf{W}}{\Rightarrow} W \quad \text{and} \quad \Lambda_W \xrightarrow{N} \Lambda_{W_n}$$

take place.

Proof. 1.Assume that $W_n \overset{\mathbf{W}}{\Rightarrow} W$ and $\Lambda_{W_n} \xrightarrow{N} \Lambda_W$. In virtue of 1.7.9. this implies $C_{W_n} \xrightarrow{N^\wedge} C_W$.

Now let f be a function in F_c and g be a mapping in $\mathbb{F}_c$. By putting $g(\{o\}) = 0$, g is extended τ_M-coutinuously to the whole of M. Thus we obtain by

$$h(a,\Phi) := f(a)g(\Phi) \quad ([a,\Phi]\in A^\wedge)$$

a mapping h in $F^\wedge{}_c$, and observe that

$$\int g(\Phi)(W_n)_{(f)}(d\Phi) = \int h(a,\Phi)C_{W_n}(d[a,\Phi])$$

$$\to \int h(a,\Phi)C_W(d[a,\Phi]) = \int g(\Phi)W_{(f)}(d\Phi) \quad , \quad \text{i. e. } (W_n)_{(f)} \Rightarrow (W)_{(f)} \ .$$

2. Let us assume, conversely, that

$$(W_n)_{(f)} \Rightarrow W_{(f)}$$

holds true for all $f \in F_c$. First, there follows for all $f \in F_c$

$$\langle \Lambda_{W_n}, f \rangle = (W_n)_{(f)}(M) \to W_{(f)}(M) = \langle \Lambda_W, f \rangle ,$$

which means $\Lambda_{W_n} \xrightarrow{N} \Lambda_W$.

To show the convergence $W_n \xrightarrow{W} W$, take an arbitrary $g \in \mathbb{F}_c$, and choose $X \in \mathcal{B}$ such that supp $g \subseteq X$, i. e. $g(\Phi)=0$ for all $\Phi \in \mathbb{A}$ such that $\Phi(X)=0$.

Now take some $f \in F_c$ with the property $f \geq 1_X$. The mapping h defined by

$$h(\Phi) := \begin{cases} g(\Phi)(\int f(a)\Phi(da))^{-1} & \text{if } \Phi(X) > 0 \\ 0 & \text{if } \Phi(X) = 0 \end{cases} \qquad (\Phi \in A)$$

is bounded and τ_A-continuous; hence follows

$$\int g(\Phi) W_n(d\Phi) = \iint h(\Phi)\, f(a)\, \Phi(da)\, W_n(d\Phi)$$

$$= \int h(\Phi)\, (W_n)_{(f)}(d\Phi) \to \int h(\Phi)\, W_{(f)}(d\Phi) = \int g(\Phi) W(d\Phi) ,$$

i. e. $W_n \overset{W}{\Rightarrow} W$. ■

For any measure H of first order $\mathcal{M}$ and any family $(H_a)_{a \in A}$ of Palm distributions of H, we obtain via

$$H^!_a := H_a((\Phi - \delta_a) \in (.)) \qquad (a \in A)$$

an $\mathcal{A}$, $\boldsymbol{P}$-measurable mapping $a \to H^!_a$ from A into **P**, calling $(H^!_a)$ a **family of reduced Palm distributions** of H.

In analogy to 1.8.1. there follows

1.8.6. An $\mathcal{A}$, $\boldsymbol{P}$-measurable mapping $a \to H_a$ is a family of reduced Palm distributions of a given measure H of first order on $\mathcal{M}$ iff for all $\mathcal{A} \otimes \mathcal{M}$-measurable mappings h from $A \times M$ into $[0, +\infty]$ there holds

$$\int h(a,\Phi) C^!_H(d[a,\Phi]) = \iint h(a,\Phi)\, H^!_a(d\Phi)\, \Lambda_H(da) .$$

For all $v \in N$ we may put

$$(Q_v)_a = \delta_{\delta_a} \; ; \quad (Q_v)^!_a = \delta_o \qquad (a \in A) .$$

Hence follows, in virtue of 1.8.3.,

1.8.7. For all $v \in N$,

$$(\Pi_v)^!_a := \Pi_v \qquad (a \in A)$$

defines a family of reduced Palm distributions of Π_v.

As a counterpart to 1.8.3. one obtains

1.8.8. For all measures H of first order on M and all $\mathcal{A}\otimes\mathcal{M}$-measurable mappings h from A×M into [0, +∞] there holds

$$\int h(a,b)\Gamma^{(2)}{}_H(d[a,b]) = \iint h(a,b)\Lambda_{H^!_a}(db)\,\Lambda_H(da) \quad .$$

Proof.

$$\int h(a,b)\Gamma^{(2)}{}_H(d[a,b]) = \iint h(a,b)\,\Phi^{(2)}(d[a,b])\,H(d\Phi)$$

$$= \iiint h(a,b)(\Phi-\delta_a)(db)\,\Phi(da)\,H(d\Phi) = \iint h(a,b)\,(\Phi-\delta_a)(db)\,C_H\,(d[a,\Phi])$$

$$= \iint h(a,b)\,\Psi(db)\,C^!{}_H(d[a,\Psi]) = \iiint h(a,b)\,\Psi(db)\,H^!_a(d\Psi)\,\Lambda_H(da)$$

$$= \iint h(a,b)\,\Lambda_{H^!_a}(db)\,\Lambda_H(da) \quad . \blacksquare$$

For all $\nu\in N$, all natural numbers k>1 and all $\mathcal{A}^{\otimes k}$-measurable mappings h from A^k into [0, +∞] we obtain

$$\int h(a_1,...a_k)\Gamma^{(k)}{}_{\Pi_\nu}(d[a_1,...a_k])$$

$$=\iint h(a_1,...a_k)\Phi^{(k)}\,(d[a_1,...a_k])\Pi_\nu(d\Phi)$$

$$=\iint h(a_1,...a_k)(\Phi-\delta_{a_k})^{(k-1)}\,(d[a_1,...a_{k-1}])\Phi(da_k)\,\Pi_\nu(d\Phi)$$

$$=\iint h(a_1,...a_k)\Phi^{(k-1)}(d[a_1,...a_{k-1}])(\Pi_\nu)^!{}_{a_k}(d\Phi)\,\nu(da_k)$$

$$=\iint h(a_1,...a_k)\Gamma^{(k-1)}{}_{\Pi_\nu}(d[a_1,...a_k])\nu(da_k) \quad .$$

By iteration, there follows

1.8.9. $\Gamma^{(k)}{}_{\Pi_\nu} = \nu^{\otimes k} \qquad (\nu\in N;\ k=1,2...)$.

By "mixing"(i. e. integrating w. r. to some measure G on $\mathcal{N}$) one obtains from 1.8.9.:

1.8.10. $\Gamma^{(k)}{}_{\mathfrak{I}(G)} = \Lambda^{(k)}{}_G \qquad (G\in \mathbf{G}_+\ ;\quad k=1,2,...)$.

1.9. The Backward Technique

The principal aim of this section is the investigation of Palm distributions of clustered measures:

1.9.1. Let τ be a clustering field on $[A, \rho_A]$ with phase space $[A', \rho_{A'}]$, and H be a measure of first order on $\mathcal{M}$ such that $\Lambda_H * J_\tau$ as well as all $J_\tau(a,(.))$, $a \in A$, belong to N'. Further, let D be a dual stochastic kernel of J_τ with respect to Λ_H. Then there holds, for $\Lambda_H * J_\tau$ almost all $b \in A'$,

$$(H_\tau)_b = \int (H^!_a)_\tau * (\tau_{(a)})_b)(.)\, D(b,da) \quad .$$

Before proving 1.9.1. we will interpret it at least for the case that H is a distribution on $\mathcal{M}$ with the property

$$H(\Phi \text{ is simple}) = 1 = H_\tau (\Phi \text{ is simple}) \quad .$$

By the initial distribution H and the stochastic branching dynamics $\Phi \to \tau_{(\Phi)}$, a random element $[\Phi_0, \Phi_1]$ in $[M \times M', \mathcal{M} \otimes \mathcal{M}']$ is generated. Assume that our model is fine enough that also "relationships" between individuals in Φ_0 and Φ_1 can be read off (this will be made precise towards the end of this section!), and denote the corresponding distribution by C.

Our aim is to determine $C(\Phi_1 \in (.) \mid \Phi_1(\{b\}) = 1)$. Denote the "father" of b in the initial population Φ_0 by a. The distribution of his location is given by the dual stochastic kernel D, namely D(b,(.)). Refining the condition, we now compute

$$C([\Phi_0, \Phi_1] \in (.) \mid \Phi_0(\{a\}) = 1 = \Phi_1(\{b\}); \text{a is father of b}) \quad .$$

This distribution has a simple structure. To begin with, $(\Phi_0 - \delta_a)$ is distributed according to $H^!_a$. The individuals in $\Phi_0 - \delta_a$ generate a random offspring Ψ, whose distribution is $(H^!_a)_\tau$. The offspring χ of the single individual a, however, is conditioned to include the individual b, hence is distributed according to $(\tau_{(a)})_b$. The distribution of the superposition $\chi + \Psi$, then, is the convolution $(H^!_a)_\tau * (\tau_{(a)})_b$. Finally, we cancel the additional condition "$\Phi_0(\{a\})=1$; a is father of b" by integrating over the distribution D(b,(.)) of the father's location and arrive at the formula in 1.9.1.

We now turn to a rigorous proof of 1.9.1., first reformulating it in terms of the Campbell measure of H_τ and in terms of the intensity kernel J_τ:

1.9.2. Under the assumptions of 1.9.1., there holds for any $\mathcal{A} \otimes \mathcal{M}'$-measurable mapping f from $A' \times M'$ into $[0, +\infty]$:

$$\iiint f(b,\Psi)((H^!_a)_\tau * (\tau_{(a)})_b)(d\Psi)\, J_\tau(a,db) \Lambda_H(da)$$
$$= \int f(b,\Psi) C_{H_\tau}(d[b,\Psi])$$
$$= \iiint f(b,\Psi)((H^!_a)_\tau * (\tau_{(a)})_b)(d\Psi) D(b,da)\, (\Lambda_H * J_\tau)(db) \quad .$$

Proof. 1. In virtue of 1.5.2. and 1.5.3., H_τ exists and has intensity measure $\Lambda_H * J_\tau$

2. Next we show that $(H^!_a)_\tau$ exists for Λ_H-almost all $a \in A$. For this purpose, let (X_n) be an ascending sequence of sets in $\mathcal{B}$ with union A. Putting $Y := M \setminus {}_\tau M$, we obtain

$$\int H_a(Y)\, \Lambda_H(da) = C_H(A \times Y)$$

$$= \lim_{n\to\infty} C_H(X_n \times Y) = \lim_{n\to\infty} \int_Y \Phi(X_n) H(d\Phi) = 0 \quad ,$$

noting that H(Y) = 0 due to 1. Hence follows for Λ_H-almost all $a \in A$:

$$(H^!_a)(\tau_{(\Phi)} \text{does not exist}) = H_a(\tau_{(\Phi)} \text{does not exist}) = 1 \; .$$

3. The $\mathcal{A}, P$-measurability of $a \to H^!_a$ yields, together with 1.4.3., the $\mathcal{A}, P'$-measurability of the (Λ_H-almost everywhere defined) mapping $a \to (H^!_a)_\tau$.

This together with 1.1.6. and 1.8.2. shows the $\mathcal{A} \otimes \mathcal{A}'$, P'-measurability of the mapping $[a,b] \to (H^!_a) * (\tau_{(a)})_b$.

4. For any $\mathcal{A}' \otimes \mathcal{M}'$- measurable mapping f from $A' \times M'$ into $[0, +\infty]$, we obtain, using 1.7.7., the following chain of equalities:

$$\int f(b,\Psi) C_{H_\tau}(d[b,\Psi]) \;=\; \iint f(b,\Psi) C_{\tau_{(\Phi)}}(d[b,\Psi]) H(d\Phi)$$

$$= \iint f(b,\Psi)\Big(\sum_{a \in \operatorname{supp}\Phi} \Phi(\{a\}) C_{\tau_{(a)}} \otimes \tau_{(\Phi - \delta_a)}\Big)(d[b,\Psi]) H(d\Phi)$$

$$= \iint f(b,\Psi)(C_{\tau_{(a)}} \otimes \tau_{(\chi)})(d[b,\Psi]) C^!_H(d[a,\chi])$$

$$= \iiint f(b,\Psi)(\tau_{(\chi)} * (\tau_{(a)})_b)(d\Psi) J_\tau(a,db) C^!_H(d[a,\chi])$$

$$= \iiiint f(b,\Psi)(\tau_{(\chi)} * (\tau_{(a)})_b)(d\Psi) J_\tau(a,db) H^!_a(d\chi) \Lambda_H(da)$$

$$= \iiint f(b,\Psi)((H^!_a)_\tau * (\tau_{(a)})_b)(d\Psi) J_\tau(a,db) \Lambda_H(da) \; ,$$

which proves the first part of our assertion.

5. The second part of the assertion follows directly from the definition of the dual stochastic kernel. ■

As an application of 1.9.1. we obtain:

1.9.3. Under the assumptions of 1.9.1. there holds:

$$\Gamma^{(2)}_{H_\tau} = \int \Gamma^{(2)}_{\tau_{(z)}}(.)\, \Lambda_H(dz) + \int J_\tau(x,(.)) \otimes J_\tau(y,(.))\, \Gamma^{(2)}_H(d[x,y]) \; .$$

Proof. According to 1.8.8. and 1.9.1. we have

$$\int h(u,v) \Gamma^{(2)}_{H_\tau}(d[u,v]) \;=\; \iint h(u,v) \Lambda_{(H_\tau)^!_u}(dv)\, \Lambda_{H_\tau}(du)$$

$$= \iiint h(u,v)(\Lambda_{(H^!_x)_\tau} + \Lambda_{(\tau_{(x)})^!_u})(dv) D(u,dx) \Lambda_{H_\tau}(du)$$

$$=\iiint h(u,v)(\Lambda_{H^!_x} * J_\tau + \Lambda_{(\tau_{(x)})^!_u})(dv)J_\tau(x,du)\,\Lambda_H(dx)$$

$$= \iiint h(u,v)J_\tau(y,dv)\,\Lambda_{H^!_x}(dy)\,J_\tau(x,du)\,\Lambda_H(dx) + \iint h(u,v)\,\Lambda_{(\tau_{(x)})^!_u}(dv)\,\Lambda_{\tau_{(x)}}(du)\,\Lambda_H(dx)$$

$$= \iiint h(u,v)J_\tau(y,dv)\,J_\tau(x,du)\Gamma^{(2)}{}_H(d[x,y]) + \iint h(u,v)\Gamma^{(2)}{}_{\tau_{(x)}}(d[u,v])\Lambda_H(dx)\ . \blacksquare$$

We are now going to apply 1.9.1. and 1.9.2. to a finer model which takes account of the relationships between individuals of two succeeding generations and thus makes the interpretation following 1.9.1. still more transparent:

We denote the direct product of $[A, \rho_A]$ and $[A', \rho_{A'}]$ by $[\underline{A}, \rho_{\underline{A}}]$, the interpretation of a pair $[a,b]$ being that a is father of b. To this interpretation, there corresponds the modified stochastic branching dynamics

$$\mathfrak{T}_{(a)} := \tau_{(a)}((\delta_a \otimes \chi) \in (.)) \quad (a \in A)\ ,$$

which has the intensity kernel

$$J_{\mathfrak{T}}(a,(.)) = \delta_a \otimes J_\tau(a,(.) \quad (a \in A)$$

Obviously, the assumptions of 1.9.1. remain valid also with the present modifications, and we obtain from 1.9.2., applied to $\mathfrak{T}$ instead of τ for all $\underline{\mathcal{A}} \otimes \underline{\mathcal{M}}$- measurable mapppings f from $\underline{A} \times \underline{M}$ into $[0, +\infty]$:

$$\int f(x,y,\Psi)C_{H_{\mathfrak{T}}}(d[x,y,\Psi])$$

$$= \int f(x,y,\Psi)((H^!_a)_{\mathfrak{T}} * (\mathfrak{T}_{(a)})_{[x,y]})(d\Psi)(\delta_a \otimes J_\tau(a,(.))(d[x,y])\Lambda_H(da)$$

$$= \int f(a,y,\Psi)((H^!_a)_{\mathfrak{T}} * (\mathfrak{T}_{(a)})_{[a,y]})(d\Psi)\ J_\tau(a,dy)\Lambda_H(da)\ .$$

It remains to determine $(\mathfrak{T}_{(a)})_{[a,y]}$ in terms of τ :

Noting that, for all $\underline{\mathcal{A}} \otimes \underline{\mathcal{M}}$ -measurable mappings g from $\underline{A} \times \underline{M}$ into $[0, +\infty]$, there holds

$$\int g(x,y,\Psi)C_{\mathfrak{T}_a}(d[x,y,\Psi]) = \iint g(x,y,\Psi)\Psi(d[x,y])\ \mathfrak{T}_{(a)}(d\Psi)$$

$$= \iint g(a,y,\delta_a \otimes \chi)\chi(dy)\tau_a(d\chi) = \iint g(a,y,\ \delta_a \otimes \chi)(\tau_{(a)})_y(d\chi)J_\tau(a,dy)$$

$$= \iint g(x,y,\ \delta_a \otimes \chi)(\tau_{(a)})_y(d\chi)J_{\mathfrak{T}}(a,d[x,y])\ ,$$

we obtain for $J_\tau(a,(.))$-almost all $y \in A'$:

$$(\mathfrak{T}_{(a)})_{[a,y]} = (\tau_{(a)})_y\ (\delta_a \otimes \chi \in (.))\ .$$

As one expects, the formula for C_{H_τ} in 1.9.2. is now recovered from the above formula for $C_{H_{\mathfrak{T}}}$ by projection onto $\mathcal{A}' \otimes \mathcal{M}'$.

One may guess from the above considerations that also an iteration of clustering leads to transparent formulae for the Palm distributions of the last generation, where now, starting from an individual a_0, at first gradually its father a_1, its grandfather a_2, etc. have to be determined, and then - given the line of ancestors- the last generation is decomposed into a_0, the brothers of a_0, the cousins of a_0, etc. For simplicity we restrict to the case that one and the same stochastic branching dynamics is acting in all generations.

1.9.4. Let τ be a clustering field on $[A, \rho_A]$ such that

$$J_\tau^{[k]}(a,(.)\in N \qquad (a\in A;\ k=1,2,...)\ .$$

Further let H be a measure on $\mathcal{M}$ of first order with the property

$$\Lambda_H * J_\tau^{[k]} \in N \qquad (k=1,2,...)\ .$$

Finally let, for all natural numbers k, D_k be a dual stochastic kernel of J_τ with respect to $\Lambda_H*(J_\tau)^{[k-1]}$. Then there holds for all natural numbers n and $\Lambda_H*(J_\tau)^{[n]}$ almost all a_0

$$(H_{\tau[n]})^!_{a_0}$$
$$= \int...\int ((H^!_{a_n})_{\tau[n]} * \mathop{\ast}_{1\le i\le n} ((\tau_{(a_i)})^!_{a_{i-1}})_{\tau[i-1]})\,(.)$$
$$D_1(a_{n-1}, da_n)...D_n(a_0, da_1)\ .$$

Proof. Due to 1.5.5. there exist all clustering powers of τ, and there holds

$$J_{\tau[k]} = J_\tau^{[k]} \qquad (k\in \mathbf{Z}_+)\ .$$

1.5.2. and 1.5.3. imply that all measures $H_{\tau[k]}$, $k\in \mathbf{Z}_+$ exist and have the intensity measure $\Lambda_H*J_\tau^{[k]}$.

2. For n=1, our assertion is implied by 1.9.1. We will now proceed by induction, assuming the validity of the assertion for an $n\in \mathbf{Z}_+$. Then we obtain in virtue of 1.9.1.:

$$(H_{\tau[n+1]})^!_{a_0} = ((H_{\tau[n]})_\tau)^!_{a_0}$$

$$= \int (((H_{\tau[n]})^!_{a_1})_\tau * (\tau_{(a_1)})^!_{a_0})(.)\ D_{n+1}(a_0,da_1)$$

$$= \int...\int (((H^!_{a_{n+1}})_{\tau[n]} * \mathop{\ast}_{2\le i\le n+1} (((\tau_{(a_i)})^!_{a_{i-1}})_{\tau[i-2]}))_\tau * (\tau_{(a_1)})^!_{a_0})(.)$$
$$D_1(a_n,da_{n+1})...D_n(a_1,da_2)\ D_{n+1}(a_0,da_1)$$

$$= \int...\int ((H^!_{a_{n+1}})_{\tau[n+1]} * \mathop{\ast}_{1\le i\le n+1} ((\tau_{(a_i)})^!_{a_{i-1}})_{\tau[i-1]})(.)$$
$$D_1(a_n,da_{n+1})...D_{n+1}(a_0,da_1)\ .\ \blacksquare$$

Continuing the interpretation of 1.9.1., the formula in 1.9.4. may be understood as follows: a_1 may be interpreted as father, a_2 as grandfather etc. of the individual a_0 in the n-th generation. The ancestors of a_0 are determined gradually "in the backward time direction": given a_0, a_1, ..., a_k, the distribution of a_{k+1} is $D_{n-k}(a_k,(.))$. The distribution of the population χ_1 of the brothers of a_0 is again $(\tau_{(a_1)})^!_{a_0}$. The population consisting of the uncles of a_0 has distribution $(\tau_{(a_2)})^!_{a_1}$, its offspring χ_2 - i.e. the cousins of a_0 - hence has distribution $((\tau_{(a_2)})^!_{a_1})_\tau$. Similarly, the other terms $((\tau_{(a_i)})^!_{a_{i+1}})_{\tau^{[i-1]}}$ may be interpreted. The first term $(H^!_{a_n})_{\tau^{[n]}}$ finally is the distribution of the population Ψ of all descendants of the initial population except the ancestor a_n of a_1. The integrand in 1.9.4. thus may be viewed as conditional distribution of the n-th generation, given a line of direct ancestors $a_0, \ldots, a_n$. The second convolution factor describes the descendants of the ancestor a_n in the n-th generation, its partial factors corresponding to the branches of the "reduced family tree" of a_0. "Reduced" means here that only such former events are considered which lead to descendants in the generation in question; extinct collateral lines are cancelled. As mentioned already, the description of the branching model may be refined in a way which grasps the complete structure of the reduced family tree.

The stochastic structure of the finite sequence a_0, a_1, ..., a_n is simply that of a time-inhomogeneous Markovian sequence with initial distribution δ_{a_0} and transition kernels D_n, D_{n-1}, ..., D_1.

We conclude this section with a generalization of 1.9.3. which is proved by induction:

1.9.5. Under the assumptions of 1.9.4., there holds for all natural numbers n

$$\Gamma^{(2)}{}_{H_{\tau^{[n]}}}$$

$$= \sum_{1\leq i\leq n} \iint (J_\tau^{[i-1]}(x,(.))\otimes J_\tau^{[i-1]}(y,(.)))\, \Gamma^{(2)}{}_{\tau_{(z)}}(d[x,y])(\Lambda_H * J_\tau^{[n-i]})(dz)$$

$$+ \int (J_\tau^{[n]}(x,(.))\otimes J_\tau^{[n]}(y,(.)))\, \Gamma^{(2)}{}_H(d[x,y]) \; .$$

2. EQUILIBRIUM DISTRIBUTIONS OF POISSON TYPE

In this chapter, above all, the branching model which is basic for all further investigations will be fixed by choosing a phase space $[A, \rho_A]$ and a stochastic branching dynamics on the space of populations $[M, \mathcal{M}]$. The moment assumption on the underlying clustering field κ means that all distributions $\kappa^{[n]}_{(a)}$ exist and are of first order. Let us emphasize that no further conditions, and in particular no continuity conditions, are imposed on κ. The intensity kernel of κ will be simply denoted by J.

As already mentioned, the main concern of this monograph is the investigation of the set $\mathbf{P}^{inv}$ of equilibrium distributions for the stochastic branching model. According to Proposition 2.1.5., a Poisson distribution Π_ν belongs to $\mathbf{P}^{inv}$ if and only if its intensity measure ν obeys the first order balance equation $\nu * J = \nu$ and, if moreover, for ν-almost all $a \in A$ the condition $\kappa_{(a)}(\chi(A) \leq 1) = 1$ is satisfied. In the "genuine branching case", hence, except δ_o no Poisson distributions belong to $\mathbf{P}^{inv}$.

Intensity measures Λ_P of equilibrium distributions P of first order will be called stable measures, forming the set N_κ^{st}. According to Proposition 2.3.1. and Theorem 2.3.6., for all $\nu \in N_\kappa^{st}$ the branching process with initial distribution Π_ν tends towards an equilibrium distribution Π^κ_ν with intensity measure ν. The infinitely divisible distributions Π^κ_ν, $\nu \in N_\kappa^{st}$, may be interpreted as model specific analoga of Poisson distributions (which is underlined by equalities like $\Pi^\kappa_\nu * \Pi^\kappa_\sigma = \Pi^\kappa_{\nu+\sigma}$) and will be called equilibrium distributions of Poisson type.

An application of the backward technique leads in Theorem 2.4.4. to an 'explicit' representation of $C^!_{\widetilde{\Pi^\kappa_\nu}}$. This constructive approach to Π^κ_ν, $\nu \in N_\kappa^{st}$, stands a first test in the derivation of a formula (Prop. 2.5.1.) for the second factorial moment measures of these distributions. Moreover, we obtain in this way in Theorem 2.5.4. a handy sufficient criterion for the stability of a given measure ν.

The measures $\nu \in N$ obeying the balance equation $\nu * J = \nu$ will be called J-invariant. In section 2.6. it will be shown that the set N_J of J-invariant measures is the "direct sum" of the set of stable and the set of "purely unstable" measures, the latter being those $\nu \in N_J$ for which a branching process with initial distribution Π_ν tends towards the trivial equilibrium distribution δ_o.

In Section 2.7. our stochastic branching model will be refined in considering also the family structure of populations (cf. Section 1.6.). The given clustering field defines a substochastic clustering field $\varkappa$ on the space $[\mathbb{A}, \rho_{\mathbb{A}}]$ of non-void populations (or "families ") Φ, which , in turn , induces a stochastic branching dynamics (" family dynamics") $\Phi \to \varkappa_{(\Phi)}$ on the space $[\mathbb{M}, \mathcal{M}]$ of family systems $\Phi = \sum_{i \in I} \delta_{\Phi_i}$. According to Proposition 2.7.18., one may associate, with each $P \in \mathbf{P}^{inv}$, a well-determined equilibrium distribution $\mathcal{G}_P$ for the family dynamics. Theorem 2.7.38. gives a simple characterization of all those equilibrium distributions for the family dynamics which are of the form $\mathcal{G}_P$, $P \in \mathbf{P}^{inv}$. In virtue of Theorem 2.7.25., a distribution $P \in \mathbf{P}^{inv}$ of first order is of Poisson type if and only if $\mathcal{G}_P$ is Poissonian. Thus, the above mentioned analogy between the distributions $\Pi^{\kappa}{}_{\nu}$ and the Poisson distributions is essentially deepened. In particular, one succeeds in this way (see Theorem 2.7.27.) to generalize the well-known Raikov- theorem for Poisson distributions to a characterization of all convolution divisors within $\mathbf{P}^{inv}$ of $\Pi^{\kappa}{}_{\nu}$, $\nu \in N_{\kappa}{}^{st}$.

As will be described in section 2.1., the question whether a measure ν in N_J is stable, cannot be decided solely by means of the kernel J. In this as well as in the following chapter it turns out, however, that many important properties of $\Pi^{\kappa}{}_{\nu}$ and quantities derived thereof are insensitive against a change of κ to κ' in the sense that they remain unchanged if only $J_{\kappa'}$ equals J_{κ} and ν is stable also with respect to the new branching dynamics κ'. According to Theorem 2.8.6. , in particular the quantity $\widetilde{\Pi^{\kappa}{}_{\nu}}(M)$ has this insensitivity property .

2.1. Equilibrium Distributions and Stable Measures

In the following, let κ always be a clustering field on some fixed phase space $[A,\rho_A]$, inducing a stochastic branching dynamics $\Phi \to \kappa_{(\Phi)}$ on the "space of populations" $[M,\mathcal{M}]$.

The intensity kernel J_{κ} of κ will, for simplicity, always be denoted by J, and it will be assumed that

$$J^{[n]}(a,(.)) \in N \qquad (a \in A\,;\, n = 1,2,...)\ .$$

This guarantees, by 1.5.5., that all clustering powers $\kappa^{[n]}$, $n \in \mathbf{Z}_+$, exist and obey

$$J^{[n]}(a,(.)) = \Lambda_{\kappa^{[n]}_{(a)}} \qquad (a \in A\ ;\ n = 1,2,...)\ .$$

A distribution $P \in \mathbf{P}$ with the property $P_{\kappa}=P$ will be called **equilibrium distribution** (of the stochastic branching dynamics); the set of all equilibrium distributions will be denoted by $\mathbf{P}^{inv}$. Obviously, $\mathbf{P}^{inv}$ is a convex subset of $\mathbf{P}$ containing δ_o (the distribution concentrated on the void population). If the convolution of an at most countable family of equilibrium distributions exists, it is again an equilibrium distribution due to 1.4.5.

Conversely, the convolution of an equilibrium distribution P_1 with a distribution P_2 is an equilibrium distribution only if P_2 has this property:

2.1.1. $P_1 \in \mathbf{P}^{inv}$, $P_2 \in \mathbf{P}$, $P_1 * P_2 \in \mathbf{P}^{inv}$ imply that $P_2 \in \mathbf{P}^{inv}$.

Proof. In virtue of 1.4.5. there holds

$$P = P_\kappa = (P_1)_\kappa * (P_2)_\kappa = P_1 * (P_2)_\kappa .$$

Uniqueness of the "convolution quotient" in **P** now implies that $(P_2)_\kappa = P_2$. ■

An infinitely divisible distribution P belongs to $\mathbf{P}^{inv}$ iff its canonical measure $\tilde{P}$ has a slightly modified invariance property. That is, there holds due to the clustering theorem 1.4.6.:

2.1.2. A distribution $P \in \mathbf{T}$ belongs $\mathbf{P}^{inv}$ iff $\tilde{P} = \tilde{P}_\kappa((.)\setminus\{o\})$. (For finite $\tilde{P}$, this amounts to $\tilde{P} = \tilde{P}_\kappa$.)

2.1.2. leads immediately to

2.1.3. For any $P \in \mathbf{P}^{inv} \cap \mathbf{T}$, and $r \in \mathbf{R}_+$, also P^r belongs to $\mathbf{P}^{inv}$.

In view of 1.5.2. there holds

2.1.4. $\Lambda_P * J = \Lambda_P \qquad (P \in \mathbf{P}^{inv})$.

If a Poisson distribution Π_ν belongs to $\mathbf{P}^{inv}$, then $\kappa_{(.)}$ coincides ν-almost everywhere with a substochastic clustering field:

2.1.5. Proposition. For $\nu \in N$, Π_ν belongs to $\mathbf{P}^{inv}$ iff $\nu * J = \nu$ and $\kappa_{(a)}(\chi(A) \leq 1) = 1$ for ν-almost all $a \in A$.

Proof. Due to 2.1.2., Π_ν belongs to $\mathbf{P}^{inv}$ iff

$$\mathcal{Q}_\nu = (\mathcal{Q}_\nu)_\kappa((.)\setminus\{o\}) = \int \kappa_{(a)}((.)\setminus\{o\})\nu(da)$$

i. e. iff

$$\text{(o)} \qquad \int \kappa_{(a)}(\chi(A) > 1)\nu(da) = 0$$

as well as

$$\text{(oo)} \qquad \nu(X) = \int \kappa_{(a)}(\chi(X) = 1)\nu(da) \qquad (X \in \mathcal{A})$$

hold true. But under (o), (oo) is equivalent to the equality $\nu * J = \nu$. ■

The following example shows that the study of equilibrium distributions cannot be reduced to those of first order:

2.1.6. Example. Choose $A := \mathbf{Z} \cup \{\infty\}$, putting $\rho_A(i,j) := |j-i|$ for $i, j \in \mathbf{Z}$ and $\rho_A(i,\infty) := \sqrt{1+i^2}$ for $i \in \mathbf{Z}$, and consider the clustering field κ defined by

$$\kappa_{(j)} := \begin{cases} \delta_{\delta_{j+1}} & \text{for } j=0,-1,-2,\ldots \\ \delta_{\delta_{j+1}} * (j^{-2}\delta_{j\delta_\infty} + (1-j^{-2})\delta_o) & \text{for } j=1,2,\ldots \end{cases},$$

$$\kappa_{(\infty)} := \delta_o \quad ,$$

i. e. κ acts as a shift on Z and, in addition, lets points in $\{1,2,\ldots\}$ generate offsprings at ∞, removing there at the same instant the offspring of the generation before. Hence $\mathbf{P}^{\text{inv}}$ is the set of all H_κ, where H is the distribution of a shift invariant random counting measure on $\mathbf{Z}$ such that H_κ exists. An example of such an H is the Poisson distribution Π_μ with intensity measure $\mu := \sum_{j\in\mathbf{Z}} \delta_j$; note that $(\Pi_\mu)_\kappa$ exists in view of

$$(\widetilde{\Pi_\mu})_\kappa(\Psi(\{\infty\})>0) = \sum_{j\geq 1} 1/j^2 < +\infty \ .$$

Thus, $\mathbf{P}^{\text{inv}} \neq \{\delta_o\}$. On the other hand we have

$$J(j,(.)) = \begin{cases} \delta_{j+1} & \text{for } j=0,-1,-2,\ldots \\ \delta_{j+1} + j^{-1}\delta_\infty & \text{for } j=1,2,\ldots \end{cases},$$

$$J(\infty,(.)) = o \ .$$

For any equilibrium distribution $P := H_\kappa$, the restriction $\Lambda_H = {}_{\mathbf{Z}}(\Lambda_P)$ of its intensity measure is translation invariant, i. e. Λ_H is of the form $r\cdot\mu$, $r\in\mathbf{R}_+\cup\{+\infty\}$. From the relation

$$\Lambda_P(\{\infty\}) = (\Lambda_H * J)(\{\infty\}) = r\cdot\sum_{j\geq 1} 1/j$$

we conclude that δ_o is the only equilibrium distribution of first order. ■

For a given $\nu\in N$, we put

$$\mathbf{P}^{\text{inv},\nu} := \{P\in\mathbf{P}^{\text{inv}} : \Lambda_P = \nu\} \ .$$

A measure $\nu\in N$ is called **stable** if $\mathbf{P}^{\text{inv},\nu}$ is nonvoid, i. e. if ν occurs as an intensity measure of some equilibrium distribution. The set of all stable measures in N is denoted by N_κ^{st}.

If for a distribution $P\in\mathbf{P}$ the clustered distributions $P_{\kappa^{[n]}}$, n=1,2,..., exist, we may form, with the given stochastic branching dynamics as a transition probability and P as initial distribution, a Markovian random sequence $\Phi_0, \Phi_1, \ldots$ with state space $[M, \mathcal{M}]$. The corresponding distribution on $\mathcal{M}^{\otimes\mathbf{Z}_+}$ will be denoted by $\vec{P}$. For all $i, j\in\mathbf{Z}_+$, $\kappa^{[j]}{}_{(\Phi_i)}$ then is a version of the conditional probability $\vec{P}(\Phi_{i+j}\in(.) \mid \Phi_i=\Phi)$, and for all $n\in\mathbf{Z}_+$ there holds $\vec{P}(\Phi_n\in(.)) = P_{\kappa^{[n]}}$. Obviously the random sequence (Φ_n) is stationary if and only if its initial distribution P belongs to $\mathbf{P}^{\text{inv}}$.

2.2. J-invariant Measures

A measure $\nu \in N$ is called **J-invariant** if $\nu*J=\nu$; the set of all J-invariant measures in N is denoted by N_J. Because of 2.1.4. we have

2.2.1. $\{o\} \subseteq N_K^{st} \subseteq N_J$.

Obviously the set N_J is $\mathcal{N}$-measurable and contains, together with ν_1 and ν_2, all linear combinations $r_1\nu_1+r_2\nu_2$, $r_1, r_2 \in \mathbf{R}_+$, and, in case of $\nu_1 \le \nu_2$, also the difference $\nu_2-\nu_1$. Moreover, N_J is closed under monotone convergence in N.

2.2.2. If (ν_n) is a sequence in N_J such that $\nu_n \uparrow \nu$ or $\nu_n \downarrow \nu$, then also ν belongs to N_J.

2.2.3. If for some $\nu \in N$ the sequence $(\nu*J^{[n]})$ is bounded in N and monotonically increasing (resp. monotonically decreasing), then there exists a measure $\sigma \in N_J$ such that $\nu*J^{[n]} \uparrow \sigma$ (resp. $\nu*J^{[n]} \downarrow \sigma$).

N_J^{ex} will denote the set of all extremal elements of the convex set N_J, i.e. the set of all $\nu \in N_J$ whose only representations as a sum of elements of N_J are of the form $\nu = r_1\nu+r_2\nu$, $r_1, r_2 \in \mathbf{R}_+$.

2.2.4. Example. We put $A := \mathbf{Z}$ and

$$\kappa_{(i)} := (1/3)\,(\delta_{\delta_{i-1}} + \delta_o + \delta_{2\delta_{i+1}}) \qquad (i \in \mathbf{Z})$$

(i. e. an individual either produces a descendant to its left or two descendants to its right or "dies out", each with probability 1/3). Hence $J(i,(.)) = (1/3)(\delta_{i-1}+2\delta_{i+1})$, i.e. J is the stochastic transition kernel of an asymmetric random walk. A measure ν in N belongs to N_J iff it obeys the difference equation

$$\nu(\{i\}) = (1/3)\nu(\{i+1\})+(2/3)\nu(\{i-1\}) \qquad (i \in \mathbf{Z}) \ .$$

Putting

$$\mu := \sum_{j \in \mathbf{Z}} \delta_j \ ; \quad \sigma := \sum_{j \in \mathbf{Z}} 2^j \delta_j \ ,$$

we observe that

$$N_J = \{r\mu+s\sigma\}_{r,s \in \mathbf{R}_+}, \quad N_J^{ex} = \{r\mu\}_{r \in \mathbf{R}_+} \cup \{s\sigma\}_{s \in \mathbf{R}_+} \ . \blacksquare$$

2.2.5. Proposition. The partially ordered set $[N_J, \le]$ is a distributive lattice. For all $\gamma, \rho \in N_J$ there hold

a) $(\gamma \wedge \rho)*J^{[n]} \downarrow \gamma \cap \rho :=$ greatest lower bound of $\{\gamma, \rho\}$ in N_J.

b) $(\gamma \vee \rho)*J^{[n]} \uparrow \gamma \cup \rho :=$ least upper bound of $\{\gamma, \rho\}$ in N_J.

c) $\gamma \cap \rho + \gamma \cup \rho = \gamma+\rho$.

Proof. 1. Let ν be some lower bound in N_J of γ and ρ.

Obviously there holds $\nu \le (\gamma \wedge \rho) \le \gamma, \rho$; hence follows $\nu \le (\gamma \wedge \rho)*J \le \gamma \wedge \rho$

and by iteration: $\nu \le (\gamma \wedge \rho)*J^{[n+1]} \le (\gamma \wedge \rho)*J^{[n]} \qquad (n=1,2,\ldots)$.

Using 2.2.3. we obtain, for some measure σ in N_J,

$$(\gamma\wedge\rho)*J^{[n]} \downarrow \sigma \geq \nu .$$

This also shows that $\sigma = \gamma\cap\rho$.

In the same way one checks the validity of b), in using, besides $(\gamma\vee\rho)*J \geq \gamma\vee\rho$, also the inequality $\gamma\vee\rho \leq \gamma+\rho$, which guarantees the boundedness of the monotonically increasing sequence $((\gamma\vee\rho)*J^{[n]})$ in N.

For all natural numbers n we obtain from $(\gamma\wedge\rho) + (\gamma\vee\rho) = \gamma+\rho$ the relation

$$(\gamma\wedge\rho)*J^{[n]} + (\gamma\vee\rho)*J^{[n]} = \gamma+\rho \quad ;$$

by using a) and b) we infer the validity of c).

Let γ, ρ and β be measures in N_J. We put

$$\sigma_1 := \rho\cup\beta - \rho\vee\beta \ ; \ \sigma_2 := \gamma\wedge\rho - \gamma\cap\rho \ ; \ \sigma_3 := \gamma\wedge\beta - \gamma\cap\beta$$

and obtain, using 1.

$$\begin{aligned} \gamma\cap(\rho\cup\beta) &\leq \gamma\wedge(\rho\cup\beta) = \gamma\wedge(\rho\vee\beta+\sigma_1) \\ &\leq \gamma\wedge(\rho\vee\beta)+\sigma_1 = (\gamma\wedge\rho)\vee(\gamma\wedge\beta)+\sigma_1 \\ &= (\gamma\cap\rho+\sigma_2)\vee(\gamma\cap\beta+\sigma_3) + \sigma_1 \leq (\gamma\cap\rho)\vee(\gamma\cap\beta) + \sigma_1 + \sigma_2 + \sigma_3 \\ &\leq (\gamma\cap\rho)\cup(\gamma\cap\beta) + \sigma_1 + \sigma_2 + \sigma_3 \ . \end{aligned}$$

Hence there holds for n=1,2,...

$$\gamma\cap(\rho\cap\beta) \leq (\gamma\cap\rho)\cup(\gamma\cap\beta) + \sigma_1*J^{[n]} + \sigma_2*J^{[n]} + \sigma_3*J^{[n]} .$$

Due to a) this implies $\gamma\cap(\rho\cup\beta) \leq (\gamma\cap\rho)\cup(\gamma\cap\beta)$.

As the converse inequality is trivially satisfied, we have proved the distributivity of the lattice $[N_J, \leq]$. ■

2.2.6. If γ, ρ_1, ρ_2,...are measures in N_J such that $\rho_n \uparrow \rho$,then $\gamma\cap\rho_n \uparrow \gamma\cap\rho$.

Proof. The sequence $(\gamma\cap\rho_n)$ is increasing, hence by 2.2.2. converges towards some $\sigma\in N_J$. Obviously there holds $\sigma \leq \gamma\cap\rho$. On the other hand we obtain for all n=1,2,...

$$\begin{aligned} (\gamma\cap\rho)-\sigma &\leq \gamma\cap\rho-\gamma\cap\rho_n \leq (\gamma\wedge\rho-\gamma\wedge\rho_n)+(\gamma\wedge\rho_n-\gamma\cap\rho_n) \\ &\leq (\rho-\rho_n) + (\gamma\wedge\rho_n - \gamma\cap\rho_n) \ . \end{aligned}$$

This yields, for all natural numbers k, n

$$(\gamma\cap\rho)-\sigma \leq (\rho-\rho_n) + (\gamma\wedge\rho_n)*J^{[k]} - \gamma\cap\rho_n \ ,$$

which, by Propositon 2.2.5. leads to

$$(\gamma\cap\rho)-\sigma \leq \rho-\rho_n \downarrow o \ . ■$$

The following sufficient criterion is an immediate consequence of 1.5.2. and 1.5.3. As to the terminology, we say, for any subset U of N, that a measure $\nu\in N$ is **U-bounded** if $\nu\leq\lambda$ for some $\lambda\in U$.

2.2.7. If the intensity measure of a measure H on $\mathcal{M}$ is N_J-bounded, then all measures $H_{\kappa^{[n]}}$, n=1,2,..., exist and obey

$$\Lambda_{H_{\kappa^{[n]}}} \leq \lambda \quad (n\in \mathbf{Z}_+) \ .$$

In view of 2.2.7., the distribution $\vec{P}$ exists for any $P\in \mathbf{P}$ with N_J-bounded intensity measure.

For a substochastic clustering field κ, Proposition 2.1.5. assures that $N_J=N_\kappa^{st}$. In general, however, not every $\nu\in N_J$ need to be stable:

2.2.8. Example. $\kappa_{(a)} := (1/2)(\delta_o+\delta_{2\delta_a})$ $(a\in A)$

defines a stochastic branching dynamics which has no equilibrium distributions apart from δ_o, hence $N_\kappa^{st}=\{o\}$. But $N_J=N$, as the intensity kernel of κ is given by $J(a,(.)) = \delta_a$, $a\in A$. ■

This example also shows that "stability" of a measure ν in N_J does not depend only on ν and J, but depends also on κ: the "trival" clustering field κ' given by

$$\kappa'_{(a)} := \delta_{\delta_a} \qquad (a\in A)$$

has the same intensity kernel as κ in Example 2.2.8., on the other hand any measure in N is stable for κ', because any $P\in \mathbf{P}$ is an equilibrium distribution for the trival branching dynamics!

A neccessary but not sufficient condition for stability of a measure ν in N_J is given by

2.2.9. Proposition. For all $\nu\in N_\kappa^{st}$ there holds

$$\lim_{c\to\infty} \sup_{n=0,1,2,\ldots} \int_{\{a:J^{[n]}(a,X)>c\}} J^{[n]}(a,X)\,\nu(da)=0 \qquad (X\in \mathcal{B}) .$$

Proof. Let P be any distribution in $\mathbf{P}^{inv,\nu}$ and $\Phi_0,\Phi_1,\ldots$ be a corresponding stationary Markovian random sequence of counting measures. By stationarity, the sequence $(\Phi_n(X))_{n\geq 0}$ is uniformly integrable with respect to $\vec{P}$; a fortiori , the sequence of conditional expectations $(E(\Phi_n(X) \mid \Phi_0=\Phi))$, i. e. the sequence $((\Phi * J^{[n]})(X))$, is uniformly integrable with respect to P, and we obtain

$$\lim_{c\to\infty} \sup_{n=0,1,\ldots} \int_{\{a:J^{[n]}(a,X)>c\}} J^{[n]}(a,X)\,\nu(da)$$

$$= \lim_{c\to\infty} \sup_{n=0,1,\ldots} \int \int_{\{a:J^{[n]}(a,X)>c\}} J^{[n]}(a,X)\,\Phi(da)\,P(d\Phi)$$

$$\leq \lim_{c\to\infty} \sup_{n=0,1,\ldots} \int_{\{\Phi:(\Phi * J^{[n]})(X)>c\}} (\Phi * J^{[n]})(X)\,P(d\Phi) = 0 \ . \ ■$$

2.3. Equilibrium Distributions of Poisson Type

The first proposition states that convergence to equilibrium holds under the stochastic branching dynamics, if the initial distribution is Poissonian with a J-invariant intensity measure:

2.3.1. Proposition. For any $\nu \in N_J$, there exists a distribution $V_{(\nu)} \in P^{inv} \cap T$ with the properties

$$(\Pi_\nu)_\kappa[n] \Rightarrow V_{(\nu)} \ ; \ \Lambda_{V_{(\nu)}} \leq \nu \ .$$

There even holds

$$\mathcal{L}_{(\Pi_\nu)_\kappa[n]}(f) \uparrow \mathcal{L}_{V_{(\nu)}}(f) \qquad (f \in F_\nu) \ .$$

Proof. 1. For all $n \in \mathbf{Z}_+$ let us put $L_n := (\Pi_\nu)_\kappa[n]$. Using 1.4.7. and 1.5.10., we get for all $f \in F$ and $n \in \mathbf{Z}_+$

$$\mathcal{L}_{L_n}(f) = \mathcal{L}_{\Pi_\nu}(-\ln \mathcal{L}_{\kappa[n]_{(.)}}(f)) = \mathcal{L}_{\Pi_{\nu * J}}(-\ln \mathcal{L}_{\kappa[n]_{(.)}}(f))$$
$$\leq \mathcal{L}_{(\Pi_\nu)_\kappa}(-\ln \mathcal{L}_{\kappa[n]_{(.)}}(f)) = \mathcal{L}_{L_{n+1}}(f) \ .$$

2. All L_n, $n \in \mathbf{Z}_+$, have intensity measure ν. Hence, by 1.2.6., the set $\{L_0, L_1, ...\}$ is τ_P-relatively compact.

3. In virtue of 1. and 2., (L_n) tends weakly towards a distribution $V_{(\nu)} \in \mathbf{P}$; since all of the L_n are infinitely divisible by 1.4.5., also $V_{(\nu)}$ belongs to $\mathbf{T}$ by 1.2.8.

In view of 1.2.7.there holds $\Lambda_{V_{(\nu)}} \leq \nu$. Finally, 1.5.11. allows to conclude that

$$L_{n+1} = (L_n)_\kappa \Rightarrow (V_{(\nu)})_\kappa \ ,$$

which shows that $V_{(\nu)} \in \mathbf{P}^{inv}$.

4. The convergence $\mathcal{L}_{L_n}(f) \to \mathcal{L}_{V_{(\nu)}}(f)$, $f \in F_\nu$, now follows from 1.3.7. ■

2.3.2. For all $\nu \in N_J$ and $r \in \mathbf{R}_+$ there holds

$$V_{(r \cdot \nu)} = (V_{(\nu)})^r \ .$$

Proof. From 2.3.1. and 1.6.3. there follows $\widetilde{(\Pi_\nu)_\kappa[n]} \xrightarrow{w} \widetilde{V_{(\nu)}}$.

1.4.6. shows that $\widetilde{(\Pi_{r \cdot \nu})_\kappa[n]} = (\widetilde{\Pi_{r\cdot\nu}})_\kappa[n]((.)\setminus\{o\}) = \int \kappa^{[n]}_{(a)}((.)\setminus\{o\}) r \cdot \nu(da) = r \cdot \widetilde{(\Pi_\nu)_\kappa[n]}$.

Letting n tend to infinity, we infer that $\widetilde{V_{(r \cdot \nu)}} = r \cdot \widetilde{V_{(\nu)}}$, which, by definition of P^r for $P \in \mathbf{T}$, yields the assertion . ■

In a similar way one obtains:

2.3.3. For all $\nu, \sigma \in N_J$ there holds

$$V_{(\nu+\sigma)} = V_{(\nu)} * V_{(\sigma)} \ .$$

Proof. From 1.4.5. and 1.1.6. we infer that

$$(\Pi_{(\nu+\sigma)})_K[n] = (\Pi_{(\nu)})_K[n] * (\Pi_\sigma)_K[n] \Rightarrow V_{(\nu)} * V_{(\sigma)} \ .$$

On the other hand, there holds

$$(\Pi_{(\nu+\sigma)})_K[n] \Rightarrow V_{(\nu+\sigma)} \ . \blacksquare$$

The assertion of 2.3.3. may be generalized as follows:

2.3.4. Proposition. If $(\nu_i)_{i\in I}$ is an at most countable family of J-invariant measures whose sum $\nu := \sum_{i\in I} \nu_i$ belongs to N, then $\nu \in N_J$ and

$$V_{(\nu)} = \mathop{\ast}_{i\in I} V_{(\nu_i)} \ .$$

Proof. 1. For finite I, the assertion follows from 2.3.3.

2. Now assume that I={1,2,...}. Because of $\sum_{i>0} \Lambda_{V_{(\nu_i)}} \leq \nu$, the convolution $\mathop{\ast}_{i>0} V_{(\nu_i)}$ exists. By 2.2.2., the measure ν belongs to N_J, and we obtain, making use of 1., of the normed algebra structure of **E**, of the algebra homomorphism $P \to {}_X P$ and of the estimate 1.1.9.:

$$_X\| V_{(\nu)} - \mathop{\ast}_{i>0} V_{(\nu_i)} \|$$

$$= {}_X\| (\mathop{\ast}_{1\leq i\leq n} V_{(\nu_i)}) * V_{(\sum_{i>n} \nu_i)} - (\mathop{\ast}_{1\leq i\leq n} V_{(\nu_i)}) * \mathop{\ast}_{i>n} V_{(\nu_i)} \|$$

$$\leq {}_X\| V_{(\sum_{i>n} \nu_i)} - \mathop{\ast}_{i\geq n} V_{(\nu_i)} \|$$

$$\leq 2\Big(\Lambda_{V_{(\sum_{i>n} \nu_i)}}(X) + \sum_{i>n} \Lambda_{V_{(\nu_i)}}(X)\Big)$$

$$\leq 4 \sum_{i>n} \nu_i(X) \xrightarrow[n\to\infty]{} 0 \ . \blacksquare$$

The construction of $V_{(\nu)}$ implies immediately:

2.3.5. The mappings $\nu \to V_{(\nu)}$ and $\nu \to \Lambda_{V_{(\nu)}}$ from N_J into **P** and N, respectively, are $\mathcal{N}, \boldsymbol{P}$ - and $\mathcal{N}, \mathcal{N}$-measurable, respectively.

We are now in the position to prove the first main result of the present chapter. It states that a J-invariant measure ν is stable (i. e. arises as intensity measure of some equilibrium distribution) if and only if the limiting distribution $V_{(\nu)}$ of a Markov chain started off with the Poisson distribution Π_ν still has intensity measure ν. Roughly speaking this happens iff an initial population distributed according to Π_ν (or an initial individual "distributed" according to $\mathcal{Q}_\nu$) does not lead to "too thick clusters" of descendants in later generations.

2.3.6. Theorem. For all $\nu \in N_J$ the following assertions are equivalent:

a) ν is stable .

b) $\Lambda_{V_{(\nu)}} = \nu$.

c) For all $X \in \mathcal{B}$, the sequence $(\Phi_n(X))$ is uniformly integrable under $\overrightarrow{\Pi_\nu}$.

d) For all $X \in \mathcal{B}$,

$$\sup_{n=1,2,\ldots} \int (\sum_{j>k} j\, \kappa^{[n]}_{(a)}(\chi(X)=j))\nu(da) \xrightarrow[k\to\infty]{} 0 \ .$$

Proof. For abbreviation we put

$$L_n := (\Pi_{(\nu)})_{\kappa^{[n]}} \quad \text{for } n \in \mathbf{Z}_+ \ ; \quad L_\infty := V_{(\nu)} .$$

1. If b) is valid, then, due to 2.3.1., $V_{(\nu)}$ is an element of $\mathbf{P}^{\mathrm{inv},\nu}$. Hence a) follows from b).

2. In view of 2.3.1. and 1.1.8., the assertion b) and c) are equivalent.

3. The convergence $L_n \Rightarrow L_\infty$, which holds by 2.3.1., implies, due to 1.6.3., the convergence $\widetilde{L_n} \xrightarrow{w} \widetilde{L_\infty}$. Hence, by 1.6.5.,

$$\Lambda_{\widetilde{L_\infty}} = \Lambda_{L_\infty} = \nu$$

is valid iff for all $X \in \mathcal{B}$ there holds

$$\sup_{n=1,2\ldots} \int (\sum_{j>k} j\, \kappa^{[n]}_{(a)}(\chi(X)=j))\, \nu(da)$$

$$= \sup_{n=1,2\ldots} \int_{\{\Psi:\Psi(X)>k\}} \Psi(X)\, \widetilde{L_n}(d\Psi) \xrightarrow[k\to\infty]{} 0 \ .$$

Thus assertions b) and d) are equivalent.

4. Let us finally assume the validity of a), i. e. $\mathbf{P}^{\mathrm{inv},\nu}$ is non-void. Let P be some distribution in $\mathbf{P}^{\mathrm{inv},\nu}$. For all natural numbers k, n there follows

$$\int (\sum_{j>k} j\, \kappa^{[n]}_{(a)}(\chi(X)=j))\, \nu(da) = \int_{\{\Psi:\Psi(X)>k\}} \Psi(X)\, \widetilde{L_n}(d\Psi)$$

$$= \mathcal{C}_{\widetilde{L_n}}(a \in X, \Psi(X)>k) = \int \mathcal{C}_{\kappa^{[n]}_{(b)}}(a \in X, \Psi(X)>k)\, \nu(db)$$

$$= \iint \mathcal{C}_{\kappa^{[n]}_{(b)}}(a \in X, \Psi(X)>k)\, \chi(db)\, P(d\chi)$$

$$\leq \iint (\mathcal{C}_{\kappa^{[n]}_{(b)}} \ast \kappa^{[n]}_{(\chi-\delta_b)})(a \in X, \Psi(X)>k)\, \chi(db)\, P(d\chi) \ .$$

Using 1.7.7., we may proceed by

$$\int \mathcal{C}_{\kappa^{[n]}_{(\chi)}}(a \in X, \Psi(X)>k)\, P(d\chi) = \mathcal{C}_{P_{\kappa^{[n]}}}(a \in X, \Psi(X)>k)$$

$$= \mathcal{C}_P(a \in X, \Psi(X)>k) \xrightarrow[k\to\infty]{} 0 \ .$$

This shows that d) is a consequence of a). ■

In virtue of Theorem 2.3.6., for all stable measures $\nu \in N_J$,

$$\Pi^{\kappa}_{\nu} := V_{(\nu)}$$

is a distinguished infinitely divisible representative in $P^{inv,\nu}$ and will be called **equilibrium distribution of Poisson type with intensity measure** ν.

If, in particular, $\kappa_{(a)}(\chi(A)\leq 1)=1$ for ν-almost all $a \in A$, then due to the clustering theorem any $(\Pi_{\nu})_{\kappa}[n]$, $n \in \mathbf{Z}_+$, and hence also Π^{κ}_{ν}, coincides with Π_{ν}.

An immediate consequence of 2.3.5. and Theorem 2.3.6. is

2.3.7. $N_{\kappa}^{st} \in \mathcal{N}$.

Observing that the intensity measure of a convolution product of distributions equals the sum of the factors' intensity measures, we get by combining Proposition 2.3.4. and Theorem 2.3.6.:

2.3.8. If $(\nu_i)_{i \in I}$ is an at most countable family of measues in N_{κ}^{st} such that $\nu := \sum_{i \in I} \nu_i$ belongs to N, then also ν belongs to N_{κ}^{st}, and

$$\Pi^{\kappa}_{\nu} = \mathop{*}_{i \in I} \Pi^{\kappa}_{\nu_i} \quad .$$

For a J-invariant measure λ which has a stable majorant σ, there holds by 2.3.3.

$$V_{(\sigma)} = V_{(\lambda)} + V_{(\sigma-\lambda)} \quad ,$$

hence

$$\sigma = \Lambda_{V_{(\sigma)}} = \Lambda_{V_{(\lambda)}} + \Lambda_{V_{(\sigma-\lambda)}} \quad .$$

Thus the summands on the right hand side must equal λ and $\sigma-\lambda$, respectively, and we obtain

2.3.9. N_{κ}^{st} is a descending subset of $[N_J, \leq]$.

Observing that for all $P \in \mathbf{T}$ and $r \in \mathbf{R}_+$ there holds

$$\Lambda_{P^r} = \Lambda_{r\tilde{P}} = r\Lambda_{\tilde{P}} = r\Lambda_P \quad ,$$

one gets immediately from 2.3.2. and Theorem 2.3.6.

2.3.10. For all $\nu \in N_{\kappa}^{st}$ and all $r \in \mathbf{R}_+$, also $r \cdot \nu$ belongs to N_{κ}^{st}, and

$$\Pi^{\kappa}_{r\nu} = (\Pi^{\kappa}_{\nu})^r \quad .$$

Finally we show the following continuity property:

2.3.11. Proposition. If for a sequence of measures (γ_n) in N_{κ}^{st} the convergence $\gamma_n \uparrow \gamma$ or $\gamma_n \downarrow \gamma$ holds, then also γ belongs to N_{κ}^{st}, and

$$\Pi^{\kappa}_{\gamma_n} \Rightarrow \Pi^{\kappa}_{\gamma} \; .$$

Proof. 1. Assume $\gamma_n \uparrow \gamma$. Putting $\nu_1 := \gamma_1$, $\nu_n := \gamma_n - \gamma_{n-1}$, $n = 2,3,\ldots$, we obtain $\gamma = \sum_{n>0} \nu_n$. As all ν_n are stable by 2.3.9., also their sum γ is stable by 2.3.8., and there holds

$$\Pi^{\kappa}_{\gamma_n} = \mathop{*}_{1\le i\le n} \Pi^{\kappa}_{\nu_i} \Rightarrow \Pi^{\kappa}_{\gamma}.$$

2. In the case $\gamma_n \downarrow \gamma$, 2.2.2. and 2.3.9. allow to conclude that γ belongs to N_κ^{st}. As to the claimed convergence, one gets for all $X \in \mathcal{B}$ and $m \in \mathbf{Z}_+$, using 1.1.9.,

$$_X\| \Pi^{\kappa}_{\gamma} - \Pi^{\kappa}_{\gamma_m} \| = {}_X\| \Pi^{\kappa}_{\gamma} - \Pi^{\kappa}_{\gamma} * \Pi^{\kappa}_{\gamma_m - \gamma} \|$$

$$\le {}_X\| \delta_o - \Pi^{\kappa}_{\gamma_m - \gamma} \| \le 2(\gamma - \gamma_m)(X) \underset{m\to\infty}{\longrightarrow} 0 \;. \blacksquare$$

2.4. An Application of the Backward Technique

For any sequence $a_0, a_1, \ldots$ of points in A, we put

$$\mathcal{K}_{(a_i)} := \bigotimes_{i\ge 0} ((\kappa_{(a_{i+1})})^!_{a_i})_{\kappa^{[i]}}$$

(provided that all factors on the right hand side exist), thus obtaining the distribution of an independent random sequence $\chi_0, \chi_1, \ldots$ of counting measures.

If ν is a J-invariant measure and D is a dual stochastic kernel of J with respect to ν, we may associate, with each $a \in A$, the distribution $q_{\nu,a}$ of a Markov chain with state space A which starts in a and has transition probability kernel D.

An inspection of 1.9.4. (putting, e. g., $H := \Pi_\nu$) shows that for ν-almost all a_0 the distribution

$$\mathcal{K}_{\nu,a_0} := \int \mathcal{K}_{(a_i)}(.) q_{\nu,a_0}(d(a_i)_{i\ge 1})$$

is well-defined. One more application of 1.9.4. leads to

2.4.1. If ν is a measure in N_J and H is a measure on $\mathcal{M}$ with $\Lambda_H = \nu$, then there holds for all $n \in \mathbf{Z}_+$ and ν-almost all $a_0 \in A$:

$$(H_{\kappa^{[n]}})^!_{a_0} = \int \Big((H^!_{a_n})_{\kappa^{[n]}} * \mathcal{K}_{(a_i)}\big(\big(\sum_{0\le i<n} \chi_i \in (.)\big)\big)\Big)(.)\, q_{\nu,a_0}(d(a_i)_{i\ge 1}) \;.$$

The special choice $H = \widetilde{\Pi_\nu} = \mathcal{Q}_\nu$ leads in view of 1.4.6. to

2.4.2. For all measures ν in N_J and all $n \in \mathbf{Z}_+$ there holds for ν-almost all $a \in A$:

$$\big(\widetilde{(\Pi_\nu)_{\kappa^{[n]}}}\big)^!_a = \mathcal{K}_{\nu,a}\big(\big(\sum_{0\le i<n} \chi_i \in (.)\big)\big)$$

We are now in the position to prove **Kallenberg's criterion for stability:**

2.4.3. Proposition. A measure $\nu \in N_J$ is stable iff $\mathcal{K}_{\nu,a}((\sum_{i\geq 0} \chi_i) \in M) = 1$ for ν-almost all a .

Proof. 1. Assume that ν is stable and choose some $P \in \mathbf{P}^{inv,\nu}$. For all natural numbers n, 2.4.1. yields for ν-almost all $a \in A$

$$\mathcal{K}_{\nu,a}((\sum_{0\leq i<n} \chi_i \in (.)) \lesssim P^!_a \ ,$$

hence

$$\mathcal{K}_{\nu,a}((\sum_{i\geq 0} \chi_i) \in M) = 1 \quad , \quad \mathcal{K}_{\nu,a}((\sum_{i\geq 0} \chi_i \in (.)) \lesssim P^!_a \quad .$$

2. Conversely, assume that $\nu \in N_J$ has the just stated property. For short, we put $L_n := (\Pi_\nu)_{\kappa^{[n]}}$, $n \in \mathbf{Z}_+$, and obtain from 2.4.2. for all $\mathcal{A}\otimes\mathcal{M}$-measurable mappings h from $A\times M$ into $[0, +\infty]$:

$$\int h(a,\chi) C^!_{\tilde{L}_n}(d[a,\chi]) = \iint h(a, \sum_{0\leq i<n} \chi_i) \mathcal{K}_{\nu,a}(d(\chi_i)) \nu(da) \ .$$

Hence follows for all $X \in \mathcal{B}$ and $k,n \in \mathbf{Z}_+$:

$$\int (\sum_{j>k} j \kappa^{[n]}_{(a)}(\chi(X) = j)) \ \nu(da) \ = \int_{\{\Psi:\Psi(X)>k\}} \Psi(X) \ \tilde{L}_n(d\Psi)$$

$$= C^!_{\tilde{L}_n}(a \in X, \chi(X) \geq k) \ \leq \int_X \mathcal{K}_{\nu,a}(\sum_{i\geq 0} \chi_i(X) \geq k) \nu(da) \xrightarrow[k\to\infty]{} 0 \ ,$$

which by Theorem 2.3.6. shows the stability of ν . ■

We are going to investigate, for stable measures ν, passage to the limit in 2.4.2.

2.4.4. Theorem. For all measures $\nu \in N_\kappa^{st}$ there holds for ν-almost all $a \in A$

$$(\widetilde{\Pi^\kappa_\nu})^!_a = \mathcal{K}_{\nu,a}((\sum_{i\geq 0} \chi_i) \in (.)) \ .$$

Proof. Let us again put $L_n := (\Pi_{(\nu)})_{\kappa^{[n]}}$, $n \in \mathbf{Z}_+$; $L_\infty := \Pi^\kappa_\nu$. The generalized Palm-Khinchin-theorem 1.7.9. ensures that, in view of

$$\tilde{L}_n \xrightarrow{w} \tilde{L}_\infty \text{ and } \Lambda_{\tilde{L}_n} = \nu = \Lambda_{\tilde{L}_\infty} \qquad (n \in \mathbf{Z}_+)$$

the convergence $C^!_{\tilde{L}_n} \xrightarrow{N^\wedge} C^!_{\tilde{L}_\infty}$ holds true. Hence we obtain for all $h \in F^\wedge_c$, using 2.4.2.

$$\int h(a,\chi) C^!_{\tilde{L}_\infty}(d[a,\chi]) = \lim_{n\to\infty} \int h(a,\chi) C^!_{\tilde{L}_n}(d[a,\chi])$$

$$= \lim_{n\to\infty} \iint h(a, \sum_{0\leq i<n} \chi_i) \ \mathcal{K}_{\nu,a}(d(\chi_i)) \nu(da) = \iint h(a, \sum_{i\geq 0} \chi_i) \mathcal{K}_{\nu,a}(d(\chi_i)) \nu(da) \ ,$$

which yields the assertion. ■

Comparing 2.4.2. and 2.4.4., we see that for all stable measures ν, $(\widetilde{(\Pi_\nu)_{\kappa[n]}})^!_a \Rightarrow$ $(\widetilde{\Pi^\kappa_\nu})^!_a$ for ν-almost all $a\in A$. On the other hand one infers from 2.4.1. and 2.4.2. that, for all $P\in \mathbf{P}^{inv,\nu}$ and $n\in \mathbf{Z}_+$, there holds

$$(\widetilde{(\Pi_\nu)_{\kappa[n]}})^!_a \precsim P^!_a \quad \nu\text{-a.e.}$$

Hence follows, by 1.1.10., the inequality

2.4.5. For all $\nu\in N_\kappa^{st}$ and all $P\in \mathbf{P}^{inv,\nu}$ there holds

$$(\widetilde{\Pi^\kappa_\nu})^!_a \precsim P^!_a \quad \nu\text{-a.e} \ .$$

One more comparison of 2.4.2. and 2.4.4. and an application of the monotone convergence theorem shows:

2.4.6. Let ν be a measure in N_κ^{st} and h be an $\mathcal{A}\otimes\mathcal{M}$-measurable mapping from $A\times M$ into $[0,+\infty]$ such that for all $a\in A$ and $\Phi_n\uparrow\Phi$ the convergence $h(a,\Phi_n)\uparrow h(a,\Phi)$ holds. Then there follows

$$\int h(a,\chi)\mathcal{C}^!_{\widetilde{(\Pi_\nu)_{\kappa[n]}}}\, d([a,\chi]) \;\uparrow\; \int h(a,\chi)\mathcal{C}^!_{\widetilde{\Pi^\kappa_\nu}}\, d([a,\chi]) \ .$$

2.5. A Sufficient Criterion for Stability

Using the results of the last section, one succeeds to compute the second factorial moment measures of equilibrium distributions of Poisson type.

2.5.1. Proposition. For all measures $\nu\in N_\kappa^{st}$ there holds

$$\Gamma^{(2)}_{\Pi^\kappa_\nu} = \nu\otimes\nu + \iint \sum_{i\geq 0} J^{[i]}(x,(.))\otimes J^{[i]}(y,(.))\, \Gamma^{(2)}_{\kappa_{(z)}}(d[x,y])\, \nu(dz) \ .$$

Proof. Again we put

$$L_n := (\Pi_\nu)_{\kappa[n]}\,, \quad n\in\mathbf{Z}_+ \quad ; \quad L_\infty := \Pi^\kappa_\nu \ .$$

For any $\mathcal{A}$-measurable mapping f from $A\times A$ into $[0,+\infty]$ we obtain, using 1.8.8. and 2.4.6 .

$$\int f(a,b)\,\Gamma^{(2)}_{\widetilde{L_n}}(d[a,b]) = \iint f(a,b)\Lambda_{(\widetilde{L_n})^!_a}(db)\,\nu(da)$$

$$= \iiint f(a,b)\,\chi(db)\,(\widetilde{L_n})^!_a(d\chi)\,\nu(da) = \iint f(a,b)\,\chi(db)\,\mathcal{C}^!_{\widetilde{L_n}}(d[a,\chi])$$

$$\uparrow \iint f(a,b)\,\chi(db)\mathcal{C}^!_{\widetilde{L_\infty}}(d[a,\chi]) = \int f(a,b)\,\Gamma^{(2)}_{\widetilde{L_\infty}}(d[a,b]) \ .$$

On the other hand there holds by 1.9.5.

$$\Gamma^{(2)}{}_{\widetilde{L_n}} = \iint\left(\sum_{0\leq i<n} J^{[i]}(x,(.)) \otimes J^{[i]}(y,(.))\right)\Gamma^{(2)}{}_{\kappa_{(z)}}(d[x,y])\nu(dz)$$

Passage to the limit $n\to\infty$ and applying 1.2.19. implies the assertion. ■

If now P is some distribution in $\mathbf{P}^{inv,\nu}$, we obtain, using 1.8.8. and 2.4.5., for all nonnegative $\mathcal{A}\otimes\mathcal{A}$-measurable mappings f:

$$\int f(a,b)\,\Gamma^{(2)}{}_{\widetilde{L_\infty}}(d[a,b]) = \iint f(a,b)\Lambda_{(\widetilde{\Pi^\kappa{}_\nu})^!{}_a}(db)\,\nu(da)$$

$$\leq \iint f(a,b)\Lambda_{P^!{}_a}(db)\nu(da) = \int f(a,b)\Gamma^{(2)}{}_P(d[a,b]) \quad ,$$

which shows, in view of 1.2.19, the estimate

2.5.2. For all $\nu\in N_\kappa{}^{st}$ and all $P\in \mathbf{P}^{inv,\nu}$ there holds

$$\Gamma^{(2)}{}_{\Pi^\kappa{}_\nu} \leq \nu\otimes\nu + \Gamma^{(2)}{}_P \ .$$

2.5.3. Example. If $\kappa_{(a)} := \delta_{\delta_a}$ for all $a\in A$, then for any $\nu\in N$ with total mass one, the distribution $P:=\mathcal{Q}_\nu$ belongs to $\mathbf{P}^{inv,\nu}$, and there holds

$$\Gamma^{(2)}{}_{\Pi^\kappa{}_\nu} = \Gamma^{(2)}{}_{\Pi_\nu} = \nu\otimes\nu \ ; \ \Gamma^{(2)}{}_P = o\otimes o \quad . \blacksquare$$

For all natural numbers k, a measure ν in N is called **stable of k-th order** if $\mathbf{P}^{inv,\nu}$ contains some distribution of k-th order. The set of all those measures ν will be denoted by $N_\kappa{}^{st(k)}$. Thus there holds

$$N_\kappa{}^{st} = N_\kappa{}^{st(1)} \supseteq N_\kappa{}^{st(2)} \supseteq \ldots$$

2.5.4. Theorem. For all measures $\nu\in N_J$, the following assertions are equivalent:

a) ν is stable of second order.

b) ν is stable and $\Pi^\kappa{}_\nu$ is of second order.

c) For all $X\in\mathcal{B}$ there holds

$$\iint\left(\sum_{i\geq 0} J^{[i]}(x,X)\, J^{[i]}(y,X)\right) \Gamma^{(2)}{}_{\kappa_{(z)}}(d[x,y])\ \nu(dz) < +\infty \quad .$$

Proof. 1. In virtue of 2.5.2., b) is a consequence of a); hence a) and b) are equivalent.

2. Due to 2.5.1., b) implies c).

3. Conversely assume that a measure $\nu\in N_J$ has property c). For all $X,Y\in\mathcal{B}$, $X\supseteq Y$, and $k\in\mathbf{Z}_+$ there follows as in the proof of 2.5.1. and with the same notation

$$\iint\left(\sum_{0\leq i<n} J^{[i]}(x,X)J^{[i]}(y,X)\right)\Gamma^{(2)}{}_{\kappa_{(z)}}(d[x,y,])\ \nu(dz)$$

$$= \Gamma^{(2)}{}_{\widetilde{L_n}}(X\times X) \geq \Gamma^{(2)}{}_{\widetilde{L_n}}(X\times Y) = \int_X \chi(Y)C^!{}_{\widetilde{L_n}}(d[a,\chi])$$

$$= \int_X\int\left(\sum_{0\leq i<n}\chi_i(Y)\right)\mathcal{K}_{\nu,a}(d(\chi_i))\ \nu(da) \quad ,$$

where we used 2.4.2. in the last equality. Hence follows

$$\mathcal{K}_{\nu,a}((\sum_{i\geq 0}\chi_i(Y))<+\infty)=1 \quad \text{for } \nu\text{-almost all } a\in X$$

and (as $X\supseteq Y$ was arbitrary) even for ν-almost all $a\in A$. But this means

$$\mathcal{K}_{\nu,a}((\sum_{i\geq 0}\chi_i)\in M)=1 \quad \text{for } \nu\text{-almost all } a\in A \quad ,$$

which guarantees stability of ν by 2.4.4. Finally, 2.5.1. now shows that $\Pi^{\kappa}{}_{\nu}$ is of second order; hence b) is a consequence of c). ■

2.5.5. Example. With the assumptions and notations of Example 2.2.4., there holds $\sigma\in N_{\kappa}{}^{st(2)}$.

Proof. We are going to apply the criterion 2.5.4. c) for second order stability. In view of

$$\Gamma^{(2)}{}_{\kappa_{(z)}}=\iiint 1_{(.)}(x,y)(\Phi-\delta_x)(dy)\Phi(dx)\kappa_{(z)}(d\Phi)=(2/3)(\delta_{z+1}\otimes\delta_{z+1}) \qquad (z\in \mathbf{Z}),$$

we obtain, for $X\in\mathcal{B}$

$$\iint\sum_{i\geq 0}J^{[i]}(x,X)J^{[i]}(y,X)\Gamma^{(2)}{}_{\kappa_{[n]}}(d[x,y])\sigma(dz)$$

$$=(2/3)\int\sum_{i\geq 0}(J^{[i]}(z+1,X))^2\sigma(dz)\ \leq\ \int(\sum_{i\geq 0}J^{[i]}(z+1,X))^2\sigma(dz)\ .$$

Now $G(z,X):=\sum_{i\geq 0}J^{[i]}(z,X)$, $z\in\mathbf{Z}$, $X\in\mathcal{B}$, is the expected number of visits in the set X of an asymmetric random walk (with transition probability J) starting in z. By transience and homogeneity, $G(z,X)\leq C<+\infty$, where C depends on X but not on z (see [Spitzer (1976), p.9]). By the Markov property there holds

$$G(z,X)\leq P_z(T_X<\infty)\cdot C$$

where $P_z(T_X<\infty)$ is the probability that the walk started in z ever hits X. The mapping $z\rightarrow P_z(T_X<\infty)$ is harmonic with respect to the kernel J on $A\backslash X$, hence follows for $z\geq \max X=:b$

$$P_z(T_X<\infty)=2^{b-z}\ .$$

Now we may proceed in our estimate as follows:

$$\int G(z+1,X)^2\sigma(dz)$$

$$\leq C^2\sigma(\{z\in\mathbf{Z}:z<b\}) \ + \int_{\{z\in\mathbf{Z}:z\geq b\}}(G(z+1,X))^2\sigma(dz)$$

$$\leq C^2\sum_{z<b}2^z+\sum_{z\geq b}(2^{b-z-1}.C)^2.2^z<+\infty\ .$$

An application of Theorem 2.5.4. now shows that σ is stable of second order. ■

2.6. Stable and Purely Unstable Components of J-invariant Measures

A measure $\nu \in N_J$ is called **purely unstable** if $V_{(\nu)}=\delta_o$, i.e. if a sequence of random populations distributed initially Poissonian with intensity measure ν converges, under the stochastic branching dynamics, towards the void population.

The set of all purely unstable measures in N_J will be denoted by N_κ^{nst}. Obviously there holds $N_\kappa^{st} \cap N_\kappa^{nst} = \{o\}$.

The same arguments which lead to 2.3.8., 2.3.9., 2.3.10. now show:

2.6.1. If $(\nu_i)_{i\in I}$ is an at most countable family of measures in N_κ^{nst} such that $\nu := \sum_{i\in I} \nu_i$ belongs to N, then also ν belongs to N_κ^{nst}.

2.6.2. N_κ^{nst} is a descending subset of $[N_J, \leq]$.

2.6.3. For all $\nu \in N_\kappa^{nst}$ and $r \in \mathbf{R}_+$, also $r \cdot \nu$ belongs to N_κ^{nst}.

We now give a criterion for pure instability:

2.6.4. Proposition. A measure $\nu \in N_J$ belongs to N_κ^{nst} iff

$$\int \kappa^{[n]}_{(a)}(\chi(X)>0)\nu(da) \downarrow 0 \qquad (X \in \mathcal{B}) \quad .$$

Proof. 1. For all $\nu \in N_J$, $X \in \mathcal{B}$ and $n \in \mathbf{Z}_+$ there holds

$$\int \kappa^{[n+1]}_{(a)}(\chi(X)>0)\nu(da) = \iint \kappa^{[n]}_{(\Phi)}(\chi(X)>0)\kappa_{(a)}(d\Phi)\nu(da)$$

$$\leq \iiint \kappa^{[n]}_{(b)}(\chi(X)>0)\Phi(db)\kappa_{(a)}(d\Phi)\nu(da)$$

$$= \iint \kappa^{[n]}_{(b)}(\chi(X)>0)J(a, db)\nu(da) = \iint \kappa^{[n]}_{(b)}(\chi(X)>0)\nu(db) \quad .$$

2. By 1.4.6. and 1.6.3., ν belongs to N^{nst} iff

$$\int \kappa^{[n]}_{(a)}((.)\setminus\{o\})\nu(da) = \widetilde{(\Pi_\nu)_{\kappa^{[n]}}} \xrightarrow{\mathbf{w}} o \quad .$$

But this latter convergence is equivalent to

$$\int \kappa^{[n]}_{(a)}(\chi(X)>0)\nu(da) \to 0 \qquad (X \in \mathcal{B}) \quad .$$

Indeed, choose for $f \in \mathbb{F}_c$ an $r \in \mathbf{R}_+$ and $X \in \mathcal{B}$ such that $f \leq r$ and $\operatorname{supp} f \subseteq X$. Then

$$\int_A \int_A f(\Phi)\kappa^{[n]}_{(a)}(d\Phi)\nu(da) \leq r \cdot \int \kappa^{[n]}_{(a)}(\chi(X)>0)\nu(da) \to 0 \quad .$$

Conversely, take for $X \in \mathcal{B}$ an $f \in \mathbb{F}_c$ such that $f \geq 1_X$. Then

$$\int \kappa^{[n]}_{(a)}(\chi(X)>0)\nu(da) \leq \int_A \int_A f(\Phi)\kappa^{[n]}_{(a)}(d\Phi)\nu(da) \to 0 \quad . \blacksquare$$

2.6.5. Example. With the assumptions and notations of 2.2.4. there holds $\mu \in N_\kappa^{nst}$.

Proof. We put

$$S := \kappa_{(0)} = (1/3)\delta_{\delta_{-1}} + (1/3)\delta_{2\delta_1} + (1/3)\delta_o$$

and obtain, using the operation $*$ of convolution of measures on $\mathcal{P}(\mathbf{Z})$, for all $a \in A$, $n \in \mathbf{Z}_+$ and $X \in \mathcal{B}$

$$\kappa^{[n]}_{(a)} = \kappa^{[n]}_{(0)}\,(\delta_a * \chi \in (.)) \quad ; \quad J^{[n]}(a,(.)) = \delta_a * (\Lambda_S)^n$$

and

$$\int \kappa^{[n]}_{(a)}(\chi(X)>0)\mu(da) \leq (\#X) \sum_{i \in \mathbf{Z}} \kappa^{[n]}_{(0)}(\chi(\{i\})>0) \quad .$$

Because of 2.6.4. it is thus sufficient to check the convergence

$$r_n := \sum_{i \in \mathbf{Z}} \kappa^{[n]}_{(0)}(\chi(\{i\})>0) \to 0 \quad .$$

For a Markovian random sequence $\Phi_0, \Phi_1, \ldots$ of counting measures which is distributed according to $\overrightarrow{\delta_{\delta_0}}$, the sequence $\Phi_0(A), \Phi_1(A), \ldots$ is obviously a Galton-Watson process starting at time zero in 1 and having offspring distribution $(1/3)(\delta_0+\delta_1+\delta_2)$. According to a well known result from the theory of critical Galton-Watson processes (see [Harris (1963), section 10.2.]), there exists a $c \in \mathbf{R}_+$ such that

$$n\,\kappa^{[n]}_{(0)}(\chi \neq o) \leq c \qquad (n=1,2,\ldots) \quad .$$

We now obtain for all $n \in \mathbf{Z}_+$ and all $\varepsilon>0$

$$\limsup_{n\to\infty} r_n \leq \limsup_{n\to\infty} \sum_{|i/n-1/3| \leq \varepsilon} \kappa^{[n]}_{(0)}(\chi(\{i\})>0) + \limsup_{n\to\infty} \sum_{|i/n-1/3|>\varepsilon} (\Lambda_S)^n(\{i\}) \quad .$$

By the law of large numbers, the second term vanishes. The first term, however, is bounded by

$$\limsup_{n\to\infty} 2\varepsilon n \kappa^{[n]}_{(0)}(\chi \neq o) \leq 2c\varepsilon \quad . ■$$

The convergence property $(\Pi_\nu)_{\kappa^{[n]}} \Rightarrow \delta_o$ which is characteristic for purely unstable measures $\nu \in N_J$ may be extended as follows:

2.6.6. Proposition. For any distribution P on $\mathcal{M}$ with N_κ^{nst} - bounded intensity measure there holds

$$P_{\kappa^{[n]}} \Rightarrow \delta_o \, .$$

Proof. For $\Lambda_P \leq \lambda \in N_\kappa^{nst}$, $X \in \mathcal{B}$ and $n \in \mathbf{Z}_+$ one obtains

$$P_{\kappa^{[n]}}(\Phi(X)>0) = \int \kappa^{[n]}_{(\Phi)}(\chi(X)>0)\,P(d\Phi)$$

$$\leq \int \langle \Phi , \kappa^{[n]}_{(.)}(\chi(X)>0)\rangle\,P(d\Phi) \leq \langle \lambda , \kappa^{[n]}_{(.)}(\chi(X)>0)\rangle \quad .$$

The last expression tends to zero by 2.6.4. ■

An immediate consequence of 2.2.5., 2.3.8., 2.3.9., 2.6.1. and 2.6.2. is

2.6.7. The sets N_κ^{st} and N_κ^{nst} are closed with respect to the lattice operations $\cup, \cap$, and there holds

$$\sigma\cap\mu = o\,,\quad \sigma\cup\mu = \sigma+\mu \qquad (\sigma\in N_\kappa^{st};\ \mu\in N_\kappa^{nst})\ .$$

However, $\sigma\in N_\kappa^{st}$ and $\mu\in N_\kappa^{nst}$ does not imply $\sigma\wedge\mu = o$:

2.6.8. Example. According to 2.5.5. and 2.6.5., the two measures μ and σ, which were introduced in Example 2.2.4., obey the conditions $\sigma\in N_\kappa^{st}$, $\mu\in N_\kappa^{nst}$. On the other hand μ and σ are equivalent measures on $\mathcal{A}$. ■

2.6.9. For all $\nu\in N_J$ there hold:

$$\nu_{st} := \Lambda_{V_{(\nu)}} = \max\{\sigma\colon \sigma\in N_\kappa^{st};\ \ \sigma\leq\nu\}\ ;\ \ V_{(\nu)} = \Pi^\kappa_{\nu_{st}}\ ;$$
$$\nu_{nst} := \nu-\nu_{st} = \max\{\gamma\colon\ \gamma\in N_\kappa^{nst}, \gamma\leq\nu\}\ .$$

Proof. Obviously ν_{st} belongs to N_κ^{st}. Because of $\nu_{st}\leq\nu$, the measure $\nu-\nu_{st}$ belongs to N_J, and there holds

$$V_{(\nu)} = V_{(\nu_{st})} * V_{(\nu_{nst})} = \Pi^\kappa_{(\nu_{st})}\ .$$

By the definition of ν_{st}, this shows that $\Lambda_{V_{(\nu_{nst})}} = o$, i. e. $\nu_{nst}\in N_\kappa^{nst}$.

If now σ is some measure in N_κ^{st} with the property $\sigma\leq\nu$, we obtain

$$V_{(\sigma)} * V_{(\nu-\sigma)} = V_{(\nu)} = \Pi^\kappa_{\nu_{st}}$$

and hence $\sigma = \Lambda_{V_{(\sigma)}} \leq \nu_{st}$.

If, finally, γ is some measure in N_κ^{nst} with the property $\gamma\leq\nu$, then the equality

$$\Pi^\kappa_{\nu_{st}} = V_{(\nu)} = V_{(\gamma)} * V_{(\nu-\gamma)} = V_{(\nu-\gamma)}$$

implies the relation $\nu_{st}\leq\nu-\gamma$; i. e. $\gamma\leq\nu_{nst}$. ■

An easy consequence of 2.6.9. is

2.6.10. Proposition. Any $\nu\in N_J$ has exactly one representation as sum of a stable and a purely unstable measure, namely $\nu = \nu_{st} + \nu_{nst}$.

Proof. Let $\nu = \nu_1 + \nu_2$ be another representation of that kind. Then, by 2.6.9., $\nu_1\leq\nu_{st}$ and $\nu_2\leq\nu_{nst}$, which implies equality in both cases. ■

By 2.6.10., N_J is the direct sum of N_κ^{st} and N_κ^{nst}; ν_{st} and ν_{nst} will be called the **stable** and the **purely unstable component** of $\nu\in N_J$, respectively.

An easy application of 2.6.10. and 2.6.7. together with 2.2.5. shows

2.6.11. The mappings $\nu\to\nu_{st}$ and $\nu\to\nu_{nst}$ from N_J onto N_κ^{st} and N_κ^{nst}, respectively, are homomorphisms with respect to the linear as well as with respect to the lattice structure.

Proof. We restrict ourselves to the proof of

$$(\nu\cup\gamma)_{st} = \nu_{st}\cup\gamma_{st} \qquad (\nu,\gamma\in N_J) \ .$$

Indeed, by 2.6.7. we get

$$\nu\cup\gamma = (\nu_{st}\cup\nu_{nst})\cup(\gamma_{st}\cup\gamma_{nst})$$
$$= (\nu_{st}\cup\gamma_{st})\cup(\nu_{nst}\cup\gamma_{nst}) = (\nu_{st}\cup\gamma_{st})+(\nu_{nst}\cup\gamma_{nst}) \ ,$$

which shows the assertion by 2.6.10. The other properties follow in a similar way. ■

2.3.5. and the definition of ν_{st} show :

2.6.12. The mapping $\nu\rightarrow\nu_{st}$ from N_J onto $N_K{}^{st}$ is $\mathcal{N}$, $\mathcal{N}$-measurable.

Combining 2.3.4., 2.3.8., 2.6.1. and 2.6.11., we obtain

2.6.13. Proposition. If $(\nu_i)_{i\in I}$ is an at most countable family of measures in N_J whose sum $\nu := \sum_{i\in I}\nu_i$ belongs to N , then also ν belongs to N_J and there holds

$$\nu_{st} = \sum_{i\in I}(\nu_i)_{st} \quad ; \quad \nu_{nst} = \sum_{i\in I}(\nu_i)_{nst} \ .$$

For all $\nu\in N_J{}^{ex}$, the measure ν_{st} is a minorant of ν in N_J, hence is of the form $r\cdot\nu$ for some $r\in\mathbf{R}_+$. If $r=0$, then ν belongs to $N_K{}^{nst}$. Otherwise, 2.3.10. allows to conclude that $\nu\in N_K{}^{st}$. Thus there holds

2.6.14. $N_J{}^{ex} \subseteq N_K{}^{st} \cup N_K{}^{nst}$.

2.7. Equilibrium Distributions for the Family Dynamics

We are now going to exploit the constructions of section 1.6. First, however, we will introduce on the space of "family systems" M a specific semi-order $\leq$ according to the "fineness of the family structure":

For $\Phi_2 = \sum_{i\in I}\delta_{\Phi_i} \in M$, we say that $\Phi_1\in M$ obeys the relation $\Phi_1 \leq \Phi_2$ iff there exists a partition $I=\bigcup_{s\in S} I_s$ of I into pairwise disjoint non-void sets I_s , $s\in S$, such that

$$\Phi_1 = \sum_{s\in S}\delta_{\sum_{i\in I_s}\Phi_i} \ .$$

Immediate consequences of this definition are:

2.7.1. $\delta_{sp\Phi} \leq \Phi \leq \sum_{a\in supp(sp\Phi)}(sp\Phi)(\{a\})\ \delta_{\delta_a} \qquad (\Phi\in M)$.

2.7.2. $\Phi_1 \leq \Phi_2$ implies

$$sp\Phi_1 = sp\Phi_2 \ ; \ \Phi_1(X) \leq \Phi_2(X) \text{ for all } X\in\mathcal{B}.$$

2.7.3. If $\mathcal{S}$ is some generating subring of $\mathcal{B}$, then $\Phi_1 = \Phi_2$ is implied by $\Phi_1 \leq \Phi_2$ together with $\Phi_1(X)=\Phi_2(X)$ for all $X\in\mathcal{S}$.

Note that the converse of 2.7.2. holds under the additional assumption that $sp\Phi_1$ and $sp\Phi_2$ are simple, but not in general. We now show the valitity of

2.7.4. The convergences $\Phi_n \xrightarrow{M} \Phi$, $\Phi_n' \xrightarrow{M} \Phi'$ together with the relations $\Phi_n' \leq \Phi_n$, n=1,2, ..., imply that $\Phi' \leq \Phi$.

Proof. By 2.7.2. there holds $sp\Phi_n = sp\Phi_n'$, $n \in \mathbf{Z}_+$; due to 1.6.6. there follows

$$sp\Phi_n = sp\Phi_n' \xrightarrow{M} sp\Phi' = sp\Phi \ .$$

Let us put, for $X \in \mathcal{B}$ and $\Psi = \sum_{i \in I} \delta_{\Psi_i} \in M$,

$$_X\Psi := \sum_{i \in I, \Psi_i(X)>0} \delta_{_X\Psi_i} \ .$$

Obviously, it suffices to show that

$$_X\Phi' \leq {}_X\Phi \text{ for all } X \in \mathcal{B}_{sp\Phi} \ .$$

1.2.4. applied to the convergent sequence $({}_X\Phi_n)$ ensures, for n large enough, the existence of indexings such that

$$_X\Phi_n = \sum_{i \in I} \delta_{\Phi_{i,n}} \ ; \ {}_X\Phi = \sum_{i \in I} \delta_{\Phi_i} \ ; \ \Phi_{i,n} \Rightarrow \Phi_i \text{ for all } i \in I, \text{ where I is a finite set} .$$

By assumption there holds $\Phi_n' \leq \Phi_n$, $n \in \mathbf{Z}_+$; hence there exist partitions $\{I_{1,n}, \ldots, I_{r_n,n}\}$ of I such that

$$_X\Phi_n' = \sum_{1 \leq s \leq r_n} \delta_{\sum_{i \in I_{s,n}} \Phi_{i,n}} \ .$$

As there are only finitely many such partitions, at least one of them occurs for infinitely many n, i. e. there exists a partition $\{I_1, \ldots, I_r\}$ and a subsequence (Φ_{n_k}) such that

$$_X\Phi_{n_k}' = \sum_{1 \leq s \leq r} \delta_{\sum_{i \in I_s} \Phi_{i,n_k}} \ .$$

Now one more application of 1.2.4. shows that the limit of the sequence $({}_X\Phi_{n_k}')$ (which, of course, coincides with ${}_X\Phi'$) equals $\sum_{1 \leq s \leq r} \delta_{\sum_{i \in I_s} \Phi_i}$; hence follows ${}_X\Phi' \leq {}_X\Phi$. ■

Any decreasing resp. increasing sequence (Φ_n) in $[M, \leq]$, has a greatest lower resp. least upper bound Φ_∞ with respect to the semi-order $\leq$. We write, in this case, $\Phi_n \downarrow \Phi_\infty$ resp. $\Phi_n \uparrow \Phi_\infty$. Obviously, in both situations, there holds $\Phi_n \xrightarrow{M} \Phi_\infty$ and $sp\Phi_n = sp\Phi_\infty$ for n=1,2,...

The semi-order $\leq$ in M induces a semi-order $\leq$ in $\mathbb{P}$; one defines that $G_1 \leq G_2$ iff there exists a distribution U on $\mathcal{M} \otimes \mathcal{M}$ with the properties

$$U(\Phi_1 \leq \Phi_2) = 1 \ ; \ U(\Phi_i \in (.)) = G_i \ \text{ for } i=1,2,$$

i. e. iff there exists a stochastic kernel K from $[\mathbb{M}, \mathcal{M}]$ to $[\mathbb{M}, \mathcal{M}]$ such that

$$G_1 * K = G_2 \quad \text{and} \quad K(\Phi_1, \Phi_1 \leq \Phi) = 1 \text{ for all } \Phi_1 \in \mathbb{M} \quad .$$

Direct consequences of this definition and the above mentioned elementary properties of the semi-order $\leq$ in $\mathbb{M}$ are:

2.7.5. If two distributions $G, H \in \mathbb{P}$ obey the relation $G \leq H$ and if $(\mathrm{sp}G)_\kappa$ and $(\mathrm{sp}H)_\kappa$ exist, there also $G_\kappa \leq H_\kappa$ holds.

2.7.6. $G \leq H$ implies that $\mathrm{sp}G = \mathrm{sp}H$ and $G_X \leq H_X$ for all $X \in \mathcal{B}$.

2.7.7. The convergences $G_n \Rightarrow G$; $H_n \Rightarrow H$ together with the relation $G_n \leq H_n$ n=1,2,..., imply that $G \leq H$.

2.7.8. $G \leq H$ and $G_X = H_X$ for all $X \in \mathcal{B}$ imply that $G = H$.

An easy consequence of 2.7.6. and 2.7.8. is

2.7.9. If G and H are distributions of first order in $\mathbb{P}$ obeying $G \leq H$ and $\Lambda_G = \Lambda_H$, then $G = H$.

Now assume that (G_n) is a decreasing sequence in $[\mathbb{P}, \leq]$. Using the stochastic kernel K figuring in the definition of the semi-order on $\mathbb{P}$, one constructs the distribution W of an almost surely decreasing sequence (Φ_n) of counting measures on $\mathcal{A}$ with the property $W(\Phi_n \in (.)) = G_n$, n=1,2,... As the greatest lower bound of (Φ_n) exists W-almost surely, we may put

$$G := W(\Phi_\infty \in (.)) \quad ,$$

thus obtaining

$$G_n \Rightarrow G , \quad \mathrm{sp}G = \mathrm{sp}G_n \quad \text{and} \quad G \leq G_n \quad \text{for } n=1,2,...$$

In this case we write $G_n \downarrow G$, because G is in fact the greatest lower bound of $\{G_1, G_2, ...\}$ in $[\mathbb{P}, \leq]$.

In the following we need the continuity assertion

2.7.10. The convergence $G_n \downarrow G$ and the existence of $(\mathrm{sp}G)_\kappa$ implies that $(G_n)_\kappa \downarrow G_\kappa$.

Proof. 1. By 2.7.5., also $((G_n)_\kappa)$ is decreasing with respect to $\leq$.

2.Now assume that $\Phi_n \downarrow \Phi_\infty$ and $(\mathrm{sp}\Phi_1) \in {}_\kappa\mathbb{M}$. Subject to these assumptions, we show the valitidy of

(o) $$\kappa(\Phi_n) \downarrow \kappa(\Phi_\infty) \quad .$$

To this purpose let $\mathrm{sp}\Phi_1 = \sum_{i \in I} \delta_{a_i}$ be some indexing of individuals in $\mathrm{sp}\Phi_1$. Obviously there exists a sequence of partitions with increasing coarseness

$$I = \bigcup_{s \in S_n} I_{n,s}$$

such that $\Phi_n = \sum_{s \in S_n} \delta_{\sum_{i \in I_{n,s}} \delta_{a_i}}$ for all n = 1,2....

Let $I = \bigcup_{s\in S_\infty} I_{\infty,s}$ be the finest partition which is coarser than all the partitions introduced above. Then there holds

$$\Phi_\infty = \sum_{s\in S_\infty} \delta_{\sum_{i\in I_{\infty,s}} \delta_{a_i}} \quad .$$

Obviously, now any family $(\chi_i)_{i\in I}$ of counting measures on $\mathcal{A}$ with the property $\sum_{i\in I}\chi_i \in M$ obeys the convergence relation

$$\sum_{s\in S_n}' \delta_{\sum_{i\in I_{n,s}} \chi_i} \downarrow \sum_{s\in S_\infty}' \delta_{\sum_{i\in I_{n,s}} \chi_i} \quad ,$$

where the symbol Σ' means that all summands of the form δ_o are to be omitted. But under $\bigotimes_{i\in I} \kappa_{(a_i)}$, the left hand side has distribution $\kappa_{(\Phi_n)}$, whereas the r. h. s. has distribution $\kappa_{(\Phi_\infty)}$, which shows assertion (o).

3. If, finally, W is a distribution on $\mathcal{M}^{\otimes \mathbf{Z}_+}$ with the properties

$$W((\Phi_n) \text{ is decreasing w. r. to } \leq) = 1 \ ; \ W((\Phi_n \in (.)) = G_n \ \text{ for } n=1,2...,$$

then 2. allows to conclude that

$$W(\kappa_{(\Phi_n)} \downarrow \kappa_{(\Phi_\infty)}) = 1 \ ,$$

hence

$$W(\kappa_{(\Phi_n)} \Rightarrow \kappa_{(\Phi_\infty)}) = 1 \ ,$$

and also

$$(G_n)_\kappa = \int \kappa_{(\Phi_n)}(.) W(d(\Phi_i)) \Rightarrow \int \kappa_{(\Phi_\infty)}(.) W(d(\Phi_i)) = G_\kappa \, .$$

We have thus identified G_κ as weak limit, hence also as greatest lower bound, of the decreasing sequence $((G_n)_\kappa)$. ■

For all $\Phi \in M$ we define

$$\Phi^x := \sum_{a\in \mathrm{supp}\Phi} \Phi(\{a\}) \delta_{\delta_a} \ ,$$

and for all distributions L on $\mathcal{M}$ we put

$$L^x := L(\Phi^x \in (.))$$

obtaining in this way a mapping $L \rightarrow L^x$ from **P** into $\mathbb{P}$.

As a counterpart to 2.7.1. we observe

2.7.11. $(\mathrm{sp}H)(\delta_\Phi \in (.)) \leq H \leq (\mathrm{sp}H)^x \qquad (H \in \mathbb{P})$.

Also note that for the clustering field

$$\tau_{(a)} := \delta_{\delta_{\delta_a}} \qquad (a \in A)$$

on $[A, \rho_A]$ with phase space $[\mathbb{A}, \rho_{\mathbb{A}}]$ there holds

$$L^x = L_\tau \qquad (L \in \mathbf{P}) \ .$$

If now L is some distribution in **P** such $\vec{L}$ exists, then we obtain by putting

$$\mathcal{G}_{L,n} := (L^x)_{\kappa[n]} \qquad (n \in \mathbf{Z}_+)$$

distributions $\mathcal{G}_{L,n}$ in $\mathbb{P}$ with the property $\mathrm{sp}\,\mathcal{G}_{L,n} = L_{\kappa[n]}$. Note that $\mathcal{G}_{L,n}$ takes into account the family structure within the n-th generation Φ_n which arose since the time zero.

Immediate consequences from this definition are

2.7.12. For all $L \in \mathbf{P}$ for which $\vec{L}$ exists, there holds

$$\mathcal{G}_{L,n+1} \leq \mathcal{G}_{L_\kappa,n} \qquad (n \in \mathbf{Z}_+) \quad .$$

2.7.13. Under the assumption of 2.7.12, the equality

$$\Lambda_{\mathcal{G}_{L,n}} = \int \kappa^{[n]}_{(a)}((.)\setminus\{o\})\Lambda_L(da) \qquad (n \in \mathbf{Z}_+)$$

is valid.

2.7.14. If L_1 and L_2 are distributions in **P** for which $\vec{L}_1$ and $\vec{L}_2$ exist, then there hold

$$\mathcal{G}_{L_1 * L_2,n} = \mathcal{G}_{L_1,n} * \mathcal{G}_{L_2,n} \qquad (n \in \mathbf{Z}_+)$$

and for $c_1, c_2 \in \mathbf{R}_+$ such that $c_1 + c_2 = 1$,

$$\mathcal{G}_{c_1 L_1 + c_2 L_2,n} = c_1 \mathcal{G}_{L_1,n} + c_2 \mathcal{G}_{L_2,n} \qquad (n \in \mathbf{Z}_+) \quad .$$

From 1.5.8. and 2.7.13. one concludes

2.7.15. If for a measure $\gamma \in N$ all $\gamma * J^{[n]}$, n=1,2,..., belong to N, then one has

$$\mathcal{G}_{\Pi_\gamma,n} = \Pi_{\widetilde{(\Pi_\gamma)_{\kappa[n]}}} \qquad (n=1,2,\ldots) \quad .$$

If now L is a distribution in P for which $\vec{L}$ exists and for which the sequence $(\mathcal{G}_{L,n})$ converges weakly towards some $G \in \mathbb{P}$, then we write $\mathcal{G}_L := G$. By 1.6.7. then holds $L_{\kappa[n]} \Rightarrow \mathrm{sp}(\mathcal{G}_L)$.

In virtue of 2.7.15. there follows

2.7.16. Proposition. For all $v \in N_\kappa^{st}$,

$$\mathcal{G}_{\Pi_v} = \Pi_{\widetilde{\Pi^\kappa_v}} \quad .$$

A quick consequence of 2.7.14. is

2.7.17. If f L_1 and L_2 are distributions in **P** for which $\mathcal{G}_{L_1}$ and $\mathcal{G}_{L_2}$ exist, then

$$\mathcal{G}_{L_1 * L_2} = \mathcal{G}_{L_1} * \mathcal{G}_{L_2}$$

and for $c_1, c_2 \in \mathbf{R}_+$ such that $c_1 + c_2 = 1$,

$$\mathcal{G}_{c_1 L_1 + c_2 L_2} = c_1 \mathcal{G}_{L_1} + c_2 \mathcal{G}_{L_2}.$$

Now let P some distribution in $\mathbf{P}^{inv}$. By 2.7.12. the sequence $(\mathcal{G}_{P,n})$ is decreasing with respect to $\leq$. Hence results $\mathcal{G}_{P,n} \downarrow \mathcal{G}_P$ and $\mathrm{sp}\,\mathcal{G}_P = P$. By 2.7.10. now follows

$$\mathcal{G}_{P,n+1} = (\mathcal{G}_{P,n})_\kappa \downarrow (\mathcal{G}_P)_\kappa ,$$

which shows the validity of

2.7.18. Propositon For all $P \in \mathbf{P}^{inv}$ the distribution $\mathcal{G}_P$ exists and obeys

$$\mathcal{G}_{P,n} \downarrow \mathcal{G}_P \ ; \ sp\mathcal{G}_P = P \ ; \ (\mathcal{G}_P)_\kappa = \mathcal{G}_P \ .$$

For $P \in \mathbf{P}^{inv}$, we call $\mathcal{G}_P$ the **family equilibrum distribution associated with** P.

The existence of $\vec{L}$ and $\mathcal{G}_L$ and the convergence $L_{\kappa[n]} \Rightarrow P$, $P \in \mathbf{P}^{inv}$, does not imply the equality $\mathcal{G}_L = \mathcal{G}_P$:

2.7.19. Example. Put $A = \{1,2\}$ and

$$\kappa_{(a)} := \begin{cases} \delta_{2\delta_2} & \text{for } a = 1 \\ \delta_{\delta_2} & \text{for } a = 2 \end{cases} .$$

$L := \delta_{\delta_1}$ obeys $L_\kappa[n] \Rightarrow P := \delta_{2\delta_2}$, on the other hand $\mathcal{G}_L = \delta_{\delta_{2\delta_2}}$ is different from $\mathcal{G}_P = \delta_{2\delta_{\delta_2}}$. ■

2.7.20. For $P_1, P_2 \in \mathbf{P}^{inv}$ there holds $\mathcal{G}_{P_1} \precsim \mathcal{G}_{P_2}$ iff $P_1 \precsim P_2$.

Proof. $P_1 \precsim P_2$ clearly implies that $P_1{}^x \precsim P_2{}^x$; this leads, for all $n \in \mathbf{Z}_+$, to $\mathcal{G}_{P_1,n} \precsim \mathcal{G}_{P_2,n}$ by 1.4.4. and finally to $\mathcal{G}_{P_1} \precsim \mathcal{G}_{P_2}$ by 1.1.10. The converse direction follows from the monotonicity of the mapping $H \to spH$ with respect to the order $\precsim$. ■

If now P is a distribution of first order in $\mathbf{P}^{inv}$ we obtain in virtue of 2.7.6. and 2.7.18.:

$$(\mathcal{G}_{P,n})_X \precsim (\mathcal{G}_{P,0})_X = P_X \qquad (X \in \mathcal{B}, n \in \mathbf{Z}_+) \ .$$

As $\mathcal{G}_{P,n} \Rightarrow \mathcal{G}_P$, we conclude by 1.1.8. that $\Lambda_{\mathcal{G}_{P,n}} \xrightarrow{N} \Lambda_{\mathcal{G}_P}$. Hence follows by 2.7.13.

$$\widetilde{(\Pi_{\Lambda_P})_{\kappa[n]}} = \int \kappa^{[n]}_{(a)}((.)\setminus\{o\})\Lambda_P(da) \xrightarrow{W} \Lambda_{\mathcal{G}_P}$$

and we recognize the validity of

2.7.21. For all $P \in \mathbf{P}^{inv}$ of first order there holds

$$\Lambda_{\mathcal{G}_P} = \widetilde{\Pi^\kappa{}_{\Lambda_P}} \ .$$

For $P \in \mathbf{P}^{inv}$ and $H \in \mathbb{P}$, the properties $spH = P$ and $H_\kappa = H$ do not imply that $H = \mathcal{G}_P$ (consider, e. g., $H = P(\delta_\Phi \in (.))$. They do, however, imply that $H \leq P^x$, leading, by 2.7.5., to $H \leq \mathcal{G}_{P,n}$ for all $n \in \mathbf{Z}_+$, and finally, by 2.7.7., to $H \leq \mathcal{G}_P$. Thus we have proved

2.7.22. Proposition. For $P \in \mathbf{P}^{inv}$ and $H \in \mathbb{P}$ the properties $H_\kappa = H$ and $spH = P$ imply that $H \leq \mathcal{G}_P$.

Combining 2.7.9., 2.7.21. and 2.7.22. we arrive immediately at

2.7.23 Theorem. For any $P \in \mathbf{P}^{inv}$ of first order, an $H \in \mathbb{P}$ coincides with $\mathcal{G}_P$ iff there holds

$$\mathrm{sp}H = P \;;\; H_\kappa = H \;;\; \Lambda_H = \widetilde{\Pi^\kappa_{\Lambda_P}} \quad .$$

If now ν is some measure in N_κ^{st}, we observe that

$$(\Pi_{\widetilde{\Pi^\kappa_\nu}})_\kappa = \Pi_{(\widetilde{\Pi^\kappa_\nu})_\kappa((.)\setminus\{o\})} = \Pi_{\widetilde{\Pi^\kappa_\nu}} \quad .$$

Moreover, 1.6.12. says that $\mathrm{sp}\Pi_{\widetilde{\Pi^\kappa_\nu}} = \Pi^\kappa_\nu$. In virtue of Theorem 2.7.23. this yields the **basic structural result:**

2.7.24. Theorem. For all measures $\nu \in N_\kappa^{st}$ there holds

$$\mathcal{G}_{\Pi^\kappa_\nu} = \Pi_{\widetilde{\Pi^\kappa_\nu}} \; .$$

Hence the representation $P = \mathrm{sp}\mathcal{G}_P$, which is valid for all $P \in \mathbf{P}^{inv}$, coincides, for $P = \Pi^\kappa_\nu$, with the standard clustering representation of the infinitely divisible distribution Π^κ_ν.

One more application of 2.7.21. shows, together with Theorem 2.7.24., the validity of

2.7.25. Theorem. A distribution $P \in \mathbf{P}^{inv}$ of first order is of Poisson type iff its associated family equilibrium distribution $\mathcal{G}_P$ is Poissonian.

If $P \in \mathbf{P}^{inv}$, however, is not of first order (and hence by definition cannot be of Poisson type),the distribution $\mathcal{G}_P$ may still be Poissonian, as the following example shows:

2.7.26. Example. Let $[A, \rho_A]$, κ and μ be as in Example 2.1.6. There it has been shown that $P := (\Pi_\mu)_\kappa$ belongs to $\mathbf{P}^{inv}$ but is not of first order. Nevertheless we observe that

$$\mathcal{G}_{P,n} = \Pi_{(\varrho_\mu)_\kappa} = \Pi_{\tilde{P}} \qquad (n \in \mathbf{Z}_+)$$

and hence also $\mathcal{G}_P = \Pi_{\tilde{P}}$, i. e. $\mathcal{G}_P$ is Poissonian . ■

Theorem 2.7.25. has an interesting consequence, stating that any equilibrium distribution which is a convolution divisor of some Poisson type equilibrium distribution, itself must be of Poisson type:

2.7.27. Theorem. Let ν be some measure in N_κ^{st}, and $\Pi^\kappa_\nu = P_1 * P_2$ for some $P_1, P_2 \in \mathbf{P}^{inv}$. Then P_1 and P_2 are of Poisson type.

Proof. In virtue of 2.7.17. and 2.7.24. there follows

$$\Pi_{\widetilde{\Pi^\kappa_\nu}} = \mathcal{G}_{P_1} * \mathcal{G}_{P_2} \quad .$$

By the generalized Raikov-theorem [MKM. Prop. 1.7.9.], the distributions $\mathcal{G}_{P_1}$ and $\mathcal{G}_{P_2}$ are Poissonian. Hence the assertion follows by Theorem 2.7.25. ■

If now for a measure $\nu \in N_{\kappa}^{st}$ there holds

$$\Pi^{\kappa}_{\nu} = P*Q \; ; \quad P \in \mathbf{P}^{inv} \quad ; \quad Q \in \mathbf{P},$$

then 2.1.1. ensures that $Q \in \mathbf{P}^{inv}$. Hence, by 2.7.27. and 2.3.9. the mapping $P \rightarrow \Lambda_P$ provides a 1-1 correspondence between the set of all divisors of Π^{κ}_{ν} which belong to $\mathbf{P}^{inv}$ and the set $\{\gamma \in N_J : \gamma \leq \nu\}$.

Now we turn once again to the special one-dimensional distributions H_X. It is immediate from the definitions that

2.7.28. $H_X(\{o\}) = (\text{sp}H)_X(\{o\}) \qquad (H \in \mathbb{P} \; ; \; X \in \mathcal{B})$.

If now L is some distribution on $\mathcal{M}$ for which $\vec{L}$ exists, then, for all $X \in \mathcal{B}$ and $n \in \mathbf{Z}_+$, the one-dimensional distribution $(\mathcal{G}_{L,n})_X$ may be interpreted as distribution of the number of forefathers in generation zero of all those individuals in the n-th generation which are situated in X.

For $P \in \mathbf{P}^{inv}$ we may imagine that the stochastic branching mechanism "began at time $-\infty$". In this case, $(\mathcal{G}_P)_X$ may, for **all** $X \in \mathcal{B}$, be interpreted as distribution of the "asymptotic fathers´ number" of those individuals in the random population at a fixed time (say, time zero) which are situated in X, because there holds

2.7.29. Proposition. $(\mathcal{G}_{P,n})_X \Rightarrow (\mathcal{G}_P)_X$ for all $P \in \mathbf{P}^{inv}$ and $X \in \mathcal{B}$.

Proof. 1. Noting that $\partial(X) \subseteq \{\Phi : \Phi \in \mathbb{A}, \Phi(\partial X) > 0\}$ for all $X \in \mathcal{B}$, the asserted convergence is clear for all $X \in \mathcal{B}_P$.

2. In virtue of 2.7.30. below, there exists, for any $X \in \mathcal{B}$ and $\varepsilon > 0$, a set $X_\varepsilon \in \mathcal{B}_P$ such that

$$P(\Phi((X \setminus X_\varepsilon) \cup (X_\varepsilon \setminus X)) > 0) < \varepsilon.$$

Hence we obtain for all $n \in \mathbf{Z}_+$

$$\| (\mathcal{G}_{P,n})_X - (\mathcal{G}_{P,n})_{X_\varepsilon} \| < \varepsilon \quad \text{and} \quad \| (\mathcal{G}_P)_X - (\mathcal{G}_P)_{X_\varepsilon} \| < \varepsilon$$

Using 1., we infer that

$$\limsup_{n \to \infty} \| (\mathcal{G}_{P,n})_X - (\mathcal{G}_P)_X \| \quad \leq \quad \varepsilon + \limsup_{n \to \infty} \| (\mathcal{G}_{P,n})_{X_\varepsilon} - (\mathcal{G}_P)_{X_\varepsilon} \| + \varepsilon \quad = \quad 2\varepsilon \; ,$$

because weak convergence of distributions on $\mathcal{P}(\mathbf{Z}_+)$ is equivalent to their convergence in variation distance. ■

We have to append:

2.7.30. For any $P \in \mathbf{P}$, $\varepsilon > 0$ and $X \in \mathcal{B}$, there exists a set $X_\varepsilon \in \mathcal{B}_P$ such that

$$P(\Phi((X \setminus X_\varepsilon) \cup (X_\varepsilon \setminus X)) > 0) < \varepsilon.$$

Proof. 1. First we show that there exists an open set U containing X and a closed subset D of X such that

(*) $$P(\Phi((U\setminus D)>0) < \varepsilon \quad .$$

Let B be some fixed bounded open subset of A which contains X. Choose $n\in \mathbf{Z}_+$ such that $P(Y) < \varepsilon/2$, where

$$Y := \{\Phi\in M : \Phi(B)>n\} \quad .$$

Being finite, the measure ν on the Polish space $[A, \rho_A]$ defined by

$$\nu(.) := \int_Y \Phi((.)\cap B)P(d\Phi)$$

is regular, hence there exists a closed set D and an open set U having the properties $D\subseteq X\subseteq U\subseteq B$ and $\nu(U\setminus D) < \varepsilon/2$. But this implies that

$$P(\Phi((U\setminus D)>0 \,;\, \Phi\in Y) < \varepsilon/2 \quad ,$$

which yields (*).

2. For D and U constructed as in 1., [MKM, 3.1.6.] ensures the existence of a set $X_\varepsilon\in \mathcal{B}_P$ such that $D\subseteq X_\varepsilon\subseteq U$. Together with (*), this completes the proof. ■

Immediate consequences of 2.7.21. and 2.7.24., respectively, are

2.7.31. Proposition. For all $P\in \mathbf{P}^{inv}$ of first order and all $X\in \mathcal{B}$ there holds

$$\sum_{m>0} m(\mathcal{G}_P)_X(\{m\}) = \widetilde{\Pi^\kappa_{\Lambda_P}}(\Psi(X)>0) \qquad (X\in \mathcal{B}).$$

2.7.32. Proposition. For all measures $\nu\in N_\kappa^{st}$ one has

$$(\mathcal{G}_{\Pi^\kappa_\nu})_X = \pi_{\widetilde{\Pi^\kappa_\nu}(\Psi(X)>0)} \qquad (X\in \mathcal{B}) \quad .$$

The distributions $(\mathcal{G}_{\Pi^\kappa_\nu})_X$, $X\in \mathcal{B}$, thus reveal, in a way, the "Poissonian infinite past" of the distribution Π^κ_ν. Under a weak additional condition the Π^κ_ν even may be characterized in this way:

2.7.33. Theorem. Let ν be a measure in N_κ^{st} with the property that

$$\Pi^\kappa_\nu\,(\Phi \text{ is simple}) = 1.$$

Then a distribution $P\in \mathbf{P}^{inv,\nu}$ coincides with Π^κ_ν iff all distributions $(\mathcal{G}_P)_X$, $X\in \mathcal{B}$, are Poissonian.

Proof. In virtue of Proposition 2.7.32, it remains to show the "if"-direction. By assumption, P and Π^κ_ν have one and the same intensity measure ν. Hence according to 2.7.31. for any $X\in \mathcal{B}$ the two distributions $(\mathcal{G}_P)_X$ and $(\mathcal{G}_{\Pi^\kappa_\nu})_X$ have the same expectation value. Now our assumptions imply that

$$(\mathcal{G}_P)_X = (\mathcal{G}_{\Pi^\kappa_\nu})_X \qquad (X\in \mathcal{B}) \quad ,$$

leading, by 2.7.28., to

(*) $$P(\Phi(X) = 0) \;=\; \Pi^\kappa_\nu\,(\Phi(X) = 0) \qquad (X\in \mathcal{B}) \quad .$$

But as Π^κ_ν obeys, by assumption, the conditon

$$\Pi^\kappa_\nu\,(\Phi \text{ is simple}) \;=\; 1 \quad ,$$

(*) yields, by a standard result from the theory of random point fields (cf. [MKM, Prop. 1.4.7.]) the equality

$$\Pi^{\kappa}_{v} = P^* := P\Big(\big(\sum_{a\in \operatorname{supp}\Phi} \min(1,\Phi(\{a\})\delta_a\big)\in(.)\Big) .$$

On the other hand we have

$$\Lambda_P = \Lambda_{\Pi^{\kappa}_{v}} = v \ ; \quad \text{i. e.} \quad \Lambda_P = \Lambda_{P^*} \in N \ ,$$

which in view of $P^* \leq P$ leads to $P = P^* = \Pi^{\kappa}_{v}$. ■

2.7.34. Let us remark that in virtue of 1.5.6., the condition P (Φ is simple) =1 (imposed on Π^{κ}_{v} in 2.7.33.) is certainly fulfilled for all $P\in \mathbf{P}^{inv}$ if for all $a\in A$ there hold

$$\kappa_{(a)}(\chi \text{ is simple}) =1 \ ; \ J(a,(.)) \text{ is "atomless".}$$

2.7.35. Putting

$$\kappa_{(a)} = \delta_{\delta_a} \qquad (a\in A)$$

one obtains

$$\mathcal{G}_P = P^x \ , \ (\mathcal{G}_P)_X = P_X \qquad (P\in \mathbf{P},\ X\in \mathcal{B}) \ ,$$

and Theorem 2.7.27. appears as generalized Raikov-theorem. Correspondingly, Theorem 2.7.33. turns out to be a generalization of a characterization due to Rényi of Poisson distributions with atomless intensity measure. This makes also clear that the assumption Π^{κ}_{v} (Φ is simple) =1, which occurs in 2.7.33., cannot be omitted.

As a generalization of a theorem of Lee [MKM, Prop. 2.2.16.] we obtain

2.7.36. Theorem. Let P be a distribution of first order in $\mathbf{P}^{inv} \cap \mathbf{T}$, $v := \Lambda_P$, and let (X_n) be an ascending sequence of sets in $\mathcal{B}$ with union A . Under these assumptions, P coincides with Π^{κ}_{v} iff all $(\mathcal{G}_P)_{X_n}$, $n\in \mathbf{Z}_+$, are Poissonian.

Proof. According to Theorem 2.7.25., P equals Π^{κ}_{v} iff $\mathcal{G}_P$ is Poissonian. In virtue of 2.7.17., $\mathcal{G}_P$ inherits from P the property of infinite divisibility. Taking into account that (X_n) is an ascending sequence of sets in $\mathcal{B}$ with union A, we may infer, by the above cited result of Lee, that $\mathcal{G}_P$ is Poissonian if all the $(\mathcal{G}_P)_{X_n}$, n=1,2,... have this property. ■

As a supplement to Proposition 2.7.22. we are going to give a characterization of all family equilibrium distributions assoicated with some equilibrium distribution. To this end we need the following lemma:

2.7.37. Let (p_n) be a sequence of distribution on $\mathcal{P}(\mathbf{Z}_+^2)$ for which $(p_n(\zeta_1\in(.))$ as well as $(p_n(\zeta_1+\zeta_2\in(.)))$ converge weakly (and hence also in variation) towards a distribution q on $\mathcal{P}(\mathbf{Z}_+)$. Then $p_n(\zeta_2>0)\to 0$ is fullfield .

Proof. Assume that $\limsup_{n\to\infty} p_n(\zeta_2>0)>0$. Then there exists a sequence $n_1<n_2<...$ for which (p_{n_m}) converges (note that (p_n) is relatively compact!) towards a distribution p and such that $(p_{n_m})(\zeta_2>0)>c$ holds for m=1,2,... and some c>0. But this implies $p(\zeta_2>0)>0$, which is a contradiction to $p(\zeta_1\in(.)) = q = p(\zeta_1+\zeta_2\in(.))$. ■

For all $\Phi = \sum_{i\in I} \delta_{a_i} \in M^{\kappa} := \bigcap_{n\geq 0} \kappa^{[n]} M$, all $X\in \mathcal{B}$ and all natural numbers m, let us put

$$\mathrm{sing}(\Phi, m, X)$$
$$:= \kappa^{[m]}{}_{(\Phi)}(\chi(X)=0) + \sum_{i\in I} \kappa^{[m]}{}_{(a_i)}(\chi(X)>0)\, \kappa^{[m]}{}_{(\Phi-\delta_{a_i})}(\chi(X)=0),$$

thus obtaining the probability that at most one individual in the family Φ has, in the m-th generation, at least one descendant situated in X. For any family system $\Phi\!\!\!\Phi = \sum_{s\in S} \delta_{\Phi_s}$ which is concentrated on M^{κ}, we put

$$\mathrm{sing}(\Phi\!\!\!\Phi, m, X) := \prod_{s\in S} \mathrm{sing}(\Phi_s, m, X).$$

2.7.38. Theorem. A distribution H on $\mathcal{M}$ is of the form $H=\mathcal{G}_P$, $P\in \mathbf{P}^{\mathrm{inv}}$, iff it has the following properties:

a) $(\mathrm{sp}H)_{\kappa}$ exists

b) $H_{\kappa}=H$

c) $\mathrm{sing}(\Phi\!\!\!\Phi, m, X) \xrightarrow[m\to\infty]{H} 1 \qquad (X\in \mathcal{B})$.

Proof. Consider two counting measures $\Phi\!\!\!\Phi_1, \Phi\!\!\!\Phi_2 \in \mathbb{M}$ with the properties

$$\sum_{i\in I} \delta_{a_i} := \mathrm{sp}\Phi\!\!\!\Phi_1 \in M^{\kappa} \quad\text{and}\quad \Phi\!\!\!\Phi_1 \leq \Phi\!\!\!\Phi_2 .$$

Then $\Phi\!\!\!\Phi_1$ and $\Phi\!\!\!\Phi_2$ have representations

$$\Phi\!\!\!\Phi_j = \sum_{s\in S_j} \delta_{\sum_{i\in I_{j,s}} \delta_{a_i}} \qquad (j=1,2) .$$

Any family $(\chi_i)_{i\in I}$ of counting measures on $\mathcal{A}$, whose sum belongs to M, now obeys the relation

$$\Phi\!\!\!\Phi_1' := \sum_{s\in S_1}{}' \delta_{\sum_{i\in I_{1,s}} \chi_i} \leq \sum_{s\in S_2}{}' \delta_{\sum_{i\in I_{2,s}} \chi_i} =: \Phi\!\!\!\Phi_2'.$$

(where the prime after the summation symbol indicates that possible summands of the form δ_o are to be omitted).

For all natural numbers m, we denote the distribution of $[\Phi\!\!\!\Phi_1', \Phi\!\!\!\Phi_2']$ under $\bigotimes_{i\in I} \kappa^{[m]}{}_{(a_i)}$ by $U_{\Phi\!\!\!\Phi_1,\Phi\!\!\!\Phi_2,m}$. These latter distributions obey

$$U_{\Phi\!\!\!\Phi_1,\Phi\!\!\!\Phi_2,m}(\zeta_1 \leq \zeta_2) = 1 ,$$
$$U_{\Phi\!\!\!\Phi_1,\Phi\!\!\!\Phi_2,m}(\zeta_j \in (.)) = \kappa^{[m]}{}_{(\Phi\!\!\!\Phi_j)} \quad \text{for } j=1,2 .$$

Moreover for $X\in \mathcal{B}$ the following inequalities hold true:

(*) $$\mathrm{sing}(\Phi\!\!\!\Phi_1, m, X) \leq U_{\Phi\!\!\!\Phi_1,\Phi\!\!\!\Phi_2,m}({}_X\zeta_1 = {}_X\zeta_2)$$

(where, for $\zeta = \sum_{s\in S} \delta_{\zeta_s}$, we put ${}_X\zeta := \sum_{s\in S, \zeta_s(X)>0} \delta_{{}_X\zeta_s}$) and

(**) $$\mathrm{sing}(\Phi\!\!\!\Phi_1, m, X) \geq U_{\Phi\!\!\!\Phi_1,(\mathrm{sp}\Phi\!\!\!\Phi_1)^X, m}(\zeta_1(X)=\zeta_2(X)) .$$

2. Now let H be a distribution in $\mathbb{P}$ with the above stated properties a), b) and c). Then P := sp$H$ is an equilibrium distribution. Due to Proposition 2.7.22. there exists a distribution U on $\mathcal{M}\otimes\mathcal{M}$ with the properties

$$U(\zeta_1 \leq \zeta_2) = 1 \quad ; \quad U(\zeta_1 \in (.)) = H \quad ; \quad U(\zeta_2 \in (.)) = \mathcal{G}_P \quad .$$

By assumption, for all natural numbers m also the distribution

$$U_m := \int U_{\Phi_1, \Phi_2, m}(.)\, U(d[\Phi_1, \Phi_2])$$

has these properties. In virtue of (*) there follows for all $X \in \mathcal{B}$ and all natural numbers m

$$U_m({}_X\zeta_1 = {}_X\zeta_2) \geq \int \text{sing}(\Phi_1, m, X)\, H(d\Phi_1) \xrightarrow[m\to\infty]{} 1 \quad ,$$

i. e. there holds

$$H({}_X\zeta_1 \in (.)) = \mathcal{G}_P\,({}_X\zeta \in (.)) \qquad (X \in \mathcal{B})$$

and hence $H = \mathcal{G}_P$.

3. Let us assume now, conversely, that $H = \mathcal{G}_P$ for some $P \in \mathbf{P}^{inv}$, and hence P = sp$H$. Property a) is now clearly fulfilled, and also property b) holds true in view of Proposition 2.7.18. To show c), we put

$$W := H([\Phi, (\text{sp}\Phi)^x] \in (.)) \quad ,$$

$$W_m := \int U_{\Phi_1, \Phi_2, m}(.)\, W(d[\Phi_1, \Phi_2]) \qquad (m=1,2,...) \quad ,$$

and observe, in view of (**), that

$$W_m\,(\zeta_1(\mathbb{X}) = \zeta_2(\mathbb{X})) \leq \int \text{sing}(\Phi_1, m, X)\, H(d\Phi) \quad .$$

Now there hold

$$W_m\,(\zeta_1(\mathbb{X}) \leq \zeta_2(\mathbb{X})) = 1 \;\; ; \;\; W_m\,(\zeta_1(\mathbb{X}) \in (.)) = H_X \quad ,$$

$$W_m\,(\zeta_2(\mathbb{X}) \in (.)) = (\mathcal{G}_{P,m})_X \quad ,$$

which by 2.7.37. and Proposition 2.7.29. allows to conclude that

$$W_m\,((\zeta_2(\mathbb{X}) - \zeta_1(\mathbb{X})) = 0) \xrightarrow[m\to\infty]{} 1 \quad .$$

Hence follows the convergence

$$\int \text{sing}(\Phi, m, X)\, H(d\Phi) \xrightarrow[m\to\infty]{} 1 \quad ,$$

which implies c). ■

2.8. Poisson Type Equilibrium Distributions with Finite Canonical Measure

A measure $\nu \in N_\kappa^{st}$ is called **of finite type** if the canonical measure of Π^κ_ν is finite. The set of all stable measures of finite type will be denoted by N_κ^{fin}.

In view of the relation

$$\nu(A) = \Lambda_{\widetilde{\Pi^\kappa_\nu}}(A) = \int \Phi(A) \widetilde{\Pi^\kappa_\nu}(d\Phi) \geq \widetilde{\Pi^\kappa_\nu}(M)$$

obviously any finite stable measure ν is of finite type. The following simple example shows that the converse is not true without further assumptions on the branching dynamics κ.

2.8.1. Example. Put $A := \{-1,0,1,...\}$; $\rho_A(i,j) := |i-j|$ for all $i, j \in A$, and define

$$\kappa_{(-1)} := \delta_{\delta_0} \; ; \; \kappa_{(0)} := \delta_{\delta_{-1}+\delta_1} \; ; \; \kappa_{(j)} := \delta_{\delta_{j+1}} \quad \text{for } j=1,2,...$$

Obviously N_J consists of all nonnegative multiples $c \cdot \sigma$, $c \in \mathbf{R}_+$, of $\sigma := \sum_{j \in A} \delta_j$.

Because of

$$(Q_\sigma)_\kappa[n] \overset{w}{\Rightarrow} \delta_{\Phi_{even}} + \delta_{\Phi_{odd}} ,$$

where

$$\Phi_{even} := \sum_{j \in A, j \text{ even}} \delta_j \; ; \; \Phi_{odd} := \sum_{j \in A, j \text{ odd}} \delta_j ,$$

σ is stable and $\widetilde{\Pi^\kappa_\sigma}(M) = 2$. ■

For all $\sigma, \nu \in N_\kappa^{fin}$ and all $c \in \mathbf{R}_+$ there holds

$$\widetilde{\Pi^\kappa_{\sigma+\nu}}(M) = \widetilde{\Pi^\kappa_\sigma}(M) + \widetilde{\Pi^\kappa_\nu}(M) < +\infty \; ; \; \widetilde{\Pi^\kappa_{c\nu}}(M) = c\widetilde{\Pi^\kappa_\sigma}(M) < +\infty .$$

Furthermore, $\gamma \in N_\kappa^{st}$, $\sigma \in N_\kappa^{fin}$ and $\gamma \leq \sigma$ allows to conclude

$$\widetilde{\Pi^\kappa_\gamma}(M) + \widetilde{\Pi^\kappa_{\sigma-\gamma}}(M) = \widetilde{\Pi^\kappa_\sigma}(M) < +\infty .$$

Hence one recognizes the validity of

2.8.2. N_κ^{fin} is a descending subset of N_κ^{st}. For all $\sigma, \nu \in N_\kappa^{fin}$ and all $c \in \mathbf{R}_+$, also $\sigma+\nu$ and $c \cdot \nu$ belong to N_κ^{fin} .

The set $\{\widetilde{\Pi^\kappa_\nu}\}_{\nu \in N_\kappa^{fin}}$ may be characterized as follows:

2.8.3. Theorem. Let ν be a measure in N and E be a measure in E_+ with $\Lambda_E = \nu$. Then the following assertions are equivalent:

a) $\nu \in N_\kappa^{fin}$ and $E = \widetilde{\Pi^\kappa_\nu}$.

b) $E_\kappa = E$; $E(\{o\}) = 0$ and $\operatorname{sing}(\Phi,m,X) \xrightarrow[m \to \infty]{E} 1 \quad (X \in \mathcal{B})$.

Proof. 1. Obviously (cf. 2.1.3.) we may restrict ourselves to the special case $E(M) = 1$.

2. Let us assume that a) holds true and, in addition, E belongs to **P**. As a canonical measure, E satisfies $E(\{o\}) = 0$, and due to 2.1.2. E belongs to $\mathbf{P}^{inv}$. Moreover, the relation

$$\Pi^{\kappa}{}_{\nu} = \mathcal{E}_E = e^{-1} \sum_{n\geq 0} (n!)^{-1} E^n$$

leads to

$$\mathcal{G}_{\Pi^{\kappa}_{\nu}} = e^{-1} \sum_{n\geq 0} (n!)^{-1} (\mathcal{G}_E)^n = \mathcal{E}_{\mathcal{G}_E} \quad .$$

On the other hand, Theorem 2.7.24. ensures that

$$\mathcal{G}_{\Pi^{\kappa}_{\nu}} = \Pi_E = \mathcal{E}_{\mathcal{Q}_E} \quad .$$

In summary, there follows

$$\mathcal{G}_E = \mathcal{Q}_E$$

and thus for all $X \in \mathcal{B}$ and all natural numbers m

$$E\,(\text{sing}(\Phi,m,X) \in (.)) = \mathcal{G}_E\,(\text{sing}(\Phi,m,X) \in (.)) \quad ,$$

which by means of Theorem 2.7.38. implies

$$\text{sing}\,(\Phi,m,X) \xrightarrow[m\to\infty]{E} 1 \qquad (X \in \mathcal{B}) \quad .$$

3. Let us assume, conversely, that a distribution $E \in \mathbf{P}^{inv,\nu}$ obeys condition b). Due to Theorem 2.7.38., $\mathcal{Q}_E$ coincides with $\mathcal{G}_E$, which implies

$$\mathcal{G}_{\mathcal{E}_E} = \mathcal{E}_{\mathcal{G}_E} = \mathcal{E}_{\mathcal{Q}_E} = \Pi_E \quad .$$

According to Theorem 2.7.25., $\mathcal{E}_E$ is thus of Poisson type, i. e. $\mathcal{E}_E = \Pi^{\kappa}{}_{\nu}$, or in other words

$$E = \widetilde{\Pi^{\kappa}{}_{\nu}} \;. \blacksquare$$

The following interesting example provides an infinite stable measure ν of finite type, which obeys, in contrast to the measure σ in Example 2.8.1., the relation

$$\int_X J^{[n]}(a,X)\nu(da) \xrightarrow[n\to\infty]{} 0 \qquad (X \in \mathcal{B}) \quad .$$

2.8.4. Example. Let us put

$$A := \bigcup_{i\geq 0} \{-i\} \times \{1,2,\ldots,2^i\},$$

and choose some metric ρ_A such that just the finite subsets of A are ρ_A-bounded. The first and second component of $[-i, j] \in A$ will be interpreted as an individual's position and age, respectively. We introduce a deterministic branching dynamics κ as follows:

a) For $[-i, j] \in A$, $j \neq 2^{i-1}$, $j \neq 2^i$, we put

$$\kappa_{(-i,j)} := \delta_{\delta_{[-i,\, j+1]}} \quad .$$

This describes transition from age j to age j+1.

b) For $i \geq 0$, $j = 2^i$ we put

$$\kappa_{(-i,j)} := \delta_o \ ,$$

i. e. an individual at position $-i$ dies after reaching the age level 2^i.

c) For $i>0$, $j=2^{i-1}$ we define

$$\kappa_{(-i,j)} := \delta_{\delta_{[-i,j+1]} + \sum_{0 \leq s<i} \delta_{[-s,1]}} \ ,$$

i. e. an individual at position $-i$, $i>0$, and "medium age" 2^{i-1} generates an offspring $\sum_{0 \leq s<i} \delta_{[-s,1]}$ and continues to live.

We are going to show that for

$$\nu := \sum_{i \geq 0} 2^{-i} \sum_{1 \leq j \leq 2^i} \delta_{[-i,j]}$$

there holds

$$\{c \cdot \nu\}_{c \geq 0} = N_J = N_\kappa^{st} = N_\kappa^{st(k)} \qquad (k \in \mathbf{Z}_+)$$

and $\widetilde{\Pi^\kappa_\nu}(M) = 1$.

1. The intensity kernel J is given by

$$J([-i,j],(.)) = \delta_{[-i,j+1]} \qquad (i>0,\ j \neq 2^{i-1},\ j \neq 2^i)$$

$$J([-i,\ 2^i],(.)) \ = o \qquad (i \geq 0)$$

$$J([-i,\ 2^{i-1}],(.)) \ = \ \delta_{[-i,2^{i-1}+1]} + \sum_{0 \leq s<i} \delta_{[-s,1]} \qquad (i>0)$$

In particular there holds

$$J^{[n]}(a,(.)) \in N \qquad (a \in A;\ n \in \mathbf{Z}_+) \ .$$

Moreover, it is clear from these relations that for any J-invariant measure σ the values

$$\alpha_i := \sigma(\{[-i,j]\}) \qquad (i \geq 0,\ j \in \{1,...2^i\})$$

do not depend on j and obey the scheme $\alpha_i = \sum_{s>i} \alpha_s$, whose solution is $\alpha_i = c \cdot 2^{-i}$, $i \geq 0$, for an arbitrary real constant c. Therefore, the measure ν is (up to a non-negative factor) the unique J-invariant measure.

2. The maximal lifespan of an individual at position $-i$ is $2^i - 1$; moreover, it follows from the relation

$$\sum_{0 \leq s<i} 2^s = 2^i - 1$$

that this individual is not survived by any of its descendants. Thus, there holds

$$\kappa^{[n]}_{(-i,j)} = \delta_o \qquad (n>2^i,\ 1 \leq j \leq 2^i) \ .$$

Particularly, this leads to the convergence

$$\int_X J^{[n]}(a,X)\nu(da) \to 0 \qquad (X \in \mathcal{B}) \ .$$

3. In this step we are going to describe, somewhat sketchy, the measure $V_{(v)}$, which turns out to be a distribution with intensity measure v ; hence v is stable (a rigorous construction and proof follows in step 4).

Let us fix a number m>0, and put

$$X := \bigcup_{0\le s\le m} \{-s\}\times\{1,2,...,2^s\} .$$

We will focus on the offspring χ in X of [−i,j] after 2^n time units ($n\ge m$). This offspring is nonvoid iff i>n, $2^{i-1}-2^n<j\le 2^{i-1}$, and it has one of 2^m different shapes $\chi_1,..., \chi_{2^m}$. For example, if $j= 2^{i-1}-2^n+1$, then $\chi=\chi_1 := \sum_{0\le s\le m} \delta_{[-s,1]}$; if $j= 2^{i-1}$, then $\chi=\chi_{2^m} := \sum_{0\le s\le m} \delta_{[-s,2^s]}$.
The total mass of the measure $(Q_v)_{\kappa}[2^n] (_X\Phi\in(.) \ ; \ \Phi(X)>0)$ is thus $\sum_{i>n} 2^{-i}\cdot 2^n = 1$.
It is rather clear that it assigns equal weight to $\chi_1,..., \chi_{2^m}$ and has intensity measure $_Xv$.
All these properties carry over in the limit to $\widetilde{V_{(v)}}(_X\Phi\in(.))$, indicating that $\widetilde{V_{(v)}} = \widetilde{\Pi^{\kappa}_{v}}$ is an equilibrium distribution.

4. Bearing in mind the considerations in 3., we are now going to construct a distribution P which by means of Theorem 2.8.3. will be proved to coincide with $\widetilde{\Pi^{\kappa}_{v}}$.

With any sequence $z=(z_i)_{i\ge1}$ in {0,1} we associate a counting measure $\Phi[z]\in M$ in the following way

$$\Phi[z] := \sum_{i\ge0} \delta_{[-i,j(i,z)]} ,$$

where $j(0,z) :=1$ and for i=0,1,...

$$j(i+1,z) := \begin{cases} 2^i+j(i,z) & \text{if } z_{i+1}=1 \\ j(i,z) & \text{if } z_{i+1}=0 \end{cases} .$$

Assume now that the sequence $(z_i)_{i\ge1}$ is random and i. i. d., each z_i taking the value 0 and 1 with probability 1/2, respectively. In this way one obtains a random counting measure, whose distribution will be denote by P. Obviously there holds

$$\Lambda_P = v.$$

If $(z_i)_{i\ge1}$ is different from the constant sequence with value 1 (which is true almost surely), then we denote the smallest value i with the property $z_i=0$ by i(z) and obtain

$$\kappa_{(\Phi[z])} = \delta_{\Phi[z']}$$

where $z'_i,..., z'_{i(z)-1} :=0$, $z'_{i(z)} :=1$ and $z'_i := z_i$ for i>i(z).

The sequence $(z'_i)_{i\ge1}$ has the same distribution as $(z_i)_{i\ge1}$. Indeed, the probability of the event $z'_1=\beta_1,..., z'_m=\beta_m$ obviously equals that of the event $z_1=\beta_1,..., z_m=\beta_m$ (namely 2^{-m}) for any $m\ge1$ and any finite sequence $[\beta_1,..., \beta_m]\in \{0,1\}^m$.

Thus we have proved that P is an equilibrium distribution. In summary we get that v is stable, and even stable of any order, because for any $X\in\mathcal{B}$ there holds

$$\Phi(X) \le \#X \quad \text{P-a.all } \Phi .$$

To show that $P=\widetilde{\Pi^{\kappa}}_{v}$, we are going to verify condition b) of Theorem 2.8.3. The only part of this condition which still needs to be checked is the convergence

$$\mathrm{sing}\,(\Phi,n,X) \xrightarrow[m\to\infty]{P} 1 \quad (X\in \mathcal{B}) \quad .$$

To this purpose, we fix an $m>0$, putting

$$X_m := \bigcup_{0\le s<m} \{-s\}\times\{1,...2^s\} \quad .$$

(Note that each bounded subset of A is covered by some suitable X_m). Now consider a fixed counting measure

$$\Phi = \Phi[z] = \sum_{i\ge 0} \delta_{[-i,j(i,z)]} \quad .$$

We may and do assume that the sequence z contains infinitely many zeros (because this is almost surely the case).The deterministic time evolution is then described recursively by

$$z^{(0)} := z \; ; \; z^{(n+1)} := (z^{(n)})' \quad (n\ge 0) \quad ,$$

and each $z^{(n)}$ corresponds to a population $\Phi^{(n)} := \Phi[z^{(n)}]$ which has exactly one individual at each position $-i$, $i\ge 0$. At the instant $\underline{n}$ when the individual at position $-m$ from the initial population Φ has reached age 2^m and dies out, all individuals in X are replaced by the offspring $\sum_{0\le s\le m} \delta_{[-s,1]}$ of a single individual b in $\Phi^{(n)}$, which shows that

$$\mathrm{sing}\,(\Phi,\underline{n}+1,X_m) = 1 \; .$$

The forefather of b in the population Φ can be happy about his descendants being alive in X_m up to his death; at this instant his position as well as the positions $\{0,-1,...,-m\}$ are filled by the offspring of another single individual. In any case, we have

$$\mathrm{sing}\,(\Phi,n,X_m) = 1 \quad (n>\underline{n}) \quad .$$

Theorem 2.7.38. now ensures that

$$P = \widetilde{\Pi^{\kappa}}_{v} \quad . \blacksquare$$

In the rest of this section, we are going to show that for any stable measure v, the quantity $\widetilde{\Pi^{\kappa}}_{v}(M)$ depends on the branching dynamics only via the intensity kernel. To this purpose, we need the following lemma:

2.8.5. Let $(z_i)_{i\in I}$ be an at most countable family of independent, $\{0,1\}$-valued random variables defined on some probability space (Ω,σ_Ω,P) and having the property

(o) $$P(\sum_{i\in I} z_i = 1) > 1-\eta$$

for some $\eta\in(0,1/2)$. Then there exists an index $i_0\in I$ such that

$$P(z_{i_0} = 1) > 1-\eta \quad \text{and} \quad P(\sum_{i\in I\setminus\{i_0\}} z_i \ge 1) < \eta(1-\eta)^{-1}.$$

Proof. 1. We may and do assume, without loss of generality, that $I=\mathbf{Z}_+$. (Indeed, a finite family could be turned into a countable one by " adding countably many zeros", which would change neither the assumption nor the conclusion). For short, we put

$$p_i := P(z_i = 1) \qquad (i \in \mathbf{Z}_+) \quad .$$

By the Borel-Cantelli lemma, there follows

$$\sum_{i \geq 1} p_i < +\infty \quad .$$

We may further assume without loss of generality that $p_1 \geq p_2 \geq \ldots$

2. Let us start with the assumption

$$(*) \qquad 0 \leq p_i \leq 1-\eta \quad \text{for all } i \in I$$

(which will actually lead to a contradiction). We show by induction that (*) implies

$$(**) \quad C_n := P(\sum_{1 \leq i \leq n} z_i = 1) = \sum_{1 \leq i \leq n} p_i \Big(\prod_{j \in \{1,\ldots,n\} \setminus \{i\}} (1-p_j) \Big) \leq 1-\eta \qquad (n \geq 1) \quad .$$

Obviously, this asssertion is true for n=1. Assuming that it holds for some natural number n, and bearing in mind the relation

$$C_{n+1} = p_{n+1} \prod_{1 \leq i \leq n} (1-p_i) + (1-p_{n+1})\, C_n$$

we obtain in the case $p_1 \geq p_2 \geq \ldots \geq p_{n+1} \geq 1/2$

$$\prod_{1 \leq i \leq n} (1-p_i) \leq 1/2 < 1-\eta$$

and therefore

$$C_{n+1} < p_{n+1}(1-\eta) + (1-p_{n+1})C_n < 1-\eta \quad .$$

In the case $p_{n+1} < 1/2$, however, there follows in view of (*) and

$$\prod_{1 \leq i \leq n} (1-p_i) = P(\sum_{1 \leq i \leq n} z_i = 0) \leq 1 - C_n$$

the estimate

$$C_{n+1} \leq p_{n+1}(1-C_n) + (1-p_{n+1})\, C_n = C_n + p_{n+1}(1-2C_n)$$
$$\leq \max(1/2\, , C_n) \leq 1-\eta \quad .$$

3. As (**) contradicts the assumed property (o), there must exist an $i_0 \in I$ with $p_{i_0} > 1-\eta$. For this index i_0 there follows

$$p_{i_0} P(\sum_{i \in I \setminus \{i_0\}} z_i \geq 1) = P(z_{i_0} = 1)\; P(\sum_{i \in I \setminus \{i_0\}} z_i \geq 1) \leq P(\sum_{i \in I} z_i \geq 2) \leq \eta$$

and thus

$$P(\sum_{i \in I \setminus \{i_0\}} z_i \geq 1) \leq \eta \cdot p_{i_0}^{-1} < \eta(1-\eta)^{-1} . \blacksquare$$

We are now in the position to characterize the total mass of the canonical measure of a Poisson type equilibrium distribution:

2.8.6. Theorem. For all $\nu \in N_{\kappa}^{st}$ there holds

$$\widetilde{\Pi^{\kappa}_{\nu}}(M) \;=\; c_{\nu}$$

$:= \inf\{c : c \in \mathbf{R}_+ ;$ for any $X \in \mathcal{B}$ and $\varepsilon > 0$ there exists, for all suitably large n, a set $S_{X,n,\varepsilon} \in \mathcal{A}$ with the properties $\nu(S_{X,n,\varepsilon}) \le c \;;\; ({}_{(A \setminus S_{X,n,\varepsilon})}\nu * J^{[n]})(X) \le \varepsilon\}$.

Proof. For abbreviation, we put $P := \widetilde{\Pi^{\kappa}_{\nu}}$.

1. For all $S \in \mathcal{A}$, $X \in \mathcal{B}$, $\varepsilon > 0$, the inequality

$$({}_{(A\setminus S)}\nu * J^{[n]})(X) \le \varepsilon$$

allows to conclude that

$$P(\Psi(X) > 0) = P_{\kappa^{[n]}}(\Psi(X) > 0)$$
$$\le ({}_{S}P)_{\kappa^{[n]}}(\Psi(X) > 0) + ({}_{(A\setminus S)}P)_{\kappa^{[n]}}(\Psi(X) > 0)$$
$$\le ({}_{S}P)_{\kappa^{[n]}}(\Psi \ne 0) + ({}_{(A\setminus S)}\nu * J^{[n]})(X)$$
$$\le P(\Psi(S) > 0) + \varepsilon \le \nu(S) + \varepsilon.$$

Hence we obtain

$$P(M) \le c_{\nu} .$$

2. In virtue of 1., our assertion holds true in the case $P(M) = +\infty$. Therefore we assume, from now on, that the measure ν is of finite type. Because of the relations

$$\widetilde{\Pi^{\kappa}_{r\gamma}}(M) = r\,\widetilde{\Pi^{\kappa}_{\gamma}}(M) \quad ; \quad c_{r\gamma} = rc_{\gamma} \quad (\text{where } 0\cdot\infty := 0)$$

which are valid for all $r \in \mathbf{R}_+$, $\gamma \in N_{\kappa}^{fin}$, we may even assume that $P(M) = 1$, i. e. $P \in \mathbf{P}$.

3. For all $X \in \mathcal{B}$, $\eta \in (0,1)$ and all natural numbers n we define

$$A_{X,n,\eta} := \{a : a \in A \;;\; \kappa^{[n]}_{(a)}(\chi(X) = 0) \le \eta\}.$$

Then there holds

$$(1-\eta) \limsup_{n\to\infty} \nu(A_{X,n,\eta}) \le 1 \quad .$$

Indeed, by 1.3.8. one has

$$(1-\eta)\,\nu(A_{X,n,\eta}) \;\le\; \langle \nu, \kappa^{[n]}_{(.)}(\chi(X) > 0)\rangle$$

$$= \widetilde{(\Pi_{\nu})_{\kappa^{[n]}}}(\Psi(X) > 0) \;=\; -\ln\,(\Pi_{\nu})_{\kappa^{[n]}}(\Phi(X) = 0)$$

$$\xrightarrow[n\to\infty]{} -\ln\,\Pi^{\kappa}_{\nu}(\Phi(X) = 0) \;=\; P(\Psi(X) > 0) \quad .$$

4. For all $X\in\mathcal{B}$, all counting measures $\Phi=\sum_{i\in I}\delta_{a_i}\in M^{\kappa}\ (=\bigcap_{m>0}\ \kappa^{[m]}M)$ and all natural numbers n we put

$$p_{\Phi,X,n}:=\mathop{*}_{i\in I}\left(\kappa^{[n]}{}_{(a_i)}(\chi(X)=0)\delta_0+\kappa^{[n]}{}_{(a_i)}(\chi(X)\neq 0)\delta_1\right) \quad ,$$

thus obtaining a distribution on $\mathcal{P}(Z_+)$ with the property

$$p_{\Phi,X,n}(\{0,1\})=\text{sing}\,(\Phi,n,X) \quad .$$

In virtue of Theorem 2.8.3. we conclude that

$$p_{\Phi,X,n}(\{0,1\})\xrightarrow[n\to\infty]{P}1 \qquad (X\in\mathcal{B}) \quad .$$

For all $X\in\mathcal{B}$ and $\eta\in(0,1)$ there follows

$$\liminf_{n\to\infty} P(p_{\Psi,X,n}(\{1\})\geq 1-\eta)$$

$$(*) \qquad \geq \liminf_{n\to\infty} P\big(p_{\Psi,X,n}(\{0,1\})\geq 1-\eta/2\ ;\ p_{\Psi,X,n}(\{0\})<\eta/2\big)$$

$$=\liminf_{n\to\infty} P\big(p_{\Psi,X,n}(\{0\})<\eta/2\big)\geq 1-(2/\eta)P(\Psi(X)=0) \quad ,$$

due to the estimate

$$P\big(p_{\Psi,X,n}(\{0\})\geq\eta/2\big)\ \leq\ (2/\eta)\int p_{\Psi,X,n}(\{0\})\,P(d\Psi)$$

$$=\ (2/\eta)\,P_{\kappa^{[n]}}(\Psi(X)=0)\ =\ (2/\eta)\,P(\Psi(X)=0) \quad .$$

Lemma 2.8.5. guarantees for any $\eta\in(0,1/2)$ and $\Psi\in M^{\kappa}$ with the property

$$p_{\Psi,X,n}(\{1\})\geq 1-\eta$$

the inequality

$$\kappa^{[n]}{}_{((A\setminus A_{X,n,\eta})\Psi)}(\chi(X)>0)\ \leq\ 2\eta \quad .$$

Combining this latter inequality with the estimate (*), we arrive, for all $X\in\mathcal{B}$ and $\eta\in(0,1/2)$, at

$$\liminf_{n\to\infty} P(\kappa^{[n]}{}_{((A\setminus A_{X,n,\eta})\Psi)}(\chi(X)>0)\leq 2\eta) \quad \geq \quad 1-(2/\eta)P(\Psi(X)=0) \quad .$$

5. Now let $X\in\mathcal{B}$ and $\varepsilon>0$ be given. Due to 4. we obtain for all $L\in\mathcal{B}$ which contain X as a subset, all $\eta\in(0,1/2)$ and all $C>0$

$$\limsup_{n\to\infty}\left(_{(A\setminus A_{L,n,\eta})}\nu * J^{[n]}\right)(X)$$

$$\leq C\limsup_{n\to\infty}\int\kappa^{[n]}{}_{(\Psi)}(\chi(L)>0)\ _{(A\setminus A_{L,n,\eta})}P(d\Psi)$$

$$+\sup_{n=1,2,\dots}\int\int_{\{\chi:\chi(X)\geq C\}}\chi(X)\,\kappa^{[n]}{}_{(\Psi)}(d\chi)\,P(d\Psi)$$

$$\leq C\cdot(2\eta+(2/\eta)P(\Psi(L)=0))+\sup_{n=1,2,\dots}\int_{\{\Psi:\Psi(X)\geq C\}}\Psi(X)\,P(d\Psi) \quad .$$

For a suitable choice of C, the second term on the right hand side is smaller than $\varepsilon/3$. Subsequently, we choose η so small that $C\cdot 2\eta$ is not larger than $\varepsilon/3$, and finally we take the set L so large that $C\cdot(2/\eta)\cdot P(\Psi(L) = 0) \leq \varepsilon/3$.

In other words: for suitable L, η there holds

$$\limsup_{n\to\infty} ({}_{(A\setminus A_{L,n,\eta})}\nu * J^{[n]})(X) \leq \varepsilon.$$

6. Combining steps 2., 3. and 5. we infer that

$$P(M) \leq c_\nu .$$

Together with step 1., this yields the assertion.■

Theorem 2.8.6. also tells that the question whether a stable measure ν is of finite type, can be decided "from the intensity kernel J alone".

If the "expected children's number" of any individual does not exceed one, then the total mass of the canonical measure of any Poisson type equilibrium distribution coincides with the total mass of its intensity measure:

2.8.7. Proposition. If J is substochastic, then there holds

$$\nu(A) = \widetilde{\Pi^\kappa_\nu}(M) \qquad (\nu \in N_\kappa^{st}) \quad .$$

Proof. 1. In any case there holds

$$\nu(A) = \int \Psi(A)\, \widetilde{\Pi^\kappa_\nu}(d\Psi) \geq \widetilde{\Pi^\kappa_\nu}(M) \quad .$$

2. If now ν is stable measure of finite type, we may choose, for any $X\in \mathcal{B}$ and $\varepsilon>0$, a sequence (S_n) of sets in $\mathcal{A}$ such that

$$\limsup_{n\to\infty} \nu(S_n) \leq \widetilde{\Pi^\kappa_\nu}(M)+\varepsilon \ ; \ \limsup_{n\to\infty} ({}_{(A\setminus S_n)}\nu * J^{[n]})(X) \leq \varepsilon \quad .$$

Due to the estimate

$$\nu(X) = (\nu * J^{[n]})(X) = ({}_{S_n}\nu * J^{[n]})(X) + ({}_{(A\setminus S_n)}\nu * J^{[n]})(X)$$
$$\leq \nu(S_n) + ({}_{(A\setminus S_n)}\nu * J^{[n]})(X),$$

this entails the inequality

$$\nu(X) \leq \widetilde{\Pi^\kappa_\nu}(M) \quad , \text{ hence } \nu(A) \leq \widetilde{\Pi^\kappa_\nu}(M) . \blacksquare$$

2.9. Finitely Fed Equilibrium Distributions

For all measures ν on $\mathcal{A}$ we put

$$\mathcal{A}_{(\nu)} := \{X : X \in \mathcal{A}, J(a,X) = 0 \text{ for } \nu\text{-almost all } a \in A \setminus X\} .$$

Thus the sets X in $\mathcal{A}_{(\nu)}$ are characterized by the property that the kernel J prohibits immigration into X from ν-almost all $a \in A \setminus X$. Notice, however, that no conditions are imposed on the values $J(a, A \setminus X)$, $a \in X$.

Straightforward estimates show that, together with X and S, also $X \cup S$ and $X \cap S$ belong to $\mathcal{A}_{(\nu)}$. Indeed, in this case, $J(a, X \cap S) = 0$ for P-a.all $a \in A \setminus X$ and for P-a.all $a \in A \setminus S$, hence also for P-a.all $A \setminus (X \cap S)$: On the other hand, $J(a,X)$ as well as $J(a,S)$ equals zero for P-a.all $a \in A \setminus (X \cup S)$, hence $J(a, X \cup S) \leq J(a,X) + (J(a,S) = 0$ for P-a.all $a \in A \setminus (X \cup S)$.

Obviously, also $\emptyset$ and A belong to $\mathcal{A}_{(\nu)}$. In summary we obtain:

2.9.1. For all measures ν on $\mathcal{A}$, the system $\mathcal{A}_{(\nu)}$ is closed with respect to finite unions as well as finite intersections, and it contains $\emptyset$ and A.

For J-invariant measures ν, the system $\mathcal{A}_{(\nu)}$ may be characterized as follows:

2.9.2. For all $\nu \in N_J$ and $X \in \mathcal{A}$, the following assertions are equivalent:

a) $X \in \mathcal{A}_{(\nu)}$.

b) ${}_X\nu \leq ({}_X\nu) * J$.

c) $({}_{A \setminus X}\nu) * J \leq {}_{A \setminus X}\nu$.

Proof. 1. For all $X \in \mathcal{A}_{(\nu)}$ and $B \in \mathcal{A}$ we have

$$({}_X\nu)(B) = \nu(X \cap B) = \int_A J(a, X \cap B)\nu(da) = \int_X J(a, X \cap B)\nu(da) \leq (({}_X\nu) * J)(B) .$$

Hence a) implies b).

2. In view of

$$({}_{A \setminus X}\nu) * J + ({}_X\nu) * J = \nu * J = \nu = {}_{A \setminus X}\nu + {}_X\nu ,$$

b) and c) are equivalent.

3. Assertion c) implies

$$\int_{A \setminus X} J(a,X)\nu(da) = (({}_{A \setminus X}\nu) * J)(X) \leq ({}_{A \setminus X}\nu)(X) = 0 ,$$

which shows the validity of a). ■

For any $P \in \mathbf{P}^{inv}$ we put for abbreviation

$$\mathcal{A}_{(P)} := \mathcal{A}_{(\Lambda_P)} .$$

The sets $X \in \mathcal{A}_{(P)}$ will be called **P-autonomous**. Obviously a set $X \in \mathcal{A}$ is P-autonomous iff there holds

$$\kappa_{(\Phi)}(\chi(X) > 0) = 0 \quad \text{for } {}_{A \setminus X}P\text{-almost all } \Phi .$$

2.9.3. For all $P \in \mathbf{P}^{inv}$ and $X \in \mathcal{A}$ the following assertions are equivalent:

a) $X \in \mathcal{A}_{(P)}$.

b) ${}_XP = {}_X((_XP)_\kappa)$.

c) ${}_XP \lesssim (_XP)_\kappa$.

d) $(_{A\setminus X}P)_\kappa \lesssim {}_{A\setminus X}P$.

Proof. 1. Put

$$H := \int(\kappa_{(_X\Phi)} \otimes \kappa_{(_{A\setminus X}\Phi)})([\chi_1,\chi_2] \in (.))\, P(d\Phi) \quad .$$

Then there hold

$$H(\chi_1 \in (.)) = (_XP)_\kappa \ ; \ H(\chi_2 \in (.)) = (_{A\setminus X}P)_\kappa \ ; \quad H((\chi_1+\chi_2) \in (.)) = P_\kappa = P \ .$$

2. Now assume that $X \in \mathcal{A}_{(P)}$. Acccording to step 1., this amounts to $H(\chi_2(X)=0) = 1$, and hence implies $H(_X(\chi_2) \in (.)) = {}_XP$, showing that ${}_XP = {}_X((_XP)_\kappa)$, i. e. the validity of b).

3. Trivially, c) is a consequence of b).

4. Assertion c) implies, in virtue of step 1., that

$${}_XP \lesssim H(_X\chi_1 \in (.)) \quad .$$

On the other hand, 1. ensures that

$${}_XP = H(_X\chi_1 + {}_X\chi_2 \in (.)) \quad ,$$

showing that $H\,(\chi_2(X) = 0) = 1$, which means that $X \in \mathcal{A}_{(P)}$.

5. Next we show that a) implies d). Indeed, exploiting a) in the second equality we obtain

$$(_{A\setminus X}P)_\kappa = \int\kappa_{(\Phi)}(.)(_{A\setminus X}P)(d\Phi) = \int\kappa_{(\Phi)}(_{A\setminus X}\Psi \in (.))(_{A\setminus X}P)(d\Phi)$$
$$\lesssim \int\kappa_{(\Phi)}(_{A\setminus X}\Psi \in (.))P(d\Phi) = {}_{A\setminus X}P \quad .$$

6. Conversely, let us assume that d) holds true. Then $\nu := \Lambda_P$ evidently has property c) of 2.9.2. (though it need not belong to N.) The very same reasoning as in step 3. of the proof of 2.9.2. now shows the validity of a) . ■

For all $X \in \mathcal{A}$ and all clustering fields τ on $[A, \rho_A]$, we define a clustering field $\tau_{[X]}$ on $[A,\rho_A]$ via

$$(\tau_{[X]})_{(a)} := \begin{cases} {}_X(\tau_{(a)}) & \text{if } a \in X \\ \delta_o & \text{otherwise} \end{cases}$$

and call it the **restriction of the clustering field τ to the set X.**

An immediate consequence of 2.9.3. is

2.9.4. Proposition. For any $P \in \mathbf{P}^{inv}$, a set $X \in \mathcal{A}$ is P-autonomous iff ${}_XP$ is an equilibrium distribution with respect to $\kappa_{[X]}$.

The following proposition introduces an important class of P-autonomous sets:

2.9.5. Proposition. Let P be an equilibrium distribution and let some $S \in \mathcal{A}$ have the properties $P(\Phi(S) < +\infty) = 1$ and

(+) $$\kappa_{(a)}(\chi(S) > 0) = 1 \qquad \text{for } \Lambda_P\text{-a.all } a \in S \ .$$

Then S is P-autonomous and obeys

$$\kappa_{(a)}(\chi(S) = 1) = 1 \qquad \text{for } \Lambda_P\text{-a.all } a \in S \ .$$

Proof. 1. Obviously there holds

$$({}_S P)_\kappa \precsim P_\kappa = P$$

and hence

$$_S((P_S)_\kappa) \precsim {}_S P \ .$$

On the other hand, (+) entails

$$_S P(\Phi(S) \in (.)) \precsim ({}_S P)_\kappa(\Phi(S) \in (.)) \ .$$

Altogether, this leads to ${}_S P = {}_S((P_S)_\kappa)$, which by 2.9.3. yields that S is P-autonomous.

2. Property (+) implies that

(++) $$\kappa_{(\Phi)}(\chi(S) \geq \Phi(S)) = 1 \qquad {}_S P\text{-a.all } \Phi \ .$$

Together with the fact that ${}_S P$ is an equilibrium distribution with respect to the restricted branching dynamics $\kappa_{[S]}$ (see 1. and Proposition 2.9.4.), (++) shows that

$$\kappa_{(\Phi)}(\chi(S) = \Phi(S)) = 1 \qquad {}_S P\text{-a.all } \Phi \ ,$$

which, combined with (+), reveals that

$$\kappa_{(a)}(\chi(S) = 1) = 1 \qquad \text{for } \Lambda_P\text{-a.all } a \in S \ . \blacksquare$$

Next we are going to show that, with any P-autonomous set X, one may associate in a natural way two equilibrium distributions $P_{[X]}$ and $P_{(X)}$, namely the limiting distributions under the branching process started off with initial distributions ${}_X P$ and ${}_{A \setminus X} P$, respectively. To this end, we need a series of simple lemmata, the first two of which follow easily from the definition of the semi-ordering $\precsim$ on **P**.

2.9.6. Let (P_n) be a sequence of distributions on $\mathcal{M}$ which increases with respect to $\precsim$. Then there exists a distribution W on $\mathcal{M}^{\otimes\{1,2,\ldots\}}$ with the properties

$$W(\Phi_n \in (.)) = P_n \quad \text{for } n=1,2,\ldots \ ; \quad W((\Phi_n) \text{ is increasing}) = 1.$$

Moreover, in this situation $\{P_1, P_2, \ldots\}$ is τ_P-relatively compact iff W-almost all (Φ_n) are bounded from above in M, i. e. increase towards a counting measure Φ_∞.
In this latter case there holds

$$P_n \Rightarrow L := W(\Phi_\infty \in (.)) \ .$$

2.9.7. Let (P_n) be a sequence of distributions on $\mathcal{M}$ which decreases with respect to $\precsim$. Then there exists a distribution U on $\mathcal{M}^{\otimes\{1,2,\ldots\}}$ such that

$$U(\Phi_n \in (.)) = P_n \quad \text{for } n=1,2,\ldots \ ; \quad U((\Phi_n) \text{ is decreasing}) = 1 \ .$$

Moreover, U-almost all (Φ_n) decrease towards a counting measure Φ_∞, and there holds

$$P_n \Rightarrow L := U(\Phi_\infty \in (.)) \ .$$

An immediate consequence of 2.9.6. together with the compactness criterion 1.2.5. is

2.9.8. A sequence (P_n) of distributions on $\mathcal{M}$, which is increasing with respect to $\lesssim$, is weakly convergent iff there exists a $P \in \mathbf{P}$ such that $(P_n) \lesssim P$ for n = 1,2,..., i. e. iff $\{P_1, P_2, \ldots\}$ is bounded in $[\mathbf{P}, \lesssim]$.

According to 2.9.3., for all $P \in \mathbf{P}^{inv}$ and $X \in \mathcal{A}_{(P)}$, the sequences $({}_X P)_{\kappa^{[n]}}$ and $({}_{A \setminus X} P)_{\kappa^{[n]}}$ are increasing and decreasing, respectively, with respect to the semi-ordering $\lesssim$. Bearing in mind that there holds

$$({}_X P)_{\kappa^{[n]}} \lesssim P_{\kappa^{[n]}} = P \qquad (n=1,2,\ldots) \quad ,$$

we conclude the existence of distributions W and U on $\mathcal{M}^{\otimes\{1,2,\ldots\}}$ with the properties stated in Lemmata 2.9.6. and 2.9.7. for the sequences $(({}_X P)_{\kappa^{[n]}})$ and $(({}_{A \setminus X} P)_{\kappa^{[n]}})$, respectively. A fortiori there follows the existence of distributions $P_{[X]}$ and $P_{(X)}$ such that the convergence relations

$$({}_X P)_{\kappa^{[n]}} \Rightarrow P_{[X]} \quad ; \quad ({}_{A \setminus X} P)_{\kappa^{[n]}} \Rightarrow P_{(X)}$$

are valid.

In the situation $\Phi_n \uparrow \Phi_\infty$, $\Phi_\infty \in {}_\kappa M$, as well as in the situation $\Phi_n \downarrow \Phi_\infty$, $\Phi_\infty \in {}_\kappa M$ there holds

$$\kappa_{(\Phi_n)} \Rightarrow \kappa_{(\Phi_\infty)} .$$

This entails, in the terminology of Lemmata 2.9.6. and 2.9.7.

$$({}_X P)_{\kappa^{[n+1]}} = \int \kappa_{(\Phi_n)}(.) W(d(\Phi_i)) \Rightarrow \int \kappa_{(\Phi_\infty)}(.) W(d(\Phi_i)) = (P_{[X]})_\kappa \quad ,$$

$$({}_{A \setminus X} P)_{\kappa^{[n+1]}} = \int \kappa_{(\Phi_n)}(.) U(d(\Phi_i)) \Rightarrow \int \kappa_{(\Phi_\infty)}(.) U(d(\Phi_i)) = (P_{(X)})_\kappa \quad ,$$

which reveals that $P_{[X]}, P_{(X)} \in \mathbf{P}^{inv}$.

Making use of the representations

$$\mathcal{L}_{({}_X P)_{\kappa^{[n]}}}(f) = \int \exp(-\langle \Phi_n, f \rangle) W(d(\Phi_i)) \qquad (n \in \mathbf{Z}_+)$$

etc. and of the monotone convergence of the sequence (Φ_n), we arrive, in summary, at

2.9.9. Proposition. For all $P \in \mathbf{P}^{inv}$ and $X \in \mathcal{A}_{(P)}$ there exist distributions $P_{[X]}$ and $P_{(X)}$ in $\mathbf{P}^{inv}$ such that for all $f \in F_b$ the convergence relations

$$\mathcal{L}_{({}_X P)_{\kappa^{[n]}}}(f) \downarrow \mathcal{L}_{P_{[X]}}(f) \quad ; \quad \mathcal{L}_{({}_{A \setminus X} P)_{\kappa^{[n]}}}(f) \uparrow \mathcal{L}_{P_{(X)}}(f)$$

hold true.

2.9.10. Example. With the assumptions and notations of Example 2.1.6., $\kappa_{[Z]}$ is a substochastic translation, and there holds

$$Z \in \mathcal{A}_{(P)} \quad ; \quad P_{[Z]} = P \quad ; \quad P_{(Z)} = \delta_o \quad . \blacksquare$$

In the following we need

2.9.11. Proposition. For all $P \in \mathbf{P}^{inv}$ and $X \in \mathcal{A}_{(P)}$ with the property

(+) $$\kappa_{(a)}(\chi(S) > 0) = 1 \quad \text{for } \Lambda_P\text{-a. all } a \in S$$

there holds

$$\mathcal{G}_{P_{[X]}} = {}_X(\mathcal{G}_P) = \mathcal{G}_{{}_X P} \quad ; \quad \mathcal{G}_{P_{(X)}} = {}_{A \setminus X}(\mathcal{G}_P) = \mathcal{G}_{{}_{A \setminus X} P} \quad .$$

For its proof we need the following lemma:

2.9.12. Let (Φ_n) be a sequence in M, $\Phi \in$ M. Then $\Phi_n \downarrow \Phi$ implies, for all $X \in \mathcal{B}$, the convergence relations

$$_X\Phi_n \xrightarrow{M} {_X\Phi} \quad \text{and} \quad {_{A\setminus X}\Phi_n} \xrightarrow{M} {_{A\setminus X}\Phi} \quad .$$

Proof. Let $\underline{n}$ be a natural number such that $_X\Phi_n$ remains constant for $n \geq \underline{n}$, i. e.

$$_X\Phi_n = {_X\Phi} = \sum_{1\leq i\leq m} \delta_{\Phi_i}$$

for some $m \in \mathbb{Z}_+$ and $\Phi_1,...\Phi_m \in$ M. Then there exist, for all $n \geq \underline{n}$, representations

$$_X\Phi_n = \sum_{1\leq i\leq m} \delta_{\Psi_{i,n}} \quad , \quad {_X\Phi} = \sum_{1\leq i\leq m} \delta_{\Psi_i} \quad ,$$

where $\Phi_i \leq \Psi_{i,n} \underset{n\to\infty}{\uparrow} \Psi_i$. Together with 1.2.4. this implies the first of the asserted convergence relations. The second one follows from the convergence $\Phi_n \xrightarrow{M} \Phi$ together with the identity $_{A\setminus X}\Phi_n + {_X\Phi_n} = \Phi_n$. ■

Proof of Proposition 2.9.11. 1. Property (+) and the definition of κ as a substochastic translation imply that

(*) $$({_X}H)_\kappa = {_X}(H_\kappa)$$

for all $H \in \mathbb{P}$ with the property $\text{sp}H = \text{P}$. Moreover, a distribution of this kind also obeys

(**) $$({_{A\setminus X}}H)_\kappa = {_{A\setminus X}}(H_\kappa) \quad .$$

2. By an induction argument, (*) implies readily, for all $n \in \mathbb{Z}_+$, the equality

$$({_X}(P^\times))_{\kappa^{[n]}} = {_X}((P^\times)_{\kappa^{[n]}}) \quad .$$

The l. h. s. of this equality equals $(({_X}P)^\times))_{\kappa^{[n]}}$, whereas the r. h. s. tends, for $n\to\infty$, weakly towards $_X(\mathcal{G}_P)$ in view of Proposition 2.7.18. and 2.9.12. Hence we arrive at

$$\mathcal{G}_{_XP} = {_X}(\mathcal{G}_P) \quad .$$

3. Replacing X by A\X and $\mathbb{X}$ by A\$\mathbb{X}$, and using (**) instead of (*), one obtains by the same arguments as in 2. the equality

$$\mathcal{G}_{_{A\setminus X}P} = {_{A\setminus X}}(\mathcal{G}_P) \quad .$$

4. The above derived equalities ensure that both $\mathcal{G}_{_XP}$ and $\mathcal{G}_{_{A\setminus X}P}$ obey conditions b) and c) of Theorem 2.7.38. In view of

$$\text{sp}\mathcal{G}_{_XP} = P_{[X]} \quad ; \quad \text{sp}\mathcal{G}_{_{A\setminus X}P} = P_{(X)} \quad ,$$

Theorem 2.7.38. now allows to conclude that

$$\mathcal{G}_{_XP} = \mathcal{G}_{P_{[X]}} \quad ; \quad \mathcal{G}_{_{A\setminus X}P} = \mathcal{G}_{P_{(X)}} \quad ■$$

It is an immediate consequence of Proposition 2.9. 11. that, subject to the assumptions stipulated there, the equality $P_{[X]} = P$ is equivalent to $P_{(X)} = \delta_o$. More generally, there holds

2.9.13. For all $P \in \mathbf{P}^{inv}$ and $X \in \mathcal{A}$, the following assertions are equivalent:

a) $({}_X P)_{\kappa^{[n]}} \Rightarrow P$.

b) $\mathcal{G}_{{}_X P} = \mathcal{G}_P$.

c) $({}_{A \setminus X}P)_{\kappa^{[n]}} \Rightarrow \delta_o$.

Proof. 1. Because of

$$sp\,\mathcal{G}_{{}_X P,n} = ({}_X P)_{\kappa^{[n]}} \quad (n=1,2,...) \ ,$$

b) implies a).

2. For all natural numbers n we put

$$H_n := \int (\kappa^{[n]}_{({}_X\Phi)} \otimes \kappa^{[n]}_{({}_{A\setminus X}\Phi)})([\chi_1,\chi_2] \in (.))\, P(d\Phi) \ ,$$

thus obtaining

$$H_n((\chi_1+\chi_2)\in(.)) = P \ ; \ H_n(\chi_1 \in (.)) = ({}_X P)_{\kappa^{[n]}} \ , \ H_n(\chi_2 \in (.)) = ({}_{A\setminus X}P)_{\kappa^{[n]}} \ .$$

Assume now that c) is valid. Then the just stated equalities imply, for all $S \in \mathcal{B}$, that $H_n((\chi_2(S) \neq 0) \longrightarrow 0$, and hence

$$(*) \qquad {}_S(({}_X P)_{\kappa^{[n]}}) \Rightarrow {}_S P \ ,$$

showing the validity of a). Conversely, a) implies that (*) holds true at least for all $S \in \mathcal{B}_P$. Together with 2.7.37., this leads to the convergence $H_n(\chi_2(S) > 0) \longrightarrow 0$, which yields c). Thus we have proved the equivalence of a) and c).

3. Now let us assume that a) - and hence also c) - is valid. From

$$sp\,\mathcal{G}_{{}_{A\setminus X}P,n} = ({}_{A\setminus X}P)_{\kappa^{[n]}} \Rightarrow \delta_o$$

one concludes that $\mathcal{G}_{{}_{A\setminus X}P}$ exists and is concentrated on the void family system, i. e. zero measure o on $\mathcal{A}$. Using the family dynamics $\varkappa$ instead of κ, one now derives, relying on the same arguments as in step 2., from

$$\mathcal{G}_{P,n} \Rightarrow \mathcal{G}_P \ , \ \mathcal{G}_{{}_{A\setminus X}P,n} \Rightarrow \delta_o$$

the equality $\mathcal{G}_{{}_X P} = \mathcal{G}_P$. ■

After these quite extensive preparations we now turn to the present section's heart of the matter. For all $P \in \mathbf{P}^{inv}$ we put

$$\mathcal{A}_{(P)}{}^{fin} := \{X : X \in \mathcal{A}_{(P)} \ ; \ P(\Phi(X) < +\infty) = 1 \ ; \ P_{[X]} = P\} \ .$$

An equilibrium distribution P will be called **finitely fed** if $\mathcal{A}_{(P)}{}^{fin}$ is non-void.

In any case, the set A is P-autonomous; this implies readily

2.9.14. Any $P \in \mathbf{P}^{inv}$ with the property $P(\Phi \text{ finite}) = 1$ is finitely fed.

The converse of the preceding assertion is not true:

2.9.15. Example. With the assumptions and notations of Example 2.8.1.,

$$P := 2^{-1}\widehat{\Pi^{\kappa}}_{V}$$

is a finitely fed equilibrium distribution with the property $P(\Phi \text{ finite}) = 0$. ■

2.9.16. For any equilibrium distribution P, the system $\mathcal{A}_{(P)}{}^{\text{fin}}$ is closed under finite unions and finite intersections.

Proof. Let $X, S \in \mathcal{A}_{(P)}{}^{\text{fin}}$ be fixed. The equality $P_{(X)} = \delta_o = P_{(S)}$ implies, for all $B \in \mathcal{B}$

$$({}_{A\setminus(X\cap S)}P)_{\kappa^{[n]}}(\Phi(B) > 0)$$
$$\leq ({}_{A\setminus X}P)_{\kappa^{[n]}}(\Phi(B) > 0) + ({}_{A\setminus S}P)_{\kappa^{[n]}}(\Phi(B) > 0) \xrightarrow[n\to\infty]{} 0$$

and hence $P_{(X\cap S)} = \delta_o$. Furthermore,

$$({}_{A\setminus(X\cup S)}P)_{\kappa^{[n]}} \preceq ({}_{A\setminus X}P)_{\kappa^{[n]}} \Rightarrow \delta_o$$

allows to conclude that $P_{(X\cup S)} = \delta_o$. Finally there follows

$$P(\Phi(X\cap S) = +\infty) \leq P(\Phi(X\cup S) = +\infty)$$
$$\leq P(\Phi(X) = +\infty) + P(\Phi(S) = +\infty) = 0 \;. ■$$

For short, we put for all $P \in \mathbf{P}^{\text{inv}}$

$$\mathcal{A}_{(P)}{}^{\text{stoch}} := \{X : X \in \mathcal{A}_{(P)}{}^{\text{fin}} \; ; \quad \kappa_{(a)}(\chi(X) = 1) = 1 \quad \text{for } \Lambda_P\text{-almost all } a \in X\} \;.$$

We are now going to prove a basic structural result:

2.9.17. Theorem. If $P \in \mathbf{P}^{\text{inv}}$ is finitely fed, then any $X \in \mathcal{A}_{(P)}{}^{\text{fin}}$ contains a set $S \in \mathcal{A}_{(P)}{}^{\text{stoch}}$.

Proof. 1. In the first three steps of the proof, let us treat the case $P(\Phi(A) < \infty) = 1$. Let S denote the set of those $a \in A$ whose progeny "almost surely survives forever", i. e. for which there holds

$$\kappa^{[n]}{}_{(a)}(\chi \neq o) = 1 \qquad (n = 1,2,\ldots) \;.$$

Clearly, S has the property

(+) $$\kappa_{(a)}(\chi(S) > 0) = 1 \quad \text{for } \Lambda_P\text{-almost all } a \in S \,,$$

which, by Proposition 2.9.5., reveals that S is P-autonomous and obeys

$$\kappa_{(a)}(\chi(S) = 1) = 1 \quad \text{for } \Lambda_P\text{-almost all } a \in S.$$

The next two steps are for showing $P_{(S)} = \delta_o$:

2. For any finite population $\Phi \in M$ there holds

(*) $$\kappa^{[n]}{}_{(\Phi)}(\chi \neq o) = 1 \qquad (n=1,2,\ldots)$$

iff $\Phi(S) > 0$. Indeed, (*) is equivalent to

$$(**) \qquad \prod_{a\in \mathrm{supp}\Phi} (1-\kappa^{[n]}_{(a)}(\chi\neq o))^{\Phi(\{a\})} = 0 \quad .$$

Due to finiteness of Φ, (**) entails that, at least for some $a\in \mathrm{supp}\Phi$, the relation $\kappa^{[n]}_{(\Phi)}(\chi\neq o) = 1$ must hold for infinitely many $n\in \mathbf{Z}_+$, and hence for all $n\in \mathbf{Z}_+$. Conversely, $\Phi(S) > 0$ clearly implies the validity of (**), and hence that of (*).

3. Let us write, for abbreviation, $P_{(S)} := H$. As a matter of fact there holds

$$\mathcal{G}_H(\Phi \text{ finite}) = 1 \quad .$$

For all $n\in \mathbf{Z}_+$, $\kappa^{[n]}$ is a substochastic clustering field on $[A, \rho_A]$ and obeys $(\mathcal{G}_H)_{\kappa^{[n]}} = \mathcal{G}_H$. The distribution H on $\mathcal{M}\otimes\mathcal{M}$ defined by

$$H(Y_1\times Y_2) := \int_{Y_1} \kappa^{[n]}_{(\Phi)}(Y_2)\, \mathcal{G}_H(d\Phi)$$

thus has the properties

$$H(\Phi_1\in (.)) = \mathcal{G}_H \quad ; \quad H(\Phi_2\in (.)) = \mathcal{G}_H \quad ; \quad H(\Phi_2(A) \leq \Phi_1(A)) = 1 \quad ,$$

which altogether implies that

$$H(\Phi_1(A) = \Phi_2(A)) = 1 \quad ,$$

i. e

$$\mathcal{G}_H\Big(\Phi\colon \kappa^{[n]}_{(\Phi)}(\chi(A) = \Phi(A)) = 1\Big) = 1 \quad ,$$

or equivalently,

$$\mathcal{G}_H\Big(\Phi\colon \kappa^{[n]}_{(\Phi)}(\chi(A) = 1) = 1 \quad \text{for } \Phi\text{-a.all } \Phi\Big) = 1 \quad .$$

This latter equality entails that

$$\kappa^{[n]}_{(\Phi)}(\chi\neq o) = 1 \quad \text{for } \Lambda_{\mathcal{G}_H}\text{-almost all } \Phi \quad ,$$

which in virtue of 2. shows that

$$\Phi(S) > 0 \quad \text{for } \Lambda_{\mathcal{G}_H}\text{-a.all } \Phi \quad .$$

On the other hand, Proposition 2.9.11. ensures that $\mathcal{G}_H(\Phi(S) = 0) = 1$, i. e.

$$\Phi(S) = 0 \qquad \text{for } \Lambda_{\mathcal{G}_H}\text{-a.all } \Phi \quad .$$

These two properties enforce that $\Lambda_{\mathcal{G}_H}$ is zero measure on $\mathcal{A}$, which implies that

$$P_{(S)} = H = \delta_o .$$

4. Let us now abandon the finiteness condition

$$P(\Phi(A) < +\infty) = 1$$

and, instead, apply the results of steps 1. to 3. to the "restricted branching dynamics" $\kappa_{[X]}$ and to the distribution $L := {}_XP$, which is an equilibrium distribution with respect to $\kappa_{[X]}$ and does, by assumption, obey the finiteness condition $L(\Phi(A) < +\infty) = 1$. Clearly the "survival set" S defined as in 1. (but now for $\kappa_{[X]}$ instead of κ) is a subset of X, and hence obeys

$$P(\Phi(S) < +\infty) = 1 \quad .$$

Because of

$$\int_{A\setminus S} J(a,S)\Lambda_P(da) \leq \int_{A\setminus X} J(a,X)\Lambda_P(da) + \int_{X\setminus S} J(a,S)\Lambda_L(da)$$

and in virtue of 1. applied to L and $\kappa_{[X]}$ there follows $S\in \mathcal{A}_{(P)}$.

For all natural numbers n there holds

$$({}_S L)_{(\kappa_{[X]})^{[n]}} \lesssim ({}_S P)_{[n]}$$

which, in virtue of step 3., leads to

$$L = {}_X P \lesssim P_{[S]} \quad .$$

This, in turn, implies for all natural numbers m the estimate

$$({}_X P)_{\kappa^{[m]}} \lesssim P_{[S]} \quad ,$$

which due to the assumption that $X \in \mathcal{A}_{(P)}{}^{\mathrm{fin}}$ yields $P \lesssim P_{[S]}$ and hence $P = P_{[S]}$.
Finally, the equality

$$\kappa_{(a)}(\chi(S) = 1) = (\kappa_{[X]})_{(a)}(\chi(S) = 1) \quad ,$$

which is valid for all $a \in S$, implies, due to step 1., that

$$\kappa_{(a)}(\chi(S) = 1) \quad \text{for } \Lambda_P\text{-a. all } a \in S \,. \blacksquare$$

For any finitely fed equilibrium distribution P, the system $\mathcal{A}_{(P)}{}^{\mathrm{stoch}}$ (which is non-void due to Theorem 2.9.17.) consists, up to Λ_P-almost-equality, of a *unique* set S:

2.9.18. Proposition. Assume that $P \in \mathbf{P}^{\mathrm{inv}}$, $S \in \mathcal{A}_{(P)}{}^{\mathrm{stoch}}$, $X \in \mathcal{A}$. Then X belongs to $\mathcal{A}_{(P)}{}^{\mathrm{stoch}}$ iff

$$\Lambda_P((X \setminus S) \cup (S \setminus X)) = 0 \quad .$$

Moreover, there exists a set $S_1 \in \mathcal{A}_{(P)}{}^{\mathrm{stoch}}$, $S_1 \subseteq S$, with the property

$$\kappa_{(a)}(\chi(S_1) = 1) = 1 \qquad \text{for all } a \in S_1 \quad .$$

Proof. 1. Assume that $X \in \mathcal{A}_{(P)}{}^{\mathrm{stoch}}$. According to 2.9.16. and Theorem 2.9.17. there exists a set $B \in \mathcal{A}_{(P)}{}^{\mathrm{stoch}}$ with the property $B \subseteq X \cap S$.

An immediate consequence of the definition of $\mathcal{A}_{(P)}{}^{\mathrm{stoch}}$ is:

$$\kappa^{[n]}{}_{(a)}(\chi(S \setminus B) > 0) = 0 \qquad \text{for } \Lambda_P\text{-almost all } a \in B \quad .$$

This implies, for all bounded measurable subsets C of $S \setminus B$, the equality

$$(({}_B P)_{\kappa^{[n]}})_C = \delta_0 \qquad (n=1,2,\dots) \quad .$$

Letting n tend to infinity, we obtain from Proposition 2.9.9. the assertion $P_C = \delta_0$. As C was arbitrary, this yields

$$P(\Phi(S \setminus B) > 0) = 0 \quad ,$$

i. e. $\Lambda_P(S \setminus H) = 0$. Analogously, there follows $\Lambda_P(X \setminus B) = 0$. In summary there results

$$\Lambda_P((S \setminus X) \cup (X \setminus S)) \leq \Lambda_P(S \setminus B) + \Lambda_P(X \setminus B) = 0 \quad .$$

2. Conversely, assume that $\Lambda_P((X \setminus S) \cup (S \setminus X)) = 0$. Then ${}_X P = {}_S P$, which implies readily that $X \in \mathcal{A}_{(P)}{}^{\mathrm{fin}}$.
Because of

$$\kappa_{(a)}\big(\chi((X \setminus S) \cup (S \setminus X)) = 0\big) = 1 \quad \text{for } \Lambda_P\text{-almost all } a \in A$$

there follows

$$\kappa_{(a)}(\chi(X) = 1) = 1 \quad \text{for } \Lambda_P\text{-almost all } a \in S$$

and hence also for Λ_P-almost all $a \in X$, which shows that $X \in \mathcal{A}_{(P)}{}^{\mathrm{stoch}}$.

3. The set $S' := \{a : a \in S, \kappa_{(a)}(\chi(S) = 1) = 1\}$ differs from S only by a Λ_P-null-set, and so does the set $S_1 := \{a : a \in S', \kappa_{(a)}(\chi(S \setminus S') = 0\}$. Due to 2., S_1 belongs to $\mathcal{A}_{(P)}{}^{stoch}$; moreover, it obeys, as required

$$\kappa_{(a)}(\chi(S_1) = 1) = 1 \quad \text{for all } a \in S_1 \quad . \blacksquare$$

As a corollary to Theorem 2.9.17. we get

2.9.19. For all finitely fed equilibrium distributions P one has $\mathcal{G}_P(\Phi \text{ finite}) = 1$.

Proof. Take some $X \in \mathcal{A}_{(P)}{}^{stoch}$ (which is non-void according to Theorem 2.9.17). Due to Proposition 2.9.11. there holds

$$\mathcal{G}_P = \mathcal{G}_{P_{[X]}} = \mathcal{G}_{{}_X P} \quad ,$$

which immediately allows to conclude that

$$\mathcal{G}_P(\Phi \text{ finite}) = 1 \quad ; \quad \mathcal{G}_P(\Psi(A) \in (.)) \lesssim P(\Phi(X) \in (.)) \quad . \blacksquare$$

The converse of 2.9.19. is not true:

2.9.20. Example. With the assumptions and notations of Example 2.8.4. there holds, on one hand,

$$\mathcal{G}_P(\Phi(A) = 1) = 1 \quad .$$

On the other hand (see step 4. of 2.8.4.), the progeny of an individual in $a \in A$ is alive up to a certain deterministic time point depending on a, and then dies altogether. In particular there holds

$$\kappa^{[n]}{}_{(a)}(\chi \neq o) \downarrow 0 \qquad (a \in A) \quad ,$$

leading, for all finite counting measures Φ, to

$$\kappa^{[n]}{}_{(\Phi)}(\chi \neq o) \leq \langle \Phi, \kappa^{[n]}{}_{(.)}(\chi \neq o) \rangle \downarrow 0 \quad ,$$

which reveals that P cannot be finitely fed. ■

With respect to any family equilibrium distribution, almost all family systems consist, on a set $X \in \mathcal{A}_{(P)}{}^{stoch}$, only of "singleton families":

2.9.21. For any $P \in \mathbf{P}^{inv}$ and $X \in \mathcal{A}_{(P)}{}^{stoch}$ there holds

$$\Phi(X) = 1 \quad \text{for } \Lambda_{\mathcal{G}_H}\text{-almost all } \Phi \quad .$$

Proof. Proposition 2.9.11. guarantees that $\delta_o = \mathcal{G}_{P_{(X)}} = {}_{A \setminus X}\mathcal{G}_P$, which reveals that

$$\Phi(X) > 0 \quad \text{for } \Lambda_{\mathcal{G}_H}\text{-almost all } \Phi \quad .$$

For $\mathcal{G}_P$-almost all $\Phi = \sum_{i \in I} \delta_{\Phi_i}$, all natural numbers m and all bounded measurable subsets B of X there holds

$$(*) \qquad \text{sing}(\Phi, m, B) \leq \prod_{i \in I} 1_{\{1\}}(\Phi_i(X)) + \kappa^{[m]}{}_{(sp\Phi)}(\chi(X \setminus B) > 0) \quad .$$

This can be seen as follows: The first summand on the right hand side accounts for the case in which all $\Phi_i(X)$, $i \in I$, equal one. In this case, the inequality is trivially satisfied. On the other hand there holds for $\Lambda_{\mathcal{G}_P}$-almost all Φ with the property $\Phi(X) > 1$:

$$\kappa^{[m]}_{(\Phi)}(\chi(X) > 1) = 1 ,$$

and hence

$$\text{sing}(\Phi, m, B) \le \kappa^{[m]}_{(\Phi)}(\chi(X\setminus B) > 0) \quad .$$

Thus, the second summand on the right hand side of (*) is, for $\mathcal{G}_P$-almost all Φ which contain a family Φ_i with $\Phi_i(X) > 1$, an upper bound of the left hand side (note that $\kappa^{[m]}_{(sp\Phi)}(\chi(X\setminus B) > 0) \ge \kappa^{[m]}_{(\Phi_i)}(\chi(X\setminus B) > 0)$ and $\text{sing}(\Phi, m, B) \le \text{sing}(\Phi_i, m, B)$).

Integrating (*) with respect to $\mathcal{G}_P$, we obtain

$$\int \text{sing}(\Phi, m, B)\, \mathcal{G}_P(d\Phi)$$

$$\le \mathcal{G}_P(\Phi(X) = 1 \text{ for all } \Phi \in \text{supp}\Phi) + \int \kappa^{[m]}_{(\Phi)}(\chi(X\setminus B) > 0)\, P(d\Phi)$$

$$= \mathcal{G}_P(\Phi(X) = 1 \text{ for all } \Phi \in \text{supp}\Phi) + P(\Phi(X\setminus B) > 0) \quad .$$

Theorem 2.7.38. now yields

$$1 \le \mathcal{G}_P(\Phi(X) = 1 \text{ for all } \Phi \in \text{supp}\Phi) + P(\Phi(X\setminus B) > 0) \quad .$$

The last term, however, can be made arbitrarily small by a suitable choice of B. ■

2.9.21 implies readily

2.9.22. For any $P \in \mathbf{P}^{inv}$ and $X \in \mathcal{A}_{(P)}{}^{stoch}$ there holds

$$\mathcal{G}_P({}_X\Psi \in (.)) = ({}_XP)^x \quad .$$

This can be interpreted as follows

For any $X \in \mathcal{A}_{(P)}{}^{stoch}$, the stochastic evolution of ${}_X\Phi$ governed by $\kappa_{[X]}$ is almost surely a stochastic translation, ${}_X\Phi$ is almost surely finite, and different individuals in ${}_X\Phi$ belong to different families according to 2.9.22.

The next theorem states, in particular, that an equilibrium distribution P is finitely fed iff there exists a set $X \in \mathcal{A}$ which is "P-almost surely finitely populated" and obeys $({}_XP)_{\kappa^{[n]}} \Rightarrow P$.

2.9.23. Theorem. For $P \in \mathbf{P}^{inv}$, let X be a set in $\mathcal{A}$ with the property $P(\Phi(X) < +\infty) = 1$. Then X contains a subset $S \in \mathcal{A}_{(P)}{}^{stoch}$ iff there holds $({}_XP)_{\kappa^{[n]}} \Rightarrow P$.

Proof. If X contains a subset $S \in \mathcal{A}_{(P)}{}^{stoch}$ then the chain of estimates

$$({}_SP)_{\kappa^{[n]}} \lesssim ({}_XP)_{\kappa^{[n]}} \lesssim P_{\kappa^{[n]}} = P \qquad (n \in \mathbf{Z}_+)$$

together with $P_{[S]} = P$ implies that $({}_XP)_{\kappa^{[n]}} \Rightarrow P$.

2. Conversely, let us assume the convergence $({}_XP)_{\kappa^{[n]}} \Rightarrow P$. For short we put $\kappa_{[X]} := \tau$. Clearly, there holds

$$({}_XP)_\tau = {}_X(({}_XP)_\kappa) \lesssim {}_X(P_\kappa) = {}_XP \quad .$$

Hence the sequence $P_n := ({}_XP)_{\tau^{[n]}}$, $n \in \mathbf{Z}_+$, is decreasing with respect to $\lesssim$, and we conclude like in the proof of Proposition 2.9.9. that (P_n) converges weakly towards some distribution L obeying $\quad {}_X L = L \lesssim {}_X P \quad ; \quad L_\tau = L \quad .$

3. By induction, we now show the validity of

(o) $$(P_m)_{\kappa^{[n]}} \underset{n\to\infty}{\Rightarrow} P \qquad (m \in \mathbf{Z}_+) \quad .$$

Because of $P_0 = {}_X P$, this convergence holds true for m=0. Let us assume that (o) is valid for some $m \in \mathbf{Z}_+$.

For all natural numbers n and all $B \in \mathcal{B}$ there follows in virtue of 2.9.13.

$$_B\| (P_{m+1})_{\kappa^{[n]}} - (P_m)_{\kappa^{[n+1]}} \| \;=\; _B\| ({}_X(P_m)_\kappa)_{\kappa^{[n]}} - ((P_m)_\kappa)_{\kappa^{[n]}} \|$$

$$= {}_B\| \int \kappa^{[n]}{}_{(_X\Phi)}(.)(P_m)_\kappa(d\Phi) - \int \kappa^{[n]}{}_{(\Phi)}(.)(P_m)_\kappa(d\Phi) \|$$

$$\leq \int {}_B\| \kappa^{[n]}{}_{(_X\Phi)}(.) - \kappa^{[n]}{}_{(\Phi)}(.) \| (P_m)_\kappa(d\Phi) \;\leq\; 2\int \kappa^{[n]}{}_{(_{A\setminus X}\Phi)}(\chi(B)>0)\,(P_m)_\kappa(d\Phi)$$

$$= 2({}_{A\setminus X}(P_m))_{\kappa^{[n]}}(\chi(B)>0) \;\leq\; 2({}_{A\setminus X}P)_{\kappa^{[n]}}(\chi(B)>0) \underset{n\to\infty}{\longrightarrow} 0 \quad .$$

For any finite sequence $X_1,\dots,X_j$ of sets in $\mathcal{B}_P$ with union B we thus obtain

$$\|((P_{m+1})_{\kappa^{[n]}})_{X_1,\dots,X_j} - ((P_m)_{\kappa^{[n+1]}})_{X_1,\dots,X_j}\|$$
$$\leq {}_B\|(P_{m+1})_{\kappa^{[n]}} - (P_m)_{\kappa^{[n+1]}}\| \underset{n\to\infty}{\longrightarrow} 0 \quad .$$

Hence our induction assumption allows to conclude that

$$((P_{m+1})_{\kappa^{[n]}})_{X_1,\dots,X_j} \underset{j\to\infty}{\Rightarrow} P_{X_1,\dots,X_j} \quad ,$$

which leads to the convergence $(P_{m+1})_{\kappa^{[n]}} \underset{n\to\infty}{\Rightarrow} P$.

4. Let us now carry the assertion of 3. over to the limit $m\to\infty$, showing that $L_{\kappa^{[n]}} \Rightarrow P$.

To this purpose, remind that (P_m) decreases with respect to $\lesssim$ towards L. Let U be a distribution on $\mathcal{M}^{\otimes \mathbf{Z}_+}$ which has the properties stated in 2.9.7. Thus there follows for all natural numbers m

$$\|P_m - L\| \leq U(\Phi_m \neq \Phi_\infty) \,.$$

Bearing in mind that

$$P_0(\Phi \text{ finite}) = {}_X P(\Phi \text{ finite}) = 1 \quad ,$$

we now obtain $\|P_m - L\| \longrightarrow 0$.

For all finite sequences $X_1,\dots,X_j$ of sets in $\mathcal{B}_P$ and all natural numbers m there results,

$$\limsup_{n\to\infty} \| (L_{\kappa^{[n]}})_{X_1,\dots,X_j} - P_{X_1,\dots,X_j} \|$$

$$\leq \limsup_{n\to\infty} \| (L_{\kappa^{[n]}})_{X_1,\dots,X_j} - ((P_m)_{\kappa^{[n]}})_{X_1,\dots,X_j} \|$$
$$+ \limsup_{n\to\infty} \| ((P_m)_{\kappa^{[n]}})_{X_1,\dots,X_j} - P_{X_1,\dots,X_j} \|$$

$$\leq \|L - P_m\| + 0 \quad ,$$

which shows that $L_{\kappa^{[n]}} \Rightarrow P$.

5. After replacing κ by its restriction τ to the set X, Theorem 2.9.17. is applicable to the equilibrium distribution L and to the set X, guaranteing the existence of a set $S\in X\cap\mathcal{A}$ with the properties

$$({}_SL)_{\tau^{[n]}} \Rightarrow L \quad ; \quad \int_{X\setminus S} J^{[n]}(a,S)\Lambda_L(da) = 0 \quad (n\in \mathbf{Z}_+) \quad ;$$

$$\kappa_{(a)}(\chi(S) = 1) = 1 \quad \text{for } \Lambda_L \text{ -almost all } a\in S \quad .$$

In view of Proposition 2.9.18. we may even choose S such that

$$\kappa_{(a)}(\chi(S) = 1) = 1 \quad \text{for all } a\in S \quad .$$

Due to Proposition 2.9.5. there follows $S\in \mathcal{A}_P$, and it only remains to show that $P_{[S]} = P$.

By assumption there holds ${}_S(({}_SL)_\tau) = {}_SL$, and hence also ${}_S(({}_SL)_\kappa) = {}_SL$, implying that ${}_SL \precsim ({}_SL)_\kappa$. Therefore, the sequence $(({}_SL)_{\kappa^{[n]}})$ is increasing with respect to $\precsim$. In addition, there holds $({}_SL)_{\kappa^{[n]}} \precsim P$ for all $n\in \mathbf{Z}_+$. Lemma 2.9.8. now shows the existence of a distribution H such that

$$({}_SL)_{\kappa^{[n]}} \Rightarrow H \precsim P \quad ;$$

due to the monotonicity of the left hand side there follows $H\in \mathbf{P}^{inv}$.

On the other hand, $({}_SL)_{\kappa^{[n]}}$ increases towards L and obeys the estimate

$$({}_SL)_{\tau^{[n]}} \precsim ({}_SL)_{\kappa^{[n]}} \qquad (n\in \mathbf{Z}_+) \quad ,$$

which together implies that

$$L \precsim H \precsim P$$

and hence

$$L_{\kappa^{[n]}} \precsim H \precsim P \qquad (n\in \mathbf{Z}_+) \quad .$$

In virtue of the convergence $L_{\kappa^{[n]}} \Rightarrow P$ (see step 4), this yields $H = P$, revealing that

$$({}_SL)_{\kappa^{[n]}} \Rightarrow P \quad . \blacksquare$$

By means of preceding result, 2.9.21. is now complemented as follows:

2.9.24. Theorem. For all $P\in \mathbf{P}^{inv}$ and $X\in\mathcal{A}$, the following assertions are equivalent:

a) P is finitely fed and $X\in \mathcal{A}_{(P)}{}^{stoch}$.

b) $\mathcal{G}_P(\Phi \text{ finite}) = 1$, $\mathcal{G}_{{}_XP} = \mathcal{G}_P$, and $\Phi(X) = 1$ for $\Lambda_{\mathcal{G}_P}$-almost all Φ .

Proof. 1. The implication a) $\Rightarrow$ b) follows immediately from 2.9.19., 2.9.13. and 2.9.21.

2. Conversely, assume the validity of b). Then P clearly obeys $P(\Phi(X) < +\infty) = 1$. Moreover, 2.9.13. ensures that $({}_XP)_{\kappa^{[n]}} \Rightarrow P$, which, in turn, by Theorem 2.9.23. implies the existence of a set $S\subseteq X$, $S\in \mathcal{A}_{(P)}{}^{stoch}$.

Due to 2.9. 21. and b) there follows

$$1 = \Phi(S) \le \Phi(X) = 1 \qquad \text{for } \Lambda_{\mathcal{G}_P} \text{ -almost all } \Phi \quad .$$

Hence $\Lambda_P(X\setminus S)$ equals zero , i. e. $X\in \mathcal{A}_{(P)}{}^{stoch}$ by Proposition 2.9.18 . ■

Proposition 2.9.11. is now complemented as follows:

2.9.25. Proposition. The conclusion of Proposition 2.9.11. holds true for any $P \in \mathbf{P}^{inv}$ and $X \in \mathcal{A}_{(P)}$ with the property $P(\Phi(X) < +\infty) = 1$.

Proof. 1. In virtue of Theorem 2.9.23. there exists a set $S \in \mathcal{A}_{(P_{[X]})}{}^{stoch}$, $S \subseteq X$. Due to Theorem 2.9.24. thre holds

$$\mathcal{G}_{S(P_{[X]})} = \mathcal{G}_{P_{[X]}} .$$

Because of $_S(P_{[X]}) = {}_SP \leq {}_XP \leq P_{[X]}$, this latter equality implies

$$\mathcal{G}_{_XP} = \mathcal{G}_{P_{[X]}} .$$

2. Relying on step 1. and 2.9.21. we obtain

$$_X(\mathcal{G}_P) = {}_X(\mathcal{G}_{P_{[X]}}) = \mathcal{G}_{P_{[X]}} .$$

3. To show the second part of the assertion, consider the distributions H_n, $n \in \mathbf{Z}_+$, defined by

$$H_n := \int (\kappa^{[n]}{}_{(_X\Phi)} \otimes \kappa^{[n]}{}_{(_{A\setminus X}\Phi)})([\Phi_1,\Phi_2] \in (.))\, P^x(d\Phi) .$$

Steps 1. and 2. imply that

$$H_n(\Phi_1(A\setminus X) > 0) \longrightarrow 0 .$$

On the other hand there holds

$$H_n({}_{A\setminus X}\Phi_1 + {}_{A\setminus X}\Phi_2 \in (.)) \Rightarrow {}_{A\setminus X}(\mathcal{G}_P) ,$$

which together yields that

$$\mathcal{G}_{_{A\setminus X}P,n} = H_n({}_{A\setminus X}\Phi_2 \in (.)) \Rightarrow {}_{A\setminus X}(\mathcal{G}_P) ,$$

i.e. $\mathcal{G}_{_{A\setminus X}P} = {}_{A\setminus X}(\mathcal{G}_P)$. The equality $\mathcal{G}_{_{A\setminus X}P} = \mathcal{G}_{P_{(X)}}$ now follows like in the last step of the proof of Proposition 2.9.11. ■

In virtue of the preceding proposition, the fact that an equilibrium distribution P is finitely fed and that a set X belongs to $\mathcal{A}_{(P)}{}^{fin}$ may be interpreted as follows:

Because of

$$\mathcal{G}_P(\Phi = {}_X\Phi) = 1 ,$$

$\mathcal{G}_P$ -almost all Φ consist only of families possessing individuals in X . If we omit all individuals outside of X, i. e. pass over to the system $_X\Phi$, then we arrive at $\mathcal{G}_P({}_X\Phi \in (.))$, which is, with respect to the restricted branching dynamics $\kappa_{[X]}$, the family equilibrium distribution associated with $_XP$. The stochastic evolution within the P-autonomous set X is not influenced from outside; the individuals in $_X\Phi$, however, are permitted to generate "emigrants" outside of X. All individuals in the population $_{A\setminus X}\Phi$ have been produced in this way - and even are "relatives" of some induviduals in the actual population $_X\Phi$. The random population $_{A\setminus X}\Phi$ developes according to κ. For its own, it is not capable to regenerate and finally collapses because of the succeeding individuals death or "emigration to infinity". (This is just what the assertion $P_{(X)} = \delta_o$ means.) Thus, the random population outside of X is stabilized only by a steady influx of descendants from $_X\Phi$.

2.10. Finitely Fed Equilibrium Distributions of Poisson Type

For equilibrium distributions of Poisson type, Proposition 2.9.9. may be refined as follows:

2.10.1. Theorem. For any $\nu \in N_\kappa^{st}$, $P := \Pi^\kappa_\nu$ and any $X \in \mathcal{A}_{(\nu)}$ with the property $P(\Phi(X) < +\infty) = 1$ there holds

$$({}_X\nu)*J^{[n]} \uparrow \gamma \in N_\kappa^{st} \quad ; \quad ({}_{A\setminus X}\nu)*J^{[n]} \downarrow \sigma \in N_\kappa^{st},$$

$$P_{[X]} = \Pi^\kappa_\gamma \text{ and } \widetilde{\Pi^\kappa_\gamma} = \tilde{P}(\Psi \in (.), \Psi(X)>0) \quad ,$$

$$P_{(X)} = \Pi^\kappa_\sigma \text{ and } \widetilde{\Pi^\kappa_\sigma} = \tilde{P}(\Psi \in (.), \Psi(X)=0)$$

and hence also

$$P = P_{[X]} * P_{(X)} \quad .$$

In particular there holds $({}_X\nu)*J^{[n]} \uparrow \nu$ for all $X \in \mathcal{A}_{(P)}^{fin}$.

Proof. Due to 2.9.2. the sequence $(({}_X\nu)*J^{[n]})$ is increasing. On the other hand, it is bounded from above by ν. By 2.2.3. and 2.3.9. now follows the convergence $({}_X\nu)*J^{[n]}\uparrow\gamma$ for some $\gamma \in N_\kappa^{st}$. In the same way one shows $({}_{A\setminus X}\nu)*J^{[n]} \downarrow \sigma$ for some $\sigma \in N_\kappa^{st}$.

Using Theorem 2.7.24. and Proposition 2.9.25. we now obtain

$$\mathcal{G}_{P_{[X]}} = {}_X(\mathcal{G}_P) = {}_X(\Pi_{\tilde{P}}) = \Pi_{({}_X\tilde{P})((.)\setminus\{o\})} = \Pi_{\tilde{P}(\Psi \in (.), \Psi(X)>0)}$$

and hence

$$P_{[X]} = \mathrm{sp}\, \mathcal{G}_{P_{[X]}} = \mathcal{E}_{\tilde{P}(\Psi \in (.), \Psi(X)>0)} \quad .$$

Analogously there follows

$$\mathcal{G}_{P_{(X)}} = \Pi_{\tilde{P}(\Psi \in (.), \Psi(X)=0)} \quad ; \quad P_{(X)} = \mathcal{E}_{\tilde{P}(\Psi \in (.), \Psi(X)=0)} \quad .$$

Bearing in mind that monotone convergence of the distributions $({}_XP)_{\kappa^{[n]}}$ towards $P_{[X]}$ implies vague convergence of the corresponding intensity measures, one arrives at $\Lambda_{P_{[X]}}=\gamma$. In the same way one derives $\Lambda_{P_{(X)}} = \sigma$. Theorems 2.7.24. and 2.7.25 now allow to conclude that

$$P_{[X]} = \Pi^\kappa_\gamma \quad ; \quad P_{(X)} = \Pi^\kappa_\sigma \quad .$$

Because of $\nu = \gamma+\sigma$, the last of the asserted equalities follows from Proposition 2.3.4. ■

In virtue of [MKM; Prop. 2.3.1. and Prop. 2.2.7.] there results immediately

2.10.2. Proposition. Under the assumptions of Theorem 2.10.1, there holds $P(\Psi(X)= 0) > 0$ and

$$P_{(X)} = P((.)|\Phi(X) = 0) \quad .$$

We are now aiming to characterize those Poisson type equilibrium distributions, which are finitely fed, in terms of their intensity measures. To this end, let us agree a measure $\sigma \in N_\kappa^{st}$ is called **finitely fed** if the set

$$N_{(\nu)}^{fin} := \{\gamma : \gamma \in N, \gamma \text{ finite}, \gamma \leq \nu, \; \gamma * J^{[n]} \xrightarrow{N} \nu\}$$

is non void. The set of all finitely fed measures $\nu \in N_\kappa^{st}$ will be denoted by N_κ^{ff}.

2.10.3. A measure $\nu \in N_\kappa^{st}$ is finitely fed iff there exists an $X \in \mathcal{A}$ such that ${}_X\nu \in N_{(\nu)}^{fin}$.

Proof. Assume that ν is finitely fed, and choose some $\gamma \in N_{(\nu)}^{fin}$.

Let f be a density function of γ with respect to ν and c be some number in (0,1). We put

$$X := \{a : a \in A, f(a) \geq c\} \quad , \quad \sigma := {}_X\gamma \; .$$

Then there results

$$\nu(X) \leq c^{-1}\gamma(A) < +\infty \; .$$

The convergence $\gamma * J^{[n]} \xrightarrow{N} \nu$ implies that $(\nu-\gamma) * J^{[n]} \xrightarrow{N} o$.

In virtue of the estimate

$$\nu - \sigma \leq (1-c)^{-1}(\nu-\gamma)$$

there follows the convergence

$$(\nu-\sigma) * J^{[n]} \xrightarrow{N} o$$

and hence

$$\sigma * J^{[n]} \xrightarrow{N} \nu \; ,$$

which reveals that $\sigma \in N_{(\nu)}^{fin}$ and, a fortiori, that ${}_X\nu \in N_{(\nu)}^{fin}$. ■

The following lemma will be used several times:

2.10.4. Let (P_n) and (H_n) be sequences of distributions of first order on $\mathcal{M}$, and P, H be distributions on $\mathcal{M}$ of first order with the properties

$$H_n \Rightarrow H \; ; \; \Lambda_{H_n} \xrightarrow{N} \Lambda_H \; ; \; P_n \precsim H_n \text{ and } \Lambda_{P_n} \leq \Lambda_H \text{ for } n = 1,2,...$$

Then there hold:

(i) The convergence $P_n \Rightarrow P$ implies that $\Lambda_{P_n} \xrightarrow{N} \Lambda_P$.

(ii) The convergence $\Lambda_{P_n} \xrightarrow{N} \Lambda_H$ implies that $P_n \Rightarrow H$.

Proof. 1. Due to 1.1.8. there holds for all $X \in \mathcal{B}$

(*) $$\sup_{n=1,2,...} \int_{\{\Phi:\Phi(X)>c\}} \Phi(X)\, H_n(d\Phi) \xrightarrow[c\to\infty]{} 0 \; .$$

Because of $P_n \precsim H_n$, also the sequence (P_n) inherits property (*).

2. According to 1.2.7. and our assumption $\Lambda_{P_n} \leq \Lambda_H$, $n \in \mathbb{Z}_+$, the set $\{P_1, P_2, ...\}$ is τ_P-relatively compact. In virtue of 1. and 1.1.8., any subsequence (P_{n_m}), which converges weakly towards some P′, obeys

$$\Lambda_{P_{n_m}} \xrightarrow{N} \Lambda_{P'} \; .$$

3. Concerning (i), assume that $P_n \Rightarrow P$. Then there follows $\Lambda_{P_n} \xrightarrow{N} \Lambda_P$ by step 2.

4. As to (ii), asume that $\Lambda_{P_n} \xrightarrow{N} \Lambda_H$, and let P_{n_m} be any subsequence of (P_n) which converges weakly towards some P'. Then 2. implies that $\Lambda_{P'} = \Lambda_H$. On the other hand, our assumptions ensure that $P' \lesssim H$, which altogether leads to $H = P'$. In summary, there follows $P_n \Rightarrow H$. ■

2.10.5. Proposition. Let P be a Poisson type equilibrium distribution whose intensity measure ν is finitely fed. Then there holds for any $\gamma \in N_{(\nu)}^{fin}$

$$\tilde{P}(M) \leq \gamma(A) \quad ; \quad (\Pi_\gamma)_{\kappa^{[n]}} \Rightarrow P \quad .$$

Proof. Putting $P_n := (\Pi_\gamma)_{\kappa^{[n]}}$, $H_n := (\Pi_\nu)_{\kappa^{[n]}}$, $n \in \mathbf{Z}_+$, $H := P$, one readily checks the condition in 2.10.4. (ii). Hence follows

$$(\Pi_\gamma)_{\kappa^{[n]}} \Rightarrow P \quad .$$

Finally the estimate

$$\widetilde{(\Pi_\gamma)_{\kappa^{[n]}}}(M) = (\mathcal{Q}_\gamma)_{\kappa^{[n]}}(\chi \neq o) \leq \gamma(A) \qquad (n \in \mathbf{Z}_+)$$

implies that $\tilde{P}(M) \leq \gamma(A)$. ■

2.10.6. Let P be a finitely fed Poisson type equilibrium distribution, and $X \in \mathcal{A}_{(P)}^{stoch}$. Then there holds

$$\tilde{P}(\Psi(X)>0) < +\infty \quad ; \quad \Psi(X) = 1 \quad \text{for } \tilde{P}\text{-almost all } \Psi.$$

Proof. By means of [MKM, Prop. 2.2.7.], $P(\Phi(X)<+\infty) = 1$ implies that $\tilde{P}(\Psi(X)>0)$ is finite. On the other hand there holds according to Theorem 2.9.24.

$$\Psi(X) = 1 \quad \text{for } \Lambda_{\mathcal{G}_P}\text{ - almost all } \Psi \quad ,$$

which by Theorem 2.7.24. implies the assertion. ■

2.10.7. Theorem. A Poisson type equilibrium distribution P is finitely fed iff its intensity measure ν has this property.

Proof. 1. Let us assume that P is finitely fed. According to Theorem 2.9.17., there exists a set $X \in \mathcal{A}_{(P)}^{stoch}$. Put $\gamma := {}_X\nu$. Then 2.10.6. together with 1.2.10. yields

$$\widetilde{{}_XP} = \mathcal{Q}_\gamma \quad ; \quad \gamma(A) = \tilde{P}(\Psi(X) > 0) < +\infty \quad .$$

Finally, Theorem 2.10.1. implies that $\gamma * J^{[n]} \uparrow \gamma$ and hence $\gamma * J^{[n]} \xrightarrow{N} \nu$.

In summary, there follows $\gamma \in N_{(\nu)}^{fin}$.

2. Conversely, assume that ν is finitely fed, and choose according to 2.10.3., a set $X \in \mathcal{A}$ such that ${}_X\nu \in \mathcal{A}_{(\nu)}^{fin}$. Putting $H_n := ({}_XP)_{\kappa^{[n]}}$, $H := P$, $n \in \mathbf{Z}_+$, we infer from 2.10.4. that

$$({}_XP)_{\kappa^{[n]}} \Rightarrow P \quad .$$

Theorem 2.9.23. now yields that P is finitely fed. ■

2.10.8. Example. With the assumptions and notations of Example 2.8.1., Π^{κ}_{ν} is finitely fed and $\{-1,0\} \in \mathcal{A}_{(\Pi^{\kappa}_{\nu})}{}^{\text{stoch}}$. ■

By Proposition 2.10.5. there holds $N_{\kappa}^{\text{ff}} \subseteq N_{\kappa}^{\text{fin}}$; on the other hand Example 2.8.4. provides a measure ν which is of finite type but not finitely fed (see Example 2.9.20. and Theorem 2.10.7.). In virtue of Proposition 2.8.7. we get, however:

2.10.9. Proposition. If J is substochastic, then $N_{\kappa}^{\text{ff}} = N_{\kappa}^{\text{fin}}$.

Proposition 2.10.5. and Theorem 2.10.7. are complemented by

2.10.10. Proposition. Let $P = \Pi^{\kappa}_{\nu}$ be a finitely fed Poisson type equilibrium distribution. Then there holds for any $X \in \mathcal{A}_{(P)}{}^{\text{stoch}}$:

a) $\tilde{P}(M) = \nu(X) < \infty$.

b) ${}_XP = \Pi_{{}_X\nu}$.

c) ${}_X\nu \uparrow \nu$; ${}_X\nu = \min N_{(\nu)}{}^{\text{fin}}$.

Proof. 1. Due to 2.10.6. there holds

$$\tilde{P}(\Psi(X)>0) < +\infty \ ; \ \ \Psi(X) = 1 \quad \text{for } \tilde{P}\text{-almost all } \Psi \ .$$

Hence follows

$$\widetilde{{}_XP} = {}_X(\tilde{P})((.)\setminus\{o\}) = \mathcal{Q}_{{}_X\nu} \ ,$$

i. e. b), and

$$\tilde{P}(M) = P(\Psi(X) > 0 = \widetilde{{}_XP}(M) = \nu(X) \ ,$$

i. e. a).

2. From Theorem 2.10.1. there follows ${}_X\nu \uparrow \nu$. Together with a), this shows that ${}_X\nu$ belongs to $N_{(\nu)}{}^{\text{fin}}$, which implies the convergence $(\nu - {}_X\nu) * J^{[n]} = ({}_{A\setminus X}\nu) * J^{[n]} \xrightarrow{N} o$.

For an arbitrary measure γ in $N_{(\nu)}{}^{\text{fin}}$, the estimate ${}_{A\setminus X}\gamma \leq {}_{A\setminus X}\nu$ now entails the convergence $({}_{A\setminus X}\gamma) * J^{[n]} \xrightarrow{N} o$, and hence $({}_X\gamma) * J^{[n]} \xrightarrow{N} \nu$, which means that ${}_X\gamma \in N_{(\nu)}{}^{\text{fin}}$. According to Proposition 2.10.3. there follows

$$\nu(X) = \tilde{P}(M) \leq \gamma(X) \ ,$$

which together with $\gamma \leq \nu$ yields ${}_X\gamma = {}_X\nu$ and, in summary, leads to $\gamma \leq {}_X\nu$. ■

The following characterization is a pendant to Theorem 2.8.3.

2.10.11. Theorem. Let ν be a measure in N, E be a measure on $\mathcal{M}$ with $\Lambda_E = \nu$ and X be a set in $\mathcal{A}$. Then the following assertions are equivalent

a) $\nu \in N_{\kappa}^{\text{ff}}$; $E = \widetilde{\Pi^{\kappa}_{\gamma}}$ and $X \in \mathcal{A}_{(\Pi^{\kappa}_{\nu})}{}^{\text{stoch}}$.

b) $E \in \mathbf{E}_+$; $E_{\kappa} = E$; $E(\Psi(X) \neq 1) = 0$; $({}_{A\setminus X}E)_{\kappa^{[n]}}(\Psi(B) > 0) \longrightarrow 0 \ \ (B \in \mathcal{B})$.

Proof. 1. Assume that a) is valid, and put $P := \Pi^{\kappa}_{\nu}$. Because of $\nu \in N_{\kappa}^{fin}$ (see Prop. 2.10.5.) there holds $E \in \mathbf{E}_{+}$, and by means of 2.1.2. there results $E_{\kappa} = E$.

Due to 2.10.6. there holds $E(\Psi(X) \neq 1) = 0$. Finally the convergence $({}_{A\setminus X}P)_{\kappa^{[n]}} \Rightarrow \delta_o$ implies that

$$\widetilde{({}_{A\setminus X}P)_{\kappa^{[n]}}} = ({}_{A\setminus X}E)_{\kappa^{[n]}}((.)\setminus\{o\}) \xrightarrow{\mathbf{w}} 0 \ ,$$

i. e.

$$({}_{A\setminus X}E)_{\kappa^{[n]}}(\Psi(B) > 0) \longrightarrow 0 \qquad (B \in \mathcal{B}) \ .$$

2. Let us now, conversely, assume that b) holds true. Because of $P := \mathcal{E}_E \in P^{inv,\nu}$ it is clear that ν is stable. By assumption there holds

$$\widetilde{({}_{A\setminus X}P)_{\kappa^{[n]}}} = ({}_{A\setminus X}E)_{\kappa^{[n]}}((.)\setminus\{o\}) \xrightarrow{\mathbf{w}} 0 \ ,$$

i. e.

$$({}_{A\setminus X}P)_{\kappa^{[n]}} \Rightarrow \delta_o \ , \quad \text{which is equivalent to}$$

(*)
$$({}_{X}P)_{\kappa^{[n]}} \Rightarrow P \ .$$

On the other hand one has

$$\widetilde{{}_{X}P} = {}_{X}E((.)\setminus\{o\}) = \widetilde{\Pi_{{}_X\nu}} \ ,$$

i.e.

(**)
$$ {}_{X}P = \Pi_{{}_X\nu} \ .$$

Because of

$$\nu(X) = E(\Psi(X) > 0) < +\infty$$

we thus conclude that

$$P(\Phi(X) < +\infty) = 1 \ .$$

(*) and (**) imply the convergence

(***)
$$(\Pi_{{}_X\nu})_{\kappa^{[n]}} \Rightarrow P \ .$$

Putting $P_n := (\Pi_{{}_X\nu})_{\kappa^{[n]}}$, $H_n := (\Pi_{\nu})_{\kappa^{[n]}}$, $n \in \mathbf{Z}_{+}$, $H := P$, one readily checks the conditions of 2.10.4. (i), which implies

$$\Lambda_{P_n} = ({}_X\nu) * J^{[n]} \xrightarrow{\mathbf{N}} \nu = \Lambda_P \ ,$$

i.e. ${}_X\nu \in N_{(\nu)}^{fin}$. Proposition 2.10.5. now ensures that $(\Pi_{{}_X\nu})_{\kappa^{[n]}} \Rightarrow \Pi^{\kappa}_{\nu}$, which together with (***) shows that $P = \Pi^{\kappa}_{\nu}$.

According to Theorem 2.9.23., X has a subset $S \in \mathcal{A}_{(P)}^{stoch}$. By 2.10.6. there follows

$$E(\Psi(X) = \Psi(S) = 1) = E(M)$$

and hence $E(\Psi(X\setminus S) \neq 0) = 0$, i. e. $\nu(X\setminus S) = 0$. Thus also the set X belongs to $\mathcal{A}_{(P)}^{stoch}$. ■

The next result deals with algebraic properties of the set N_{κ}^{ff}:

2.10.12. Proposition. N_κ^{ff} is a descending subset of N_κ^{st}. If the sum ν of an at most countable family $(\nu_i)_{i\in I}$ of measures in N_κ^{ff} is of finite type, then it even belongs to N_κ^{ff}. Together with σ, any multiple $r\cdot\sigma$, $r\in\mathbf{R}_+$, belongs to N_κ^{ff}.

Proof. 1. Assume that $\rho\in N_\kappa^{st}$, $\nu\in N_\kappa^{ff}$, $\rho\le\nu$. According to 2.10.3. there exists a set $X\in\mathcal{A}$ with the property ${}_X\nu\in N_{(\nu)}^{fin}$. The convergence $({}_{A\setminus X}\nu)*J^{[n]}\longrightarrow o$ implies that $({}_{A\setminus X}\rho)*J^{[n]}\longrightarrow o$, which leads to $({}_X\rho)*J^{[n]}\longrightarrow\rho$. Moreover, there holds $\rho(X)\le\nu(X)<+\infty$, which shows that ${}_X\rho\in N_{(\rho)}^{fin}$, i.e. $\rho\in N_\kappa^{ff}$.

2. Assume that $\nu_i\in N_\kappa^{ff}$, $\gamma_i\in N_{(\nu_i)}^{fin}$, and $\sum_{i\in I}\nu_i := \nu\in N_\kappa^{fin}$.

For $\gamma := \sum_{i\in I}\gamma_i$ we obtain in virtue of Proposition 2.10.5.

$$\gamma\le\nu \;;\quad \gamma(A) = \sum_{i\in I}\gamma_i(A) = \sum_{i\in I}\widetilde{\Pi^\kappa_{\nu_i}}(M) = \widetilde{\Pi^\kappa_\nu}(M)<+\infty \;.$$

Due to 1.1.4., the sequence $(\gamma*J^{[n]})$ is τ_N-relatively compact. Let (σ_n) be a subsequence converging vaguely towards some $\sigma\in N$. Then there holds for all finite subsets S of I:

$$\sum_{i\in S}\nu_i\le\sigma\le\nu \;,$$

which reveals that $\sigma=\nu$. Hence the whole sequence $(\gamma*J^{[n]})$ converges vaguely towards ν, i.e. $\nu\in N_\kappa^{ff}$.

3. For all $r\in\mathbf{R}_+$ and $\sigma\in N_\kappa^{ff}$, by means of the estimate $r\cdot\sigma\le([r]+1)\cdot\sigma$, steps 1. and 2. allow to conclude that $r\cdot\sigma\in N_\kappa^{ff}$. ■

Now we give a sufficient first-order-criterion for an equilibrium distribution to be finitely fed:

2.10.13. Any equilibrium distribution P of first order whose intensity measure ν is finitely fed, also has this property. In this case there holds $\mathcal{A}_{(\Pi^\kappa_\nu)}^{stoch}\subseteq\mathcal{A}_{(P)}^{stoch}$.

Proof. According to Theorems 2.10.7. and 2.9.17., $\mathcal{A}_{(\Pi^\kappa_\nu)}^{stoch}$ is non-void. Any $X\in\mathcal{A}_{(\Pi^\kappa_\nu)}^{stoch}$ belongs to $\mathcal{A}_{(\nu)}$; moreover $\nu(X)$ is finite by Proposition 2.10.10. and hence, a fortiori, $P(\Phi(X)<+\infty)=1$. Therefore 2.10.1. implies that $({}_X\nu)*J^{[n]}\uparrow\nu$. In view of the estimate

$$({}_XP)_{\kappa^{[n]}}\preceq P \qquad (n=1,2,\ldots) \;,$$

2.10.4. (ii) now readily implies that $({}_XP)_{\kappa^{[n]}}\Rightarrow P$, which altogether shows that $X\in\mathcal{A}_{(P)}^{fin}$. Finally, by assumption there holds $\kappa_{(a)}(\chi(X)=1)=1$ for ν-almost all $a\in X$, which yields that X even belongs to $\mathcal{A}_{(P)}^{stoch}$. ■

Concludingly we state a "local structural result":

2.10.14. Proposition. For any $\nu \in N_{\kappa}^{ff}$ and $S \in \mathcal{A}_{(\Pi\kappa_{\nu})}^{stoch}$, the mapping $P \rightarrow {}_{S}P$ establishes a 1-1 corespondence between $\mathbf{P}^{inv,\nu}$ and the set $\{L : L \in \mathbf{P}, \Lambda_L = {}_{S}\nu , L_{\kappa_{[S]}} = L\}$ (i.e.the set of all equilibrium distributions L with respect to the "stochastic translation on S" $\kappa_{[S]}$). Its inverse mapping maps L into the weak limit of the sequence $(L_{\kappa^{[n]}})$.

Proof: 1. By Theorem 2.10.1. there holds $({}_{S}\nu) * J^{[n]} \uparrow \nu$.

2. Due to Proposition 2.9.4., for any $P \in \mathbf{P}^{inv}$ the distribution ${}_{S}P$ actually *is* an equilibrium distribution with respect to $\kappa_{[S]}$. By means of 2.10.13. there follows $({}_{S}P)_{\kappa^{[n]}} \Rightarrow P$.

3. On the other hand, for any equilibrium distribution L with respect to $\kappa_{[S]}$, which has intensity measure ${}_{S}\nu$, the sequence $(L_{\kappa^{[n]}})$ is increasing w. r. to $\preceq$ and weakly convergent towards some $P \in \mathbf{P}^{inv}$, whose intensity measure is ν due to 1. and 2.10.4. Finally, monotonicity of convergence together with ${}_{S}(L_{\kappa^{[n]}}) = L$ implies that ${}_{S}P = L$. ■

2.11. A Characterization of the Equilibrium Distributions of Poisson Type via Random Marking

In this section it is shown that, among the equilibrium distributions of first order, those of Poisson type are characterized by a certain commutativity relation between clustering and "thinning", i. e. removing individuals independently of each other with a certain probability c. "Thinning with probability $c \in [0, 1]$" is modelled by the substochastic translation $\vartheta_c := \tau(cI)$, which, according to 1.5.7., is associated with the substochastic kernel

$$a \rightarrow c\delta_a = c \cdot I(a,(.)) \quad .$$

Thus, there holds

$$(\vartheta_c)_{(a)} = c\delta_{\delta_a} + (1-c)\delta_o \qquad (a \in A) \quad ,$$

i. e. an individual at site a "dies" with probability (1−c) and stays at its place with "survival probability" c.

An immediate consequence of 1.5.8. is the following classical property of arbitrary Poisson distributions:

2.11.1. $(\Pi_{\nu})_{\vartheta_c} = \Pi_{c\nu} \qquad (\nu \in N;\ 0 \leq c \leq 1)$.

In a similar way, Poisson distributions even may be characterized (cf. [Fichtner (1975), Satz 2]):

2.11.2. Let m be an arbitrary natural number different from 1, and $c_1,...,c_m \in [0,1)$ be such that $\sum_{1 \le i \le m} c_m = 1$. Then a distribution L of first order is Poissonian iff

$$L = \mathop{*}_{1 \le i \le m} L_{\vartheta_{c_i}} .$$

For any kernel K from $[A, \mathcal{A}]$ to $[A, \mathcal{A}]$ there holds

$$K*(cI) = cK = (cI)*K \qquad (0 \le c \le 1) .$$

This implies

2.11.3. For all substochastic kernels K from $[A, \mathcal{A}]$ to $[A, \mathcal{A}]$ there holds

$$\tau(K) \circ \vartheta_c = \tau(cK) = \vartheta_c \circ \tau(K) \qquad (0 \le c \le 1) .$$

An immediate consequence of the definition of the semi-order $\precsim$ in **P** is

2.11.4. $P_{\vartheta_c} \precsim P \qquad (P \in \mathbf{P} \; ; \; 0 \le c \le 1)$.

In the sequel we still need the continuity statement

2.11.5. For all $c \in [0,1]$, the convergence $P_n \Rightarrow P$ implies that $(P_n)_{\vartheta_c} \Rightarrow P_{\vartheta_c}$.

Proof. By means of 1.2.4. there results the continuity of the mapping $\Phi \to (\vartheta_c)_{(\Phi)}$ with respect to the vague topology on M and the weak topology on **P**. This however implies readily our assertion. ■

The commutativity relation 2.11.3. between "thinnings" ϑ_c and arbitrary substochastic translations does not carry over to arbitrary clustering fields τ on $[A, \rho_A]$. We thus cannot expect that together with P also P_{ϑ_c} belongs to $\mathbf{P}^{inv}$. This excludes a straightforward generalization of 2.11.1. to equilibrium distributions of Poisson type. It turns out, however, that a more sophisticated generalization is possible, which is based on

2.11.6. Theorem. For any $P \in \mathbf{P}^{inv}$ and $c \in [0,1]$ there exists a $P_c \in \mathbf{P}^{inv}$ with the properties

$$\mathcal{G}_{P_c} = (\mathcal{G}_P)_{\vartheta_c} \quad ; \quad (P_{\vartheta_c})_{\kappa^{[n]}} \underset{n \to \infty}{\Rightarrow} P_c .$$

Proof. 1. First we show by means of Theorem 2.7.38. that

$$H := (\mathcal{G}_P)_{\vartheta_c}$$

is of the form $H = \mathcal{G}_L$, $L \in \mathbf{P}^{inv}$.

Because of $H \precsim \mathcal{G}_P$ there holds sp$H \precsim$ sp $\mathcal{G}_P = P$; hence results the existence of $(\mathrm{sp}H)_\kappa$.

In virtue of 2.11.3. we now obtain

$$H_\kappa = ((\mathcal{G}_P)_\kappa)_{\vartheta_c} = (\mathcal{G}_P)_{\vartheta_c} = H .$$

The relation $\Phi_1 \le \Phi_2$ obviously implies, for all $X \in \mathcal{B}$ and all natural numbers m, the estimate

$$\mathrm{sing}\,(\Phi_2, m, X) \le \mathrm{sing}\,(\Phi_1, m, X) .$$

Hence the convergence

$$\text{sing}(\Phi_1, m, X) \xrightarrow[m\to\infty]{\mathcal{G}_P} 1 \quad ,$$

which holds due to Theorem 2.7.38., implies the stochastic convergence

$$\text{sing}(\Phi_2, m, X) \xrightarrow[m\to\infty]{H} 1 \quad ,$$

Thus H has all characteristic properties required in Theorem 2.7.38. , and consequently there exists an $L \in \mathbf{P}^{\text{inv}}$ such that

$$(\mathcal{G}_P)_{\vartheta_c} = \mathcal{G}_L \quad .$$

2. Putting now for abbreviation $H := P_{\vartheta_c}$, we obtain by means of 2.11.3. for all natural numbers n

$$\mathcal{G}_{H,n} = (H^x)_{\kappa[n]} = ((P^x)_{\kappa[n]})_{\vartheta_c} = (\mathcal{G}_{P,n})_{\vartheta_c} \quad ,$$

which due to 2.11.5. and $\mathcal{G}_{P,n} \Rightarrow \mathcal{G}_P$ allows to conclude that

$$\mathcal{G}_{H,n} \Rightarrow (\mathcal{G}_P)_{\vartheta_c} = \mathcal{G}_L \quad .$$

This, in turn, implies that

$$H_{\kappa[n]} = \text{sp}\, \mathcal{G}_{H,n} \Rightarrow \text{sp}\, \mathcal{G}_L = L \quad . \blacksquare$$

For any equilibrium distribution P of Poisson type, 2.11.1. together with Theorem 2.7.24. yields for all $c \in [0,1]$

$$(\mathcal{G}_P)_{\vartheta_c} = (\Pi_{\tilde{P}})_{\vartheta_c} = \Pi_{c\tilde{P}} \quad .$$

By Theorem 2.11.6. this leads to $P_c = \text{sp}\Pi_{c\tilde{P}}$; hence we arrive at the desired analogue of 2.11.1. for equilibrium distributions of Poisson type:

2.11.7. Proposition. $(\Pi^{\kappa}_{v})_c = \Pi^{\kappa}_{cv} \quad (v \in N_{\kappa}^{st} \ ; \ 0 \le c \le 1) \quad .$

By means of 2.11.2. we now get a first characterization theorem for Poisson type equilibrium distributions.

2.11.8. Theorem. Let m be an arbitrary natural number different from 1 , and $c_1, \ldots c_m \in [0,1)$ be such that $\sum_{1\le i\le m} c_m = 1$. Then a distribution $P \in \mathbf{P}^{\text{inv}}$ of first order is of Poisson type iff it has the property

(+) $$P = \mathop{\ast}_{1\le i\le m} P_{c_i} \quad .$$

Proof. Due to Proposition 2.11.7., any Poisson type equilibrium distribution P has property (+). Conversely, by Theorem 2.11.6., property (+) implies that

$$\mathcal{G}_P = \mathop{\ast}_{1\le i\le m} (\mathcal{G}_P)_{\vartheta_{c_i}} \quad ,$$

which by 2.11.2. reveals that $\mathcal{G}_P$ is Poissonian. Hence P is of Poisson type according to Theorem 2.7.25. ■

The preceding result will now be deepened towards a characterization of Poisson type equilibrium disributions by means of "random marking".

For the rest of this section, let m be a fixed natural number different from 1. The points $a \in A$ will be "marked" by numbers from $\{1,...,m\}$, which is modelled in using, as a new phase space, the direct product $[A_m, \rho_{A_m}]$ of $[A, \rho_A]$ with the discretely metrized "mark space "$\{1,...,m\}$. Objects referring to the phase space $[A_m, \rho_{A_m}]$ will be denoted by subscript m.

For all $\mathbf{c} = [c_1,...c_m] \in \mathbf{R}_+{}^m$ with the property $c := \sum_{1 \le i \le m} c_i \le 1$, we put

$$(\vartheta_{\mathbf{c}})_{(a)} := \sum_{1 \le i \le m} c_i \delta_{\delta_{[a,i]}} + (1-c)\delta_o \qquad (a \in A) \quad ,$$

thus obtaining a stochastic clustering field $\vartheta_{\mathbf{c}}$ on $[A, \rho_A]$ with phase space $[A_m, \rho_{A_m}]$, which models a kind of "random marking" combined with thinning: An individual situated at site a either dies with probability 1–c or gets the mark i with probability c_i. Obviously, $\vartheta_{\mathbf{c}}$ a is stochastic clustering field iff the "survival probability" c equals 1.

Any counting measure on $\mathcal{A}_m$ can be interpreted as an m-tuple of counting measures on $\mathcal{A}$; more precisely

$$[\Phi_1,...\Phi_m] \rightarrow \sum_{i \le i \le m} \Phi_i \otimes \delta_i$$

is a bijective and in both directions measurable mapping from $[M^m, \mathcal{M}^{\otimes m}]$ onto $[M_m, \mathcal{M}_m]$, whose inverse mapping is given by

$$\Phi \rightarrow [\Phi((.)\times\{1\}),..., \Phi((.)\times\{m\})] \quad .$$

In the just described sense, the "marginal distributions" of P coincide with the distributions $P_{\vartheta_{c_i}}$, $1 \le i \le m$. More generally there holds:

2.11.9. For any $P \in \mathbf{P}$, $S \subseteq \{1,...m\}$, $\mathbf{c} \in \mathbf{R}_+{}^m$ obeying $\sum_{1 \le i \le m} c_i \le 1$, there holds with the notation $c_S := \sum_{i \in S} c_i$:

$$P_{\vartheta_{\mathbf{c}}}\Big(\big(\sum_{i \in S} \Phi((.)\times\{i\})\big) \in (.)\Big) = P_{\vartheta_{c_S}} \quad .$$

If now ν is some measure in N, and $\mathbf{c}$ obeys the just stated assumptions , then there results by means of 1.5.8.

$$(\Pi_\nu)_{\vartheta_{\mathbf{c}}} = \Pi_\gamma \quad \text{, where } \gamma = \nu * J_{\vartheta_{\mathbf{c}}} \quad .$$

There holds, however

$$\nu * J_{\vartheta_{\mathbf{c}}} = \sum_{1 \le i \le m} c_i(\nu \otimes \delta_i)$$

and hence

$$(\Pi_\nu)_{\vartheta_{\mathbf{c}}} = \mathop{\ast}_{1 \le i \le m} \Pi_{c_i(\nu \otimes \delta_i)} \quad .$$

Thus we obtain, as a refinement of 2.11.1.:

2.11.10. For all $\nu \in N$ and all $\mathbf{c} \in \mathbf{R}_+{}^m$ with the property $\sum_{1 \le i \le m} c_i \le 1$ there holds

$$(\Pi_\nu)_{\vartheta_{\mathbf{c}}} = \Big(\bigotimes_{1 \le i \le m} \Pi_{c_i \nu}\Big)\Big(\big(\sum_{1 \le i \le m} \Phi_i \otimes \delta_i\big) \in (.)\Big) \quad .$$

In other words: $(\Pi_\nu)_{\vartheta_{\mathbf{c}}}$ may be identified with $\bigotimes_{1 \le i \le m} \Pi_{c_i \nu}$.

With any clustering field ω on $[A, \rho_A]$, we associate via

$$(\omega_m)_{(a,i)} := \omega_{(a)}((\chi\otimes\delta_i)\in(.)) \qquad ([a,i]\in A_m)$$

a clustering field ω_m on $[A_m, \rho_{A_m}]$. Note that, according to ω_m, the original branching dynamics is refined only in the sense that an individual's mark devolves upon any of its descendants.

Obviously there holds for all $n\in\mathbf{Z}_+$

$$(\omega_m)^{[n]} = (\omega^{[n]})_m =: \omega^{[n]}{}_m \ ,$$

provided only that the clustering powers $\omega^{[n]}$, $n\in\mathbf{Z}_+$, exist.

If K is some substochastic kernel from $[A, \mathcal{A}]$ to $[A, \mathcal{A}]$, then we obtain, in putting

$$K_m([a,i],(.)) := \sum_{1\le i\le m} K(a,(.))\otimes\delta_i \qquad ([a,i]\in A_m)$$

a substochastic kernel K_m from $[A_m, \mathcal{A}_m]$ to $[A_m, \mathcal{A}_m]$ which obeys

$$(\tau(K))_m = \tau(K_m) \ .$$

On the analogy of 2.11.3. there follows

2.11.11. For all $c\in\mathbf{R}_+{}^m$ with the property $\sum_{1\le i\le m} c_i \le 1$ and all substochastic kernels K from $[A, \mathcal{A}]$ to $[A, \mathcal{A}]$, there holds

$$(\tau(K))_m\circ\vartheta_c = \tau(K_m)\circ\vartheta_c = \tau(H) = \vartheta_c\circ\tau(K) \ ,$$

where we put

$$H(a,(.)) := \sum_{1\le i\le m} c_i K(a,(.))\otimes\delta_i \ .$$

In particular, any substochastic clustering field ω on $[A, \rho_A]$ obeys

$$\omega_m\circ\vartheta_c = \vartheta_c\circ\omega \ .$$

The same argument as in the proof of 2.11.5. yields

2.11.12. For all $c\in\mathbf{R}_+{}^m$ with the property $\sum_{1\le i\le m} c_i \le 1$, the weak convergence $P_n \Rightarrow P$ in $\mathbf{P}$ implies that $(P_n)_{\vartheta_c} \Rightarrow P_{\vartheta_c}$.

Before we are going to refine Theorem 2.11.6., the following remark is necessary:

2.11.13. The mapping

$$\sum_{i\in I} (\delta_{\Phi_i})\otimes\delta_{j_i} \rightarrow \sum_{i\in I} \delta_{\Phi_i\otimes\delta_{j_i}}$$

provides a natural embedding of A_m into A_m.

Note that the image of A_m under this embedding contains exactly those systems $\Phi := \sum_{i\in I}\delta_{\chi_i}$, whose families $\chi_i\in M_m\setminus\{o\}$ consist, for all $i\in I$, only of individuals with one and the same mark j_i.

After these preparations we now get as a refinement of Theorem 2.11.6:

2.11.14. Theorem For any $P \in \mathbf{P}^{inv}$ and $c \in \mathbf{R}_+^m$ with the property $c := \sum_{1 \le i \le m} c_i \le 1$, there exists a distribution P_c on $\mathcal{M}_m$ with the properties

$$(P_c)_{\kappa_m} = P_c \;\; ; \;\; (P_{\vartheta_c})_{\kappa^{[n]}_m} \underset{n \to \infty}{\Rightarrow} P_c \;\; ; \;\; \mathcal{G}_{P_c} = (\mathcal{G}_P)_{\vartheta_c} \; .$$

Proof. 1. Paralleling the proof of Theorem 2.11.6. we first show by means of Theorem 2.7.38. that $H := (\mathcal{G}_P)_{\vartheta_c}$ is of the form $H = \mathcal{G}_L$, $L_{\kappa_m} = L$.

Under H, almost all realizations Φ are of the form

(*) $$\Phi = \sum_{i \in I} \delta_{\Phi_i} \otimes \delta_{j_i} \quad .$$

Thus, due to 2.11.9., under H the system $\sum_{i \in I} \delta_{\Phi_i}$ has the distribution $(\mathcal{G}_P)_{\vartheta_c}$. Therefore the estimate

$$(\mathcal{G}_P)_{\vartheta_c} \lesssim \mathcal{G}_P$$

yields that

$$(\sum_{i \in I} \Phi_i) \in {}_{\kappa}M \quad \text{for } H\text{-almost all } \Phi \, .$$

A fortiori, $(\kappa_m)_{(sp\Phi)}$ exists for H-almost all Φ.

Next, we show that H is cluster-invariant with respect to (κ_m) (which is the family dynamics associated with the "refined branching dynamics" κ_m.) Noting that H is concentrated on family system of the form (*), that for all such family systems there holds $(\dot{\kappa}_m)_{(\Phi)} = (\kappa_m)_{(\Phi)}$, (where we made use of the embedding 2.11.13.) and that κ is a substochastic clustering field on $[A, \rho_A]$, we get by means of the commutativity assertion 2.11.11.

$$H_{\kappa_m} = ((\mathcal{G}_P)_{\vartheta_c})_{\kappa_m} = ((\mathcal{G}_P)_{\vartheta_c})_{\kappa_m}$$
$$= ((\mathcal{G}_P)_{\kappa})_{\vartheta_c} = (\mathcal{G}_P)_{\vartheta_c} = H \quad .$$

For all $X \in \mathcal{B}$, all natural numbers n and all family systems

$$\Phi = \sum_{i \in I} \delta_{\Phi_i} \otimes \delta_{j_i} \quad \text{with} \quad sp \sum_{i \in I} \delta_{\Phi_i} \in {}_{\kappa^{[n]}}M$$

we obtain the equality

$$\text{sing}\,(\Phi, n, X \times \{1,...,m\}) = \text{sing}\,(\sum_{i \in I} \delta_{\Phi_i}, n, X) \quad .$$

Bearing in mind that, under H, $\sum_{i \in I} \delta_{\Phi_i}$ has distribution $(\mathcal{G}_P)_{\vartheta_c}$, which equals $\mathcal{G}_{P_c}$ by Theorem 2.11.6., we first get from Theorem 2.7.38.

$$\text{sing}\,(\Psi, n, X) \xrightarrow[n \to \infty]{\mathcal{G}_{P_c}} 1$$

and hence

$$\text{sing}\,(\Phi, n, X \times \{1,...m\}) \xrightarrow[n \to \infty]{H} 1 \quad .$$

In summary, one more application of Theorem 2.7.38. yields that $H = \mathcal{G}_L$ for some $L \in \mathbf{P}_m$ with the property $L = L_{\kappa_m}$.

2. Putting $H := P_{\vartheta_c}$, one obtains by means of 2.11.11. for all natural numbers n

$$\mathcal{G}_{H,n} = (H^x)_{\kappa_m^{[n]}} = (H^x)_{\kappa^{[n]}_m}$$

$$= ((P^x)_{\vartheta_c})_{\kappa^{[n]}_m} = = (P^x{}_{\kappa^{[n]}})_{\vartheta_c} ,$$

which by 2.11.12. leads to

$$\mathcal{G}_{H,n} \Rightarrow (\mathcal{G}_P)_{\vartheta_c} = \mathcal{G}_L$$

and, consequently, implies

$$H_{\kappa^{[n]}_m} = \mathrm{sp}\, \mathcal{G}_{H,n} \Rightarrow \mathrm{sp}\mathcal{G}_L = L \quad . \blacksquare$$

For any equilibrium distribution P of Poisson type, 2.11.10. together with Theorem 2.7.24. yields

$$(\mathcal{G}_P)_{\vartheta_c} = (\Pi_{\tilde{P}})_{\vartheta_c} = \Pi_W \quad ,$$

where $W := \sum_{1\le i\le m} c_i \tilde{P}((\Psi \otimes \delta_i) \in (.))$.

By Theorem 2.11.14. this leads to $P_c = \mathrm{sp}\Pi_W$; hence we get the following analogue of 2.11.10. for equilibrium distributions of Poisson type:

2.11.15. Proposition. For all $\nu \in N_\kappa^{st}$ and all $c \in \mathbf{R}_+^m$ with the property $\sum_{1\le i\le m} c_i \le 1$, there holds

$$(\Pi^\kappa{}_\nu)_c = (\bigotimes_{1\le i\le m} \Pi^\kappa{}_{c_i\nu})((\sum_{1\le i\le m} \Phi_i \otimes \delta_i) \in (.)) \quad .$$

Now we prove a characterization theorem for Poisson type equilibrium distributions in terms of "random marking":

2.11.16. Theorem. Let $c \in [0,1)^m$ obey $\sum_{1\le i\le m} c_i = 1$. Then a distribution $P \in \mathbf{P}^{inv}$ of first order is of Poisson type iff it has the property

$$(*) \qquad P_c = (\bigotimes_{1\le i\le m} P_{c_i})((\sum_{1\le i\le m} \Phi_i \otimes \delta_i) \in (.)) \quad .$$

Proof. In view of Propositions 2.11.7. and 2.11.15., any Poisson type equilibrium distribution P has property (*). Conversely, by 2.11.9., (*) implies that $P = \mathop{*}_{1\le i\le m} P_{c_i}$, which due to Theorem 2.11.8. yields that P is of Poisson type. ■

3. EQUILIBRIUM DISTRIBUTIONS OF COX TYPE

In the preceding chapter we constructed, for all stable J-invariant measures ν, a distinguished representative $\Pi^{\kappa}{}_{\nu}$ of the set $\mathbf{P}^{inv,\nu} := \{P: P\in \mathbf{P}^{inv,\nu}, \Lambda_P = \nu\}$, namely the equilibrium distribution of Poisson type with intensity measure ν. In this chapter we are going to introduce a whole class of distributions in $\mathbf{P}^{inv,\nu}$, by taking as initial distribution for the branching process not merely the Poisson distribution Π_{ν}, but any Cox distribution $\mathfrak{S}(Q)$, where Q belongs to the set

$$\mathbf{V}^{inv,\nu} := \{V: V\in \mathbf{V}, V(\gamma * J \in (.)) = V, \Lambda_V = \nu\}$$

(i. e. the set of all "J-invariant" distributions on $\mathcal{N}$ with intensity measure ν). In the special case $Q=\delta_{\nu}$, this J-invariace of Q coincides with the J-invariance of ν, i. e. the validity of the balance equation $\nu * J=\nu$. It turns out (Proposition 3.7.1.) that $(\mathfrak{S}(Q))_{\kappa^{[n]}}$ tends, for $n\to\infty$, weakly towards a distribution $V_{(Q)}$ in $\mathbf{P}^{inv}$, which, by Theorem 3.7.7., has intensity measure ν.

The distributions $V_{(Q)}$ may be interpreted as model specific analoga of Cox distributions and will be called equilibrium distributions of Cox type. This interpretation is also justified by Theorem 3.7.11., which states that the first order equilibrium distributions of Cox type are exactly those whose associated family equilibrium distribution is a Cox distribution.

By Theorem 3.7.10., $Q\to V_{(Q)}$ is, for all stable measures ν, a 1-1 mapping from $\mathbf{V}^{inv,\nu}$ into the set $\mathbf{P}^{inv,\nu}$. Later on, we will see in Chapter 4 that, under a certain regularity condition on the measure ν, the whole set $\mathbf{P}^{inv,\nu}$ consists of distributions of the form $V_{(Q)}$, $Q\in \mathbf{V}^{inv,\nu}$.

According to Proposition 3.8.2. and Theorem 3.7.8., any $Q\in \mathbf{V}^{inv,\nu}$ is concentrated on $N_{\kappa}{}^{st}$, provided that the measure $\nu\in N_{\kappa}{}^{st}$ obeys the condition

$$\int |J^{[n]}(a,X)-J^{[n+1]}(a,X)|\, \nu(da) \xrightarrow[n\to\infty]{} 0 \qquad (X\in \mathcal{B}) \quad .$$

Thus, in this case there holds

$$V_{(Q)} = \int \Pi^{\kappa}{}_{\gamma}(.)\, Q(d\gamma) \ ,$$

i.e. the equilibrium distribution of Cox type is a mixture of equilibrium distributions of Poisson type. This is not true in the general case, in which, however, a corresponding statement always holds true in a suitable space-time picture. There, the concept of an equilibrium distribution P has to be replaced by that of a "κ-process", i. e. the distribution

$\underline{P}$ of a Markovian random sequence $(\Phi_n)_{n\in Z}$ governed by the stochastic branching dynamics, or equivalently, by the corresponding "entrance law" $(P_n)_{n\in Z}$.

The results of the preceding chapter will be exploited for the construction of "Poisson type entrance laws " $(\Pi^{\kappa}{}_{(v_{i+n})})_{n\in Z}$. This constructive approach to the distributions $V_{(Q)}$ proves useful in deriving the stability criterion in Theorem 3.7.8. and the formula for the second factorial moment measure of $V_{(Q)}$ (Theorem 3.7.12.), and it helps to abandon the first order-assumption on Q. In this latter context, the concept of a J-invariant distribution Q with stable intensity measure generalizes to that of a "stable J-invariant distribution" (not necessarly of first order).

3.1. κ-processes

A distribution $\underline{P}$ on the σ-algebra $\mathcal{F} := \mathcal{M}^{\otimes Z}$ is termed **κ-process** if it is the distribution of a Markovian random sequence $(\Phi_n)_{n\in Z}$, whose marginal distributions $(P_n)_{n\in Z}$ are all concentrated on ${}_{\kappa}M$ and have the property

(*) $$(P_n)_{\kappa} = P_{n+1} \qquad (n\in Z) \quad .$$

In view of the Kolmogorov extension theorem, there is an obvious 1-1 correspondence between the set $\underline{P}^{\kappa}$ of all κ-processes $\underline{P}$ and the set P^{entr} of all **entrance laws** $(P_n)_{n\in Z}$ (i.e. sequences $(P_n)_{n\in Z}$ in P with property (*)), which is given by

$$P_n = \underline{P}(\Phi_n \in (.)) \qquad (n\in Z) \quad .$$

For any sequence $(\Phi_n) \in M^Z$ and $m\in Z$, we define

$$\theta^m(\Phi_n) := (\Phi_{n+m})_{n\in Z} \quad .$$

For any distribution $\underline{H}$ on $\mathcal{F}$, and $m\in Z$, we put

$$\theta^m\underline{H} := \underline{H}((\Phi_{n+m})_{n\in Z} \in (.)) \quad .$$

$\underline{H}$ is called **stationary** if $\theta^m\underline{H} = \underline{H}$ for all $m\in Z$.

For any κ-process $\underline{P}$ and $m\in Z$, $\theta^m\underline{P}$ remains to be a κ-process and corresponds to the entrance law $(P_{n+m})_{n\in Z}$.

Any equilibrium distribution $P\in P^{inv}$ induces a stationary κ-process $\underline{P}$ (corresponding to the "constant entrance law" $P_i=P$, $i\in Z$), and vice versa. On the other hand, there may exist nontrivial entrance laws even if $P^{inv} = \{\delta_o\}$:

3.1.1. Example. Let, for some "survival probability" $c\in(0,1)$, κ be the "thinning" ϑ_c introduced in 2.11. Then $P^{inv} = \{\delta_o\}$; on the other hand we obtain, in view of 1.5.8., for all distributions Q on $\mathcal{N}$ via

(o) $$P_i := \int \Pi_{c^i\gamma}(.)Q(d\gamma) \qquad (i\in Z)$$

an entrance law $(P_i) \in \mathbb{P}^{entr}$. Using results of [MKM, Chapter 7] it can even be shown that (o) provides a 1-1 correspondence between V and $\mathbb{P}^{entr}$. ■

A κ-process $\underline{P}$ as well as the corresponding entrance law (P_n) will be called **of k-th order** and **infinitely divisible**, respectively, if all distributions P_n, $n \in \mathbf{Z}$, have this property. The sequence $(\Lambda_{P_n})_{n\in \mathbf{Z}}$ will be referred to as **intensity sequence** of $\underline{P}$.

The elementary assertions 2.1.2. and 2.1.4. generalize to entrance laws as follows:

3.1.2. A sequence $(P_i)_{i\in \mathbf{Z}}$ of infinitely divisible distributions belongs to $\mathbb{P}^{entr}$ iff there holds

$$(\widetilde{P_i})_\kappa((.)\setminus\{o\}) = \widetilde{P_{i+1}} \qquad (i \in \mathbf{Z}) \quad .$$

3.1.3. $$\Lambda_{P_{n+1}} = \Lambda_{P_n} * J \qquad ((P_i) \in \mathbb{P}^{entr};\ n \in \mathbf{Z}) \quad .$$

A κ-process $\underline{P}$ is called **Poissonian** resp. **Coxian** if the corresponding entrance law $(P_i)_{i\in \mathbf{Z}}$ is a sequence of Poisson resp. Cox distributions on $\mathcal{M}$. The following proposition, which is an easy generalization of 2.1.5., characterizes the situation in which Poissonian κ-processes exist:

3.1.4. Proposition. For a sequence $(\nu_i)_{i\in Z}$ of measures in N, (Π_{ν_i}) belongs to $\mathbb{P}^{entr}$ iff for all $i \in Z$ the following properties hold true

$$\nu_i * J = \nu_{i+1} \quad ; \quad \kappa_{(a)}(\chi(A) \le 1) = 1 \quad \text{for } \nu_i\text{-almost all } a \in A.$$

Now let $\underline{P}$ be some κ-process. For each $m \in \mathbf{Z}$, the sequence $(\mathcal{G}_{P_{m-n},n})_{n\in \mathbf{Z}}$ is, in virtue of 2.7.12., decreasing with respect to $\le$. Hence there exists a distribution $\mathcal{G}_m^{\underline{P}} \in \mathbb{P}$ such that

$$\mathcal{G}_{P_{m-n},n} \downarrow \mathcal{G}_m^{\underline{P}} \quad ; \quad \text{sp}\, \mathcal{G}_m^{\underline{P}} = P_m \; .$$

For all $m \in \mathbf{Z}$ and $n \in \mathbf{Z}_+$ there holds

$$(\mathcal{G}_{P_{m-n},n})_\kappa = \mathcal{G}_{P_{m-n},n+1} = \mathcal{G}_{P_{(m+1)-(n+1)},n+1} \; .$$

This yields in virtue of 2.7.10.:

$$(\mathcal{G}_m^{\underline{P}})_\kappa = \mathcal{G}_{m+1}^{\underline{P}} \, ,$$

that is, $(\mathcal{G}_m^{\underline{P}})_{m\in \mathbf{Z}}$ is an entrance law for the family dynamics $\Phi \to \kappa_{(\Phi)}$, and will be called **family entrance law** associated with $\underline{P}$. The corresponding κ-process will be denoted by $\underline{\mathcal{G}}^{\underline{P}}$ and will be called **family process** associated with $\underline{P}$. In summary, one recognizes the validity of

3.1.5. Proposition. For any κ-process $\underline{P}$, there exists a κ-process $\underline{\mathcal{G}}^{\underline{P}}$ (corresponding to a κ-entrance law $(\mathcal{G}_m^{\underline{P}})_{m\in \mathbf{Z}}$) such that for all $m \in \mathbf{Z}$ there hold:

$$\mathcal{G}_{P_{m-n},n} \downarrow \mathcal{G}_m^{\underline{P}} \; ; \; \text{sp}\, \mathcal{G}_m^{\underline{P}} = P_m \, .$$

An immediate generalization of Proposition 2.7.22. is

3.1.6. Proposition. For any κ-process $\underline{P}$ and any sequence $(H_m)_{m\in Z}$ of distributions in $\mathbb{P}$, the properties

$$(H_m)_\kappa = H_{m+1} \ ; \ \mathrm{sp}H_m = P_m \qquad (m\in \mathbf{Z})$$

imply that $H_m \leq \mathcal{G}_m{}^{\underline{P}}$ for all $m\in \mathbf{Z}$.

Only plain modifications in the proofs of Proposition 2.7.29. and Theorem 2.7.38. lead to the following generalizations:

3.1.7. Proposition. For any κ-process $\underline{P}$, $X \in \mathcal{B}$ and $m \in \mathbf{Z}$ there holds

$$(\mathcal{G}_{P_{m-n},n})_X \underset{n\to\infty}{\Rightarrow} (\mathcal{G}_m{}^{\underline{P}})_X \ .$$

3.1.8. Theorem. A κ-process $\underline{\mathcal{H}}$ is a family process associated with some κ-process $\underline{P}$ iff the entrance law $(\mathcal{H}_i)$ corresponding to $\underline{\mathcal{H}}$ obeys the following conditions:

a) $(\mathrm{sp}\ \mathcal{H}_i)_\kappa$ exists for all $i\in \mathbf{Z}$.

b) $\int \mathrm{sing}(\Phi,n,X)\ \mathcal{H}_{m-n}(d\Phi) \underset{n\to\infty}{\longrightarrow} 1$ for all $m \in \mathbf{Z}$ and $X \in \mathcal{B}$.

The **convolution** of two distributions $\underline{P}$, $\underline{P}'$ on $\mathcal{F}$ is defined by

$$\underline{P}*\underline{P}' := (\underline{P}\otimes\underline{P}')\,((\Phi_i + \Phi'_i)_{i\in Z}\in(.)).$$

If, more generally, $(\underline{P}_s)_{s\in S}$ is an at most countable family of distributions $P_s((\Phi_{i,s})_{i\in Z}\in(.))$ on $\mathcal{F}$, then we put

$$\mathop{*}_{s\in S} \underline{P}_s := (\bigotimes_{s\in S} \underline{P}_s)((\sum_{s\in S} \Phi_{i,s})_{i\in Z}\in(.))$$

provided there holds

$$(\bigotimes_{s\in S} \underline{P}_s)((\sum_{s\in S} \Phi_{i,s})\in M \text{ for all } i\in \mathbf{Z}) = 1 \ .$$

An obvious consequence of 1.4.5. is

3.1.9. If the convolution $\underline{P}$ of an at most countable family $(\underline{P}_s)_{s\in S}$ of κ-process exists, then $\underline{P}$ is again a κ-process, obeying

$$P_i = \mathop{*}_{s\in S} P_{s,i} \qquad (i\in \mathbf{Z}) \ .$$

For any sequence $(v_i)_{i\in Z}$ of measures in N we put

$$\mathbb{P}^{entr,(v_i)} := \{(P_i)\in \mathbb{P}^{entr} : \Lambda_{P_i} = v_i \ \text{ for all } i\in \mathbf{Z}\} \ .$$

The corresponding set of κ-processes will be denoted by $\underline{\mathbb{P}}^{\kappa,(v_i)}$. A sequence $(v_i)_{i\in Z}$ of measures in N is called **stable** if $\mathbb{P}^{entr,(v_i)}$ is non-empty, i. e. if (v_i) occurs as intensity sequence of some κ-process of first order. The set of all stable sequences of measures will be denoted by $N_\kappa{}^{entr,st}$.

3.2. J-processes and κ-processes of Cox Type

Any sequence $(v_i)_{i\in Z}$ of measures in N with the property

$$v_i * J = v_{i+1} \quad (i\in \mathbf{Z})$$

will be called a **J-sequence**; the set of J-sequences will be denoted by N_J^{entr}.

In virtue of 3.1.3. there holds

3.2.1. $N_\kappa^{entr,st} \subseteq N_J^{entr}$.

For $(v_i)\in N_J^{entr}$ and $m\in \mathbf{Z}$, clearly also the shifted sequence $(v_{i+m})_{i\in Z}$ is in N_J^{entr}. Together with (v_i) and (γ_i), all linear combinations $(r_1 v_i + r_2 \gamma_i)$ with nonnegative coefficients r_1, r_2 belong to N_J^{entr}. For (v_i), $(\gamma_i)\in N_J^{entr}$ we write $(v_i) \le (\gamma_i)$ if this inequality holds component - wise in N. In this case, also $(\gamma_i - v_i)_{i\in Z}$ belongs to N_J^{entr}.

As $\gamma \to \gamma * J$ is an $\mathcal{N}$, $\mathcal{N}$- measurable mapping from a subset of N into N, there holds

3.2.2. $N_J^{entr} \in \mathcal{N}^{\otimes Z}$.

For abbreviation we put $\mathcal{N}_J^{entr} := N_J^{entr} \cap \mathcal{N}^{\otimes Z}$. As an analogue to Proposition 2.3.1. there holds

3.2.3. Proposition. For any $(v_i)\in N_J^{entr}$, the sequence $((\Pi_{v_{-n}})_{\kappa[n]}$ tends weakly towards an infinitely divisible distribution $V_{(v_i)}$ with the property $\Lambda_{V_{(v_i)}} \le v_0$.

There even holds

$$\mathcal{L}_{(\Pi_{v_{-n}})_{\kappa[n]}}(f) \uparrow \mathcal{L}_{V_{(v_i)}}(f) \qquad (f\in F_{v_0}) \ .$$

The sequence $(V_{(v_{i+m})})_{m\in Z}$ belongs to P^{entr}.

For $v\in N_J$, the constant sequence $v_i := v$, $i\in \mathbf{Z}$, belongs to N_J^{entr}. In this case there holds $V_{(v_i)} = V_{(v)}$ (i. e. the notation is consistent with 2.3.1.).

The assertions 2.3.2. to 2.3.5. extend as follows:

3.2.4. For all $(v_i)\in N_J^{entr}$ and $r\in \mathbf{R}_+$ there holds

$$V_{(rv_i)} = (V_{(v_i)})^r .$$

3.2.5. For all (v_i), $(\sigma_i)\in N_J^{entr}$ there holds

$$V_{(\gamma_i+\sigma_i)} = V_{(\gamma_i)} * V_{(\sigma_i)} .$$

3.2.6. If $(v_{i,s})_{s\in S}$ is an at most countable family in N_J^{entr} such that $\sum_{s\in S} v_{i,s} =: v_i \in N$ for all $i\in \mathbf{Z}$, then $(v_i)_{i\in Z}\in N_J^{entr}$ and

$$V_{(v_i)} = \mathop{\ast}_{s\in S} V_{(v_{i,s})} .$$

3.2.7. The mappings $(\nu_i) \to V_{(\nu_i)}$ and $(\nu_i) \to \Lambda_{V_{(\nu_i)}}$ from N_J^{entr} into **P** and from N_J^{entr} into N, respectively, are $\mathcal{N}_J^{entr}$, $\boldsymbol{P}$ - and $\mathcal{N}_J^{entr}$, $\mathcal{N}$-measurable.

Any distribution $\underline{Q}$ on $\mathcal{N}_J^{entr}$ will be called a **J-process**; any sequence $(Q_n)_{n\in\mathbf{Z}}$ of distributions on $\mathcal{N}$ with the property

$$Q_{i+1} = Q_i(\gamma * J \in (.)) \qquad (i\in\mathbf{Z})$$

will be called a **J-entrance law**. There is an obvious 1-1 correpondence between the set $\underline{\mathbf{Y}}_J$ of J-processes $\underline{Q}$ and the set $\mathbf{V}^{entr}$ of J-entrance laws (Q_i), which is given by

$$Q_i = \underline{Q}(\gamma_i \in (.)) \qquad (i\in\mathbf{Z}) \quad .$$

The sequence $(\Lambda_{Q_i})_{i\in\mathbf{Z}}$ will be addressed as **intensity sequence** of $\underline{Q}$; obviously it has the property

$$\Lambda_{Q_i} * J = \Lambda_{Q_{i+1}} \qquad (i\in\mathbf{Z}) \quad .$$

For any sequence $(\nu_i)_{i\in\mathbf{Z}}$ of measures in N, let us put

$$\mathbf{V}^{entr,(\nu_i)} := \{(Q_i)\in \mathbf{V}^{entr} : \Lambda_{Q_i} = \nu_i \quad \text{for all } i\in\mathbf{Z}\} \quad .$$

The corresponding set of J-processes will be denoted by $\underline{\mathbf{Y}}_J^{(\nu_i)}$.

A J-process $\underline{Q}$ is called **of k-th order** if all Q_i , $i\in\mathbf{Z}$, have this property.

For any J-process $\underline{Q}$ and $m\in\mathbf{Z}$, let us put

$$\theta^m \underline{Q} \;:=\; \underline{Q}((\gamma_{i+m})_{i\in\mathbf{Z}} \in (.)) \; .$$

Clearly, $\theta^m \underline{Q}$ remains to be a J-process and corresponds to the J-entrance law $(Q_{i+m})_{i\in\mathbf{Z}}$. In the case $\theta\underline{Q} = \underline{Q}$, the J-process $\underline{Q}$ is called **stationary**. For any J-process $\underline{Q}$ we define relying on 3.2.7.

$$V_m^{\underline{Q}} := \int V_{(\gamma_{i+m})}(.)\underline{Q}(d(\gamma_i)) \qquad (m\in\mathbf{Z}) \quad .$$

By Proposition 3.2.3. there follows, for each $m\in\mathbf{Z}_+$,

$$(V_m^{\underline{Q}})_\kappa = \int (V_{(\gamma_{i+m})})_\kappa(.)\underline{Q}(d(\gamma_i)) = \int (V_{(\gamma_{i+m+1})})(.)\underline{Q}(d(\gamma_i)) = V_{m+1}^{\underline{Q}} \; .$$

Furthermore, one observes that

$$\Lambda_{V_m^{\underline{Q}}} = \int \Lambda_{V_{(\gamma_{i+m})}}(.)\underline{Q}(d(\gamma_i)) \le \int \gamma_m(.)\underline{Q}(d(\gamma_i)) = \Lambda_{Q_m} \quad .$$

Finally, noting that $F_b \subseteq \bigcap_{\gamma\in N} F_\gamma$, one concludes from Proposition 3.2.3. for all $f\in F_b$

$$\mathcal{L}_{(\mathfrak{I}(Q_{m-n}))_\kappa[n]}(f) = \int \mathcal{L}_{(\Pi_\gamma)_\kappa[n]}(f)\, Q_{m-n}(d\gamma)$$

$$= \int \mathcal{L}_{(\Pi_{\gamma_{m-n}})_\kappa[n]}(f)\, \underline{Q}(d(\gamma_i)) \underset{n\to\infty}{\uparrow} \mathcal{L}_{V_m^{\underline{Q}}}(f) \quad .$$

Summarizing, we recognize the validity of

3.2.8. For all J-processes $\underline{Q}$ there holds

$$(V_m^{\underline{Q}})_{m\in Z} \in P^{entr} \quad ,$$

$$\Lambda_{V_m^{\underline{Q}}} \leq \Lambda_{Q_m} \qquad (m\in \mathbf{Z}) \quad ,$$

$$\mathcal{L}_{(\mathfrak{Z}(Q_{m-n})_{\kappa^{[n]}}}(f) \underset{n\to\infty}{\uparrow} \mathcal{L}_{V_m^{\underline{Q}}}(f) \qquad (f\in F_b,\ m\in \mathbf{Z}) \quad .$$

The κ-process corresponding to the entrance law $(V_m^{\underline{Q}})_{m\in Z}$ will be called **κ-process of Cox type induced by the J-process $\underline{Q}$** and will be denoted by $\underline{V}^{\underline{Q}}$.

An immediate consequence of this definition is:

3.2.9. $\underline{V}^{c\underline{Q}_1+(1-c)\underline{Q}_2} = c\underline{V}^{\underline{Q}_1} + (1-c)\underline{V}^{\underline{Q}_2} \qquad (0\leq c\leq 1\ ;\ \underline{Q}_1, \underline{Q}_2\in \underline{V}_J)$.

3.2.10. Let $(\underline{Q}_s)_{s\in S}$ be an at most countable family of J-processes with the property

$$\gamma_i := \sum_{s\in S} \gamma_{i,s} \in N \qquad \bigotimes_{s\in S} \underline{Q}_s \text{ -almost surely} \qquad (i\in \mathbf{Z}) \quad .$$

Then

$$\underline{Q} := \mathop{\ast}_{s\in S} \underline{Q}_s := \bigotimes_{s\in S} \underline{Q}_s\left(\left(\sum_{s\in S}\gamma_{i,s}\right)_{i\in Z}\in (.)\right)$$

is a J-process in virtue of 3.2.6.,and there holds

$$\underline{V}^{\underline{Q}} = \mathop{\ast}_{s\in S} \underline{V}^{\underline{Q}_s} \quad .$$

For two J-processes $\underline{Q}^{(1)}, \underline{Q}^{(2)}$ which obey $\underline{Q}^{(1)} \lesssim \underline{Q}^{(2)}$, (i. e. this relation holds component-wise for the corresponding J-entrance laws) there holds $\mathfrak{Z}(Q^{(1)}_{-n}) \lesssim \mathfrak{Z}(Q^{(2)}_{-n})$ and hence also $\mathfrak{Z}(Q^{(1)}_{-n})_{\kappa^{[n]}} \lesssim \mathfrak{Z}(Q^{(2)}_{-n})_{\kappa^{[n]}}$, $n\in \mathbf{Z}_+$. Applying 1.1.10. and 3.2.8., one arrives at

3.2.11. Proposition. For all J-processes $\underline{Q}^{(1)}, \underline{Q}^{(2)}$ with the property $\underline{Q}^{(1)} \lesssim \underline{Q}^{(2)}$, also $V_m^{\underline{Q}^{(1)}} \lesssim V_m^{\underline{Q}^{(2)}}$, $m\in \mathbf{Z}$, holds true.

If the clustering field κ is substochastic and $\underline{Q}$ is a J-process, then $V_m^{\underline{Q}}$ is just the Cox distribution $\mathfrak{Z}(Q_m)$.

3.3. κ-processes of Poisson Type

Theorem 2.3.6. admits a generalization to the non-stationary case:

3.3.1. Theorem. For all $(\nu_i)\in N_J^{entr}$ the following assertions are equivalent:

a) (ν_i) is stable .

b) $\Lambda_{V_{(\nu_{i+m})}} = \nu_m$ for all $m\in \mathbf{Z}$.

c) For all $X\in \mathcal{B}$ and all $m\in \mathbf{Z}$,

$$\sup_{n=1,2,\dots} \int_{\{\Phi:\Phi(X)>c\}} \Phi(X)(\Pi_{\nu_{-n+m}})_{\kappa^{[n]}}(d\Phi) \underset{c\to\infty}{\longrightarrow} 0 .$$

d) For all $X\in \mathcal{B}$ and all $m\in \mathbf{Z}$,

$$\sup_{n=1,2,\dots} \int (\sum_{j>c} j\kappa_{(a)}^{[n]}(\chi(X)=j))\nu_{-n+m}(da) \underset{c\to\infty}{\longrightarrow} 0 .$$

Proof. This follows along the same lines as 2.3.6.., replacing, for fixed $m\in\mathbf{Z}$, $(\Pi_\nu)_{\kappa^{[n]}}$ by $(\Pi_{\nu_{-n+m}})_{\kappa^{[n]}}$ for all $n\in \mathbf{Z}_+$, $V_{(\nu)}$ by $V_{(\nu_{i+m})}$, the invariant distribution $P\in \mathbf{P}^{inv,\nu}$ by an entrance law $(P_i)\in \mathbf{P}^{entr,(\nu_i)}$, and using Proposition 3.2.3. instead of 2.3.1.■

In virtue of Theorem 3.3.1., for all sequences $(\nu_i)\in N_\kappa^{entr,st}$, the sequence

$$\Pi^\kappa_{(\nu_{i+m})} := V_{(\nu_{i+m})} \qquad (m\in \mathbf{Z})$$

is a distinguished representative in $\mathbf{P}^{entr,(\nu_i)}$ and will be called **entrance law of Poisson type with intensity sequence** (ν_i). The corresponding κ-process will be termed **κ-process of Poisson type with intensity sequence** (ν_i) and will be denoted by $\underline{\Pi}^\kappa_{(\nu_i)}$.

For a constant stable sequence $(\nu_i) = (\nu)$, Theorem 3.3.1. states that $\Lambda_{V_{(\nu)}} = \nu$, hence the measure ν is stable , and $\Pi^\kappa_{(\nu_{i+m})}$ coincides , for all $m\in\mathbf{Z}$, with Π^κ_ν. In summary, we obtain:

3.3.2. A constant sequence $(\nu)_{i\in\mathbf{Z}}$ of measure in N is stable iff ν is a stable measure. In this case, the κ-process of Poisson type with intensity sequence (ν) is stationary, its corresponding equilibrium distribution being of Poisson type with intensity measure ν.

An immediate consequence of Theorem 3.3.1. and 3.2.7. is

3.3.3. $N_\kappa^{entr,st} \in \mathcal{N}_J^{entr}$.

3.3.1. together with 3.2.6. implies

3.3.4. If for any s from an at most countable set S, $(\nu_{i,s})_{i\in\mathbf{Z}}$ is an element of $N_{\kappa}^{entr,st}$ such that $\nu_i := \sum_{s\in S} \nu_{i,s}$ belongs to N for all $i\in\mathbf{Z}$, then also $(\nu_i)_{i\in\mathbf{Z}}$ is a stable sequence, and there holds

$$\Pi^{\kappa}_{(\nu_{i+m})} = \mathop{*}_{s\in S} \Pi^{\kappa}_{(\nu_{i+m,s})} \qquad (m\in\mathbf{Z}) \ .$$

As an analogue to 2.3.9. there results

3.3.5. $N_{\kappa}^{entr,st}$ is a descending subset of $[N_J^{entr}, \leq]$.

In virtue of 3.2.4. there follows

3.3.6. For all $(\nu_i)\in N_{\kappa}^{entr,st}$ and $r\in\mathbf{R}_+$, $(r\cdot\nu_i)$ is stable and there holds

$$\Pi^{\kappa}_{(r\nu_{i+m})} = (\Pi^{\kappa}_{(\nu_{i+m})})^{r} \qquad (m\in\mathbf{Z}) \ .$$

A fundamental property of the entrance laws of Poisson type is given by

3.3.7. Theorem. For all $(\nu_i)\in N_{\kappa}^{entr,st}$ there holds

$$\int \left(\left(\widetilde{\Pi^{\kappa}_{(\nu_{i-n})}}\right)^!_a\right)_{\kappa^{[n]}}(\chi(X)>0)\, J^{[n]}(a,X)\nu_{-n}(da) \underset{n\to\infty}{\longrightarrow} 0 \qquad (X\in\mathcal{B}) \ .$$

This is a direct consequence of the following more general assertion:

3.3.8. Theorem. For all $(\nu_i)\in N_{\kappa}^{entr,st}$ and any sequence $P_1, P_2,\ldots$ of distributions in T with $\Lambda_{P_n} = \nu_{-n}$, n=1,2,..., the convergence

$$(*) \qquad (P_n)_{\kappa^{[n]}} \Rightarrow \Pi^{\kappa}_{(\nu_i)}$$

holds true iff

$$(**) \qquad \int ((P_n)^!_a)_{\kappa^{[n]}}(\chi(X)>0)\, J^{[n]}(a,X)\, \nu_{-n}(da) \underset{n\to\infty}{\longrightarrow} 0 \qquad (X\in\mathcal{B}) \ .$$

Proof. 1. For abbreviation, we put

$$L_n := (P_n)_{\kappa^{[n]}} \quad \text{for } n = 1,2,\ldots \ ; \quad L_\infty := \Pi^{\kappa}_{(\nu_i)} \ .$$

Bearing in mind that

$$\Lambda_{\widetilde{L_n}} = \Lambda_{L_n} = \nu_0 = \Lambda_{L_\infty} = \Lambda_{\widetilde{L_\infty}} \ ,$$

one infers from 1.6.5. and the generalized Palm-Khinchin-theorem 1.8.5. that the convergence (*) holds true iff

$$(\widetilde{L_n})_{(f)} \Rightarrow (\widetilde{L_\infty})_{(f)} \qquad \text{for all } f\in F_c \ .$$

In view of 1.9.1. this is case iff for all $f\in F_c$

$$(o) \quad \iint(((P_n)^!_b)_{\kappa^{[n]}} * (\kappa^{[n]}_{(b)})_a)(.)\, f(a) J^{[n]}(b,da)\, \nu_{-n}(db) \Rightarrow \int(\Pi^{\kappa}_{(\nu_i)})_a(.)f(a)\nu_0(da) = (\widetilde{L_\infty})_{(f)}$$

Correspondingly, the convergence $(\Pi^{\kappa}_{\nu_{-n}})_{\kappa^{[n]}} \Rightarrow \Pi^{\kappa}_{(\nu_i)}$ ensures that

$$(oo) \qquad \iint(\kappa_{(b)}^{[n]})_a(.)f(a)J^{[n]}(b,da)\, \nu_{-n}(db) \Rightarrow (L_\infty)_{(f)} \quad \text{for all } f\in F_c \ .$$

If now (**) holds true, then the validity of (o) follows, for any $f \in F_c$, from (oo). Thus (*) is fulfilled.

2. Conversely, assume the convergence (*). If f is some function in F_c with the property $\langle v_0, f\rangle = 1$, then there results for all $X \in \mathcal{B}_{(\widetilde{L_\infty})_{(f)}}$

$$\iint ((\widetilde{P_n})^!_b)_{\kappa^{[n]}} * (\kappa^{[n]}_{(b)})_a)_X(.)\, f(a)\, J^{[n]}(b,da)\, v_{-n}(db) \Rightarrow ((\widetilde{L_\infty})_{(f)})_X$$

and

$$\iint ((\kappa^{[n]}_{(b)})_a)_X(.)\, f(a)\, J^{[n]}(b,da)\, v_{-n}(db) \Rightarrow ((\widetilde{L_\infty})_{(f)})_X ,$$

which, by 2.7.37., implies that

$$\text{(ooo)} \qquad \iint ((P_n)^!_b)_{\kappa^{[n]}}(\chi(X)>0) f(a) J^{[n]}(b,da)\, v_{-n}(db) \longrightarrow 0 \ .$$

Any set in $\mathcal{B}$ is included in a set in $\mathcal{B}_{(\widetilde{L_\infty})_{(f)}}$. Hence (ooo) is valid for $X \in \mathcal{B}$ and all $f \in F_c$ with the property $\langle v_0, f\rangle = 1$. Putting finally $f := (v_0(X))^{-1} 1_X$, we obtain (**), at least for $X \in \mathcal{B}$ with $v_0(X) > 0$. But for $X \in \mathcal{B}$ with $v_0(X) = 0$, (**) is trivially satisfied. ■

As a generalization of 2.7.21. one obtains

3.3.9. For all κ-processes $\underline{P}$ of first order with intensity sequence (v_i) there holds

$$\Lambda_{\mathcal{G}_m \underline{P}} = \widetilde{\Pi^\kappa_{(v_{i+m})}} \qquad (m \in \mathbf{Z}) \ .$$

Proof. Based on 2.7.6. and Proposition 3.1.6. there follows

$$(\mathcal{G}_{P_{-n+m},n})_X \preceq (\mathcal{G}_{P_m,0})_X = (P_m)_X \quad (X \in \mathcal{B};\, m \in \mathbf{Z};\, n \in \mathbf{Z}_+) \ .$$

Since for all $m \in \mathbf{Z}$ there holds

$$\mathcal{G}_{P_{-n+m},n} \underset{n\to\infty}{\Rightarrow} \mathcal{G}_m \underline{P} \ ,$$

we conclude by 1.1.8. that

$$\Lambda_{\mathcal{G}_{-n+m},n} \xrightarrow{N} \Lambda_{\mathcal{G}_m \underline{P}} \ .$$

Hence follows, by 2.7.13., that

$$(\widetilde{\Pi_{v_{-n+m}}})_{\kappa^{[n]}} = \int \kappa_{(a)}^{[n]}(\chi((.)\setminus\{o\}))\, v_{-n+m}(da)$$

$$\underset{n\to\infty}{\xrightarrow{W}} \Lambda_{\mathcal{G}_m \underline{P}} \ . \ ■$$

A direct consequence of 2.7.9., 3.1.6. and 3.3.9. is

3.3.10 Theorem. For any κ-process $\underline{P}$ of first order with intensity sequence (v_i), a sequence $(H_m)_{m\in\mathbf{Z}}$ of distributions in $\mathbb{P}$ coincides with $(\mathcal{G}_m \underline{P})$ iff there hold

$$\mathrm{sp} H_m = P_m \ ; \ (H_m)_\kappa = H_{m+1} \ ; \ \Lambda_H = \widetilde{\Pi^\kappa_{(v_{i+m})}} \qquad (m \in \mathbf{Z}) \ .$$

For all $(v_i)\in N_\kappa^{entr,st}$ and all $m\in\mathbf{Z}$ we obtain

$$(\Pi_{\widetilde{\Pi^\kappa(v_{i+m})}})_\kappa = \Pi_{\widetilde{(\Pi^\kappa(v_{i+m}))_\kappa((.)\setminus\{o\})}} = \Pi_{\widetilde{\Pi^\kappa(v_{i+(m+1)})}} .$$

Furthermore, 1.6.12. says that

$$sp\Pi_{\widetilde{\Pi^\kappa(v_{i+m})}} = \Pi^\kappa(v_{i+m}) .$$

Combined with Theorem 3.3.10. this yields

3.3.11. Theorem For all $(v_i)\in N_\kappa^{entr,st}$ there holds

$$\mathcal{G}_m^{\Pi^\kappa(v_i)} = \Pi_{\widetilde{\Pi^\kappa(v_{i+m})}} \qquad (m\in\mathbf{Z}) .$$

Together with Theorem 3.3.11., one more application of 3.3.9. leads to

3.3.12. Theorem. A κ-process $\underline{P}$ of first order is of Poisson type iff its associated family process is Poissonian.

3.4. Stable J-processes

A J-process $\underline{Q}$ is called **stable** if it is concentrated on the set $N_\kappa^{entr,st}$. In a similar way as we did with J-invariant measures, we will introduce "stable components" of J-sequences and of J-processes. To this end, we obtain first, on the analogy of 2.6.9

3.4.1. For all $(v_i)\in N_\kappa^{entr,st}$,

$$(v^{st}_m) := \left(\Lambda V_{(v_{m+i})}\right)_{m\in\mathbf{Z}}$$

is the largest element of the set $\{(\gamma_i) : \{(\gamma_i)\in N_\kappa^{entr,st},\ (\gamma_i)\le(v_i)\}$ with respect to the semi-order $\le$ in N_J^{entr}, and there holds

$$V_{(v_{i+m})} = \Pi^\kappa(v^{st}_{i+m}) \qquad (m\in\mathbf{Z}) .$$

We will refer to the sequence $(v^{st}_m)_{m\in\mathbf{Z}}$ as **stable component of the J-sequence** $(v_i)_{i\in\mathbf{Z}}$.

There exist sequences $(v_i)\in N_J^{entr}$ which are not stable but for which $v^{st}_0 = v_0$:

3.4.2. Example. Put $A := \mathbf{Z}$; $\rho_A(i,j) := |i-j|$ for all $i,j\in\mathbf{Z}$ and

$$\kappa_{(j)} := \begin{cases} (1/2)\delta_{2\delta_{j+1}} + (1/2)\delta_o & \text{for } j\le -2 \\ \delta_o & \text{for } j\ge -1 \end{cases} .$$

Then there results

$$J(j,(.)) = \begin{cases} \delta_{j+1} & \text{for } j \leq -2 \\ o & \text{for } j \geq -1 \end{cases} ,$$

hence

$$\nu_i := \begin{cases} \delta_i & \text{for } i \leq -1 \\ o & \text{for } i \geq 0 \end{cases}$$

defines a sequence $(\nu_i) \in \mathcal{N}_J^{entr}$ with the properties

$$\nu^{st}_i = o \quad \text{for all } i \in \mathbf{Z} \quad ; \quad \nu^{st}_0 = o = \nu_0 \ . \ ■$$

3.4.3. For all $m \in \mathbf{Z}$, the mapping $(\nu_i) \to \nu^{st}_m$ from N_J^{entr} into N is $\mathcal{N}_J^{entr}$, $\mathcal{N}$-measurable.

For any J-process $\underline{Q}$, let us define its **stable component** $\underline{Q}^{st}$ by

$$\underline{Q}^{st} := \underline{Q}((\gamma^{st}_i) \in (.)) \ .$$

Evidently, $\underline{Q}^{st}$ is indeed a stable J-process, with corresponding J-entrance law $Q^{st}_i := \underline{Q}((\gamma^{st}_i) \in (.))$, $i \in \mathbf{Z}$. A straightforward consequence of 3.4.1. is:

3.4.4. For all J-processes $\underline{Q}$ there hold

$$\underline{Q}^{st} \precsim \underline{Q} \quad \text{and} \quad \underline{Y}^{\underline{Q}} = \underline{Y}^{\underline{Q}^{st}}$$

(i. e., $\underline{Q}$ and $\underline{Q}^{st}$ induce the same κ-process of Cox type).

The following theorem shows that the family process associated with a κ-process of Cox type is Coxian.

3.4.5. Theorem. Let $\underline{P}$ be a κ-process of Cox type induced by a stable J-process $\underline{Q}$. Then there holds for all $m \in \mathbf{Z}$:

$$\mathcal{G}_m^{\underline{P}} = \mathfrak{I}(S_m) \ , \quad \text{where } S_m := \underline{Q}\big(\widetilde{\Pi^{\kappa}_{(\gamma_{i+m})}} \in (.)\big) \ .$$

Proof. Using Theorem 3.3.11. and an appropriate generalization of 2.7.17. we observe that

$$\mathfrak{I}(S_m) = \int \Pi_{\widetilde{\Pi^{\kappa}_{(\gamma_{i+m})}}}(.)\, \underline{Q}(d(\gamma_i)) = \int \mathcal{G}_{\Pi^{\kappa}_{(\gamma_{i+m})}}(.)\, \underline{Q}(d(\gamma_i)) \ . ■$$

Now we are in the position to prove

3.4.6. Theorem. Any κ-process $\underline{P}$ of Cox type is induced by a unique stable J-process $\underline{Q}$ (which will be called **intensity sequence distribution** of $\underline{P}$).

Proof. The fact that $\underline{P}$ is induced by some stable J-processes $\underline{Q}$ follows directly from 3.4.4. As to uniqueness of $\underline{Q}$, note that the corresponding J-entrance law (Q_i) obeys, in virtue of Theorem 3.3.1., for all $m \in \mathbf{Z}$

$$Q_m = \underline{Q}(\gamma_m \in (.)) = \underline{Q}(\Lambda_{\Pi^{\kappa}_{(\gamma_{i+m})}} \in (.)) \ .$$

But this latter distribution is uniquely determined by $\underline{P}$ due to Theorem 3.4.5. and 1.2.14. ■

In the sequel we need the following lemma:

3.4.7. For any J-process $\underline{Q}$ of first order with intensity sequence $(\nu_i)_{i\in \mathbf{Z}}$ there holds for all $m \in \mathbf{Z}$

$$\widetilde{V_{(\nu_{i+m})}} = \int \widetilde{V_{(\gamma_{i+m})}}(.)\underline{Q}(d(\gamma_i))$$

Proof. For all $(\gamma_i) \in N_J^{entr}$ there holds

$$\text{(o)} \qquad \int \kappa^{[n]}_{(a)}((.)\setminus\{o\})\,\gamma_{-n+m}(da) = \widetilde{(\Pi_{\gamma_{-n+m}})_{\kappa^{[n]}}} \xrightarrow{\mathbf{w}} V_{(\gamma_{i+m})}$$

Hence follows

$$\text{(oo)} \qquad \int \kappa^{[n]}_{(a)}((.)\setminus\{o\})\nu_{-n+m}(da) = \iint \kappa^{[n]}_{(a)}((.)\setminus\{o\})\gamma_{-n+m}(da)\underline{Q}(d(\gamma_i)) \xrightarrow{\mathbf{w}} \int \widetilde{V_{(\gamma_{i+m})}}\,\underline{Q}(d\gamma_i)) \ .$$

On the other hand, (o) applied to (ν_i) yields the convergence of the l.h.s. of (oo) towards $\widetilde{V_{(\nu_{i+m})}}$, thus showing the desired equality. ■

3.4.8. Theorem. Let $\underline{P}$ be a κ-process of Cox type, induced by some J-processes $\underline{Q}$ of first order.Then the following sequences coincide

a) the intensity sequence of $\underline{P}$,

b) the stable component of the intensity sequence of $\underline{Q}$,

c) the intensity sequence of the stable component of $\underline{Q}$.

Proof. Putting $\nu_i := \Lambda_{Q_i}$, $i \in \mathbf{Z}$, we compute, using 3.4.7., for all $m \in \mathbf{Z}$

$$\begin{aligned} \nu^{st}{}_m &= \Lambda_{V_{(\nu_{i+m})}} = \Lambda_{\widetilde{V_{(\nu_{i+m})}}} = \int \Lambda_{\widetilde{V_{(\gamma_{i+m})}}}(.)\underline{Q}(d(\gamma_i)) \\ &= \int \Lambda_{V_{(\gamma_{i+m})}}(.)\underline{Q}(d(\gamma_i)) = \Lambda_{V_m \underline{Q}} = \Lambda_{P_m} . \end{aligned}$$

On the other hand there holds

$$\begin{aligned} \int \Lambda_{V_{(\gamma_{i+m})}}(.)\underline{Q}(d(\gamma_i)) &= \int \gamma^{st}{}_{i+m}(.)\underline{Q}(d(\gamma_i)) \\ &= \int \gamma_{i+m}(.)\underline{Q}^{st}(d(\gamma_i)) = \Lambda_{Q_m^{st}} \ . ■ \end{aligned}$$

Because of $\underline{Q}^{st} \lesssim \underline{Q}$, a J-process $\underline{Q}$ of first order is stable iff its intensity sequence coincides with that of $\underline{Q}^{st}$. In virtue of Theorem 3.4.8. we thus obtain

3.4.9. Theorem. A J-process $\underline{Q}$ of first order is stable iff its intensity sequence has this property.

Combining Theorems 3.4.6., 3.4.8. and 3.4.9. we arrive at

3.4.10. Theorem. For all stable sequences $(\nu_i) \in N_\kappa^{entr,st}$, $\underline{Q} \to \underline{V}\underline{Q}$ is a 1-1 mapping from the set $\underline{V}_J^{(\nu_i)}$ of all J-processes with intensity sequence (ν_i) into the set $\underline{P}^{(\nu_i)}$ of κ-processes with intensity sequence (ν_i).

3.5. A Characterization of the κ-processes of Cox Type

Let $\underset{\sim}{P}$ be a κ-process of Cox type with intensity sequence distribution $\underset{\sim}{Q}$. According to Theorem 3.4.5., for any $m \in \mathbf{Z}$, $\mathcal{G}_m^{\underset{\sim}{P}}$ is a Cox distribution $\mathfrak{J}(S_m)$, where

$$S_m \; := \; \underset{\sim}{Q}\left(\widetilde{\Pi^\kappa_{(\gamma_{i+m})}} \in (.)\right) \quad .$$

Note that $S_m(\Lambda_W \in (.)) = Q_m$, and a fortiori

$$S_m(\Lambda_W \in N) \; = \; 1 \qquad (m \in \mathbf{Z}_+) \quad .$$

The following theorem gives a converse statement:

3.5.1. Theorem. A κ-process $\underset{\sim}{P}$ is of Cox type iff its associated family entrance law $(\mathcal{G}_m^{\underset{\sim}{P}})_{m \in \mathbf{Z}}$ is a sequence of Cox distributions $(\mathfrak{J}(S_m))_{m \in \mathbf{Z}}$, where the distributions S_m on $\mathcal{N}$ have the property

$$S(\Lambda_W \in N) = 1 \quad .$$

Proof. 1. As mentioned above, any κ-process $\underset{\sim}{P}$ of Cox type has the asserted properties.

2. Let us assume, conversely, that, for some κ-process $\underset{\sim}{P}$ there holds $\mathcal{G}_m^{\underset{\sim}{P}} = \mathfrak{J}(S_m)$, where $S_m \in V$, $S_m(\Lambda_W \in N) = 1$, $m \in \mathbf{Z}$. Observing, for $m \in \mathbf{Z}$

$$\begin{aligned} \mathcal{G}_{m+1}^{\underset{\sim}{P}} \; &= \; (\mathcal{G}_m^{\underset{\sim}{P}})_\kappa \; = \; \left(\int \Pi_W(.)\, S_m(dW)\right)_\kappa \\ &= \; \int (\Pi_W)_\kappa(.)\, S_m(dW) \; = \; \int \Pi_{W_\kappa((.)\setminus\{o\})}(.)\, S_m(dW) \end{aligned}$$

and noting that S_m is uniquely determined by $\mathfrak{J}(S_m)$, we conclude that

$$S_m\left(W_\kappa((.)\setminus\{o\}) \in (.)\right) \; = \; S_{m+1} \quad .$$

Hence $(S_m)_{m \in \mathbf{Z}}$ is, on the "family level", a J_κ-entrance law, corresponding to a J_κ-process $\underset{\sim}{S}$. We put

$$\underset{\sim}{U} := \underset{\sim}{S}((\mathcal{E}_{W_i})_{i \in \mathbf{Z}} \in (.)),$$

thus obtaining a distribution $\underset{\sim}{U}$ on $P^{\otimes \mathbf{Z}}$ which is concentrated on the set of all infinitely divisible entrance laws $(H_i) \in P^{entr}$ of first order. Furthermore, there holds for all $m \in \mathbf{Z}$

$$\underset{\sim}{U}(\widetilde{H_m} \in (.)) = S_m \quad ;$$

$$P_m = \mathrm{sp}\, \mathcal{G}_m^P = \int (\mathrm{sp}\Pi_{\widetilde{H_m}})(.)\, \underset{\sim}{U}(d(H_i)) = \int H_m(.)\underset{\sim}{U}(d(H_i))$$

and hence

(o) $$\mathcal{G}_m^{\underset{\sim}{P}} \; = \; \int \mathcal{G}_m^{\underset{\sim}{H}}(.)\, \underset{\sim}{U}(d(H_i)) \qquad (m \in \mathbf{Z}) \quad .$$

On the other hand, the equality $\mathcal{G}_m^{\underset{\sim}{P}} = \mathfrak{J}(S_m)$ together with the definition of $\underset{\sim}{U}$ leads to

(oo) $$\mathcal{G}_m^P \; = \; \int (\Pi_{\widetilde{H_m}})(.)\, \underset{\sim}{U}(d(H_i)) \qquad (m \in \mathbf{Z}) \quad .$$

For any infinitely divisible entrance (H_i) of first order, the sequence $(\Pi_{\tilde{H}_m})_{m\in Z}$ is an entrance law for the family dymanics and obeys

$$\mathrm{sp}\,\Pi_{\tilde{H}_m} = H_m \qquad (m\in Z) \quad .$$

Hence we infer from Proposition 3.1.6. that

$$\Pi_{\tilde{H}_m} \leq \mathcal{G}_m^H \qquad (m\in Z) \; .$$

In virtue of (o) and (oo) there follows for $\underline{U}$-almost all (H_i)

$$\mathcal{G}_m^H = \Pi_{\tilde{H}_m} \qquad (m\in Z) \; ,$$

which, by Theorem 3.3.12. yields, for $\underline{U}$-almost all (H_i)

$$H_m = \Pi^\kappa_{(\Lambda_{H_{i+m}})} \qquad (m\in Z) \; .$$

Putting $Q_m := \underline{U}(\Lambda_{H_m}\in(.))$, $m\in Z$, (which defines a J-entrance law due to $\underline{U}(\Lambda_{H_m}*J\in(.)) = \underline{U}(\Lambda_{(H_m)_\kappa}\in(.)) = \underline{U}(\Lambda_{H_{m+1}}\in(.))$), we obtain

$$P_m = \int H_m(.)\,\underline{U}(d(H_i)) = \int \Pi^\kappa_{(\gamma_{i+m})}(.)\,\underline{Q}(d(\gamma_i)) \; ,$$

i. e. $\underline{P}$ is of Cox type . ■

If a κ-process $\underline{P}$ is of first order and has the property $\mathcal{G}_m^{\underline{P}} = \mathfrak{I}(S_m)$, $m\in Z$, where (S_m) is a sequence in $\mathbf{V}$, then 3.3.9. and the equality $\Lambda_{\mathcal{G}_m^{\underline{P}}} = \int W(.)\,S_m(dW)$ show that $\Lambda_{P_m} = \int \Lambda_W(.) S_m(dW)$ and, a fortiori, $S_m(\Lambda_W\in N) = 1$, $m\in Z$. This yields the following version of Theorem 3.5.1. for κ-processes of first order:

3.5.2. Theorem. A κ-processes of first order is of Cox type iff its associated family process is Coxian.

3.6. Once Again Applying the Backward Technique

We are now going to extend the results of section 2.4. and 2.5. to J-sequences and to κ-processes of Poisson type.

Let us fix some sequence $(v_i)_{i\in Z}\in N_J^{entr}$, and let us associate, with each $i\in Z$, a dual stochastic kernel D_i of J with respect to v_{i-1}. For each $a\in A$, let $q_{(v_i),a}$ denote the distribution of a time-inhomogeneous Markov chain with state space A, starting at time zero in point a and having transition probabilities $D_0, D_{-1}, \ldots$

1.9.4. tells us that for ν_0-almost all a_0 the distribution

$$\mathcal{K}_{(\nu_i),a_0} := \int \mathcal{K}_{(a_j)_{j\in Z_+}}(.)\, q_{(\nu_i),a_0}(d(a_i)_{i\geq 1})$$

is well-defined (where $\mathcal{K}_{(a_j)}$ was defined at the beginning of section 2.4.). One more application of 1.9.4. leads to

3.6.1. For $(\nu_i)\in N_J^{entr}$, $n\in \mathbf{Z}_+$, and any measure H on $\mathcal{M}$ with $\Lambda_H = \nu_{-n}$, there holds for ν_0-almost $a\in A$:

$$(H_{\kappa^{[n]}})^!_{a_0} = \int (H^!_{a_n})_{\kappa^{[n]}} * \mathcal{K}_{(a_j)}((\sum_{0\leq j<n} \chi_j)\in(.))\, q_{(\nu_i),a_0}(d(a_i))$$

Choosing $H := \widetilde{\Pi_{-\nu_n}} = \mathcal{Q}_{-\nu_n}$, we obtain from 3.6.1.

3.6.2. For $(\nu_i)\in N_J^{entr}$ and $n\in \mathbf{Z}_+$, there holds for ν_0-almost $a_0\in A$:

$$(\widetilde{(\Pi_{\nu_{-n}})_{\kappa^{[n]}}})^!_a = \mathcal{K}_{(\nu_i),a}((\sum_{0\leq j<n} \chi_j)\in(.)) \quad .$$

As a generalization of Proposition 2.4.3. we get

3.6.3. Proposition. For $(\nu_i)\in N_J^{entr}$, the equality $\nu_0=\nu^{st}_0$ holds true iff for ν_0-almost all $a\in A$

(o)
$$\mathcal{K}_{(\nu_i),a}((\sum_{j\geq 0} \chi_j)\in M) = 1$$

is fulfilled.

Proof. 1. Assume that $\nu_0=\nu^{st}_0$. For all $n\in \mathbf{Z}_+$, 3.6.1. yields for ν_0-almost all $a \in A$

$$\mathcal{K}_{(\nu_i),a}((\sum_{0\leq j<n} \chi_j)\in(.)) \leq ((V_{(\nu_{i-n})})_{\kappa^{[n]}})^!_a = (V_{(\nu_i)})^!_a$$

(the last equality follows from Proposition 3.2.3.). Hence (o) is fulfilled ν_0-almost everywhere.

2. Conversely, assume that for some $(\nu_i)\in N_J^{entr}$ the relation (o) is valid ν_0-almost everywhere. Putting $L_n := (\Pi_{\nu_{-n}})_{\kappa^{[n]}}$, $n\in \mathbf{Z}_+$, we obtain, exploiting 3.6.2., in a similar way as in the 2nd step of the proof of 2.4.3. for all $X\in \mathcal{B}$ and $m\in \mathbf{Z}_+$:

$$\sup_{n=1,2,\ldots} \int (\sum_{j>m} j\kappa^{[n]}(a)(\chi(X)=j))\, \nu_{-n}(da)$$

$$= \sup_{n=1,2,\ldots} \int_{\{\Psi:\Psi(X)>m\}} \Psi(X)\, \widetilde{L_n}(d\Psi)$$

$$\leq \int_X \mathcal{K}_{(\nu_i),a}(\sum_{j\geq 0} \chi_j(X)\geq m)\, \nu_0(da) \underset{m\to\infty}{\longrightarrow} 0 .$$

Hence we conclude, due to $\widetilde{L_n} \xrightarrow{w} \widetilde{V_{(\nu_i)}}$, that $\Lambda_{\widetilde{V_{(\nu_i)}}} = \nu_0$, i. e. $\nu = \Lambda_{V_{(\nu_i)}} = \nu^{st}_0$. ■

On the analogy of Theorem 2.4.4., 2.4.5. and 2.4.6. we now obtain

6.4. Theorem. For $(\nu_i)\in N_\kappa^{entr,st}$ there holds for ν_0-almost all $a\in A$

$$(\widetilde{\Pi^\kappa_{(\nu_i)}})^!_a = \mathcal{K}_{(\nu_i),a}(\sum_{j\geq 0} \chi_j \in (.))$$

3.6.5. For $(\nu_i)\in N_\kappa^{entr,st}$, $(P_i)\in \mathbf{P}^{entr,(\nu_i)}$ and $m\in \mathbf{Z}$ there holds

$$(\Pi^\kappa_{(\nu_{i+m})})^!_a \lesssim (P_m)^!_a \quad \text{for } \nu_m\text{-almost all } a\in A .$$

3.6.6. Let $(\nu_i)\in N_\kappa^{entr,st}$ and h be an $\mathcal{A}\otimes\mathcal{M}$-measurable mapping from $A\times M$ into $[0,+\infty]$ such that for all $a\in A$ and $\Phi_n\uparrow\Phi$ the convergence $h(a,\Phi_n)\uparrow h(a,\Phi)$ holds. Then there follows

$$\int h(a,\chi)\mathcal{C}^!_{\widetilde{(\Pi_{\nu_{-n}})_{\kappa[n]}}}(d[a,\chi]) \;\uparrow\; \int h(a,\chi)\mathcal{C}^!_{\widetilde{\Pi^\kappa_{(\nu_i)}}}(d[a,\chi]) .$$

Generalizing Proposition 2.5.1. we get

3.6.7. Proposition. For all $(\nu_i)\in N_\kappa^{st}$ there holds

$$\Gamma^{(2)}_{\Pi^\kappa_{(\nu_i)}} = \nu_0\otimes\nu_0 + \sum_{i\geq 0}\iint J^{[i]}(x,(.))\otimes J^{[i]}(y,(.))\, \Gamma^{(2)}_{\kappa_{(z)}}(d[x,y])\nu_{-(i+1)}(dz) .$$

Proof. Putting

$$L_n := (\Pi_{\nu_{-n}})_{\kappa[n]} \;, n\in \mathbf{Z}_+ \;;\; L_\infty := \Pi^\kappa_{(\nu_i)} ,$$

we obtain like in the proof of 2.5.1. (now using 3.6.6. instead of 2.4.6.) for all $\mathcal{A}\otimes\mathcal{A}$-measurable mappings f from $A\times A$ into $[0,+\infty]$:

$$\int f(a,b)\, \Gamma^{(2)}_{\widetilde{L_n}}(d[a,b]) \uparrow \int f(a,b)\, \Gamma^{(2)}_{\widetilde{L_\infty}}(d[a,b]).$$

On the other hand there holds by 1.9.5. for all $n\in \mathbf{Z}_+$

$$\Gamma^{(2)}_{\widetilde{L_n}} = \sum_{0\leq i<n}\iint J^{[i]}(x,(.))\otimes J^{[i]}(y,(.))\, \Gamma^{(2)}_{\kappa_{(z)}}(d[x,y])\nu(dz) .$$

Passage to the limit $n\to\infty$ and applying 1.2.19. yields the assertion. ■

If now $(P_i)_{i\in\mathbf{Z}}$ is some entrance law in $\mathbf{P}^{entr,(\nu_i)}$, then we obtain, using 1.8.8. and 3.6.5., the estimate

$$\Gamma^{(2)}_{\widetilde{\Pi^\kappa_{(\nu_{i+m})}}} \lesssim \Gamma^{(2)}_{P_m} \qquad (m\in \mathbf{Z}) ,$$

which shows in view of 1.2.19. :

3.6.8. For all $(\nu_i)\in N_\kappa^{entr,st}$ and all $(P_i)\in \mathbf{P}^{entr,(\nu_i)}$ there holds

$$\Gamma^{(2)}_{\Pi^\kappa_{(\nu_{i+m})}} \lesssim \nu_m\otimes\nu_m + \Gamma^{(2)}_{P_m} \qquad (m\in \mathbf{Z}) .$$

For all $k\in \mathbf{Z}_+$ a sequence (ν_i) of measure in N is called **stable of k-th order** if $\mathbf{P}^{entr,(\nu_i)}$ contains some sequence of distributions of k-th order.

Paralleling the arguments in the proof of Theorem 2.5.4., one arrives at the following second order stability criterion:

3.6.9. Theorem. For all $(v_i) \in N_\kappa^{entr}$, the following assertions are equivalent:

a) (v_i) is stable of second order.

b) (v_i) is stable and $\Pi^\kappa_{(v_{i+m})}$ is of second order for all $m \in \mathbf{Z}$.

c) For all $X \in \mathcal{B}$ and all $m \in \mathbf{Z}$ there holds

$$\sum_{i \geq 0} \iint J^{[i]}(x,X) J^{[i]}(y,X)\, \Gamma^{(2)}_{\kappa_{(z)}}(d[x,y])\, v_{m-i}(dz) < +\infty \quad .$$

3.7. J-invariant Distributions and Equilibrium Distributions of Cox Type

A distribution Q on $\mathcal{N}$ is called **J-invariant** if

$$Q(\gamma * J \in (.)) = Q \quad .$$

The set of J-invariant distributions will be denoted by V^{inv}; for $v \in N$, $V^{inv,v}$ will stand for the set $\{Q: Q \in V^{inv}, \Lambda_Q = v\}$.

Clearly, a distribution Q is J-invariant iff the constant sequence $(Q)_{i\in\mathbf{Z}}$ is a J-entrance law. There is an obvious 1-1-correspondence between the set V^{inv} and the set of all stationary J-processes $\underline{Q}$, which is given by

$$Q = \underline{Q}(\gamma_m \in (.)) \qquad (m \in \mathbf{Z}) \quad .$$

In the sequel, we will freely use the symbol $\underline{Q}$ for the stationary J-process corresponding to a J-invariant distribution Q.

Concepts and results of sections 3.2., 3.4. and 3.5. may now be specialized to the "stationary case":

For $Q \in V^{inv}$, $V_{(Q)} := V_0^{\underline{Q}}$ is, in virtue of 3.2.8., an equilibrium distribution and will be termed **equilibrium distribution of Cox type induced by Q.**

3.2.8. to 3.2.11. yield

3.7.1. Proposition. For all $Q \in V^{inv}$ there holds

$$\Lambda_{V_{(Q)}} \leq \Lambda_Q \; ;$$

$$\mathcal{L}_{(\mathfrak{I}(Q))_{\kappa^{[n]}}}(f) \underset{n\to\infty}{\uparrow} \mathcal{L}_{V_{(Q)}}(f) \qquad (f \in F_b) \quad .$$

3.7.2. $V_{(cQ_1+(1-c)Q_2)} = cV_{(Q_1)} + (1-c)V_{(Q_2)} \qquad (0 \leq c \leq 1 \; ; Q_1, Q_2 \in V^{inv})$.

3.7.3. If the convolution Q of an at most countable family $(Q_s)_{s\in S}$ of J-invariant distributions exists, then Q is also a J-invariant distribution, and there holds

$$V_{(Q)} = \mathop{\ast}_{s\in S} V_{(Q_s)} \quad .$$

3.7.4. Proposition. For all J-invariant distributions Q_1, Q_2 with the property $Q_1 \precsim Q_2$, also $V_{(Q_1)} \precsim V_{(Q_2)}$ holds true.

If the clustering field κ is substochastic and Q is a J-invariant distribution, then $V_{(Q)}$ is just the Cox distribution $\mathfrak{Z}(Q)$.

For a J-invariant distribution Q, its **stable component** Q_{st} is defined by

$$Q_{st} := \underline{Q}(\gamma^{st}{}_0 \in (.)) \ .$$

Obviously, Q_{st} is again a J-invariant distribution and corresponds to the stable component $\underline{Q}^{st}$ of the J-process $\underline{Q}$. A J-invariant distribution is called **stable** if $Q = Q_{st}$; note that this is equivalent to the stability of $\underline{Q}$.

Specializing 3.4.4. and Theorem 3.4.6., respectively, one gets

3.7.5. For all J-invariant distribution Q there hold

$$Q \precsim Q_{st} \quad \text{and} \quad V_{(Q)} = V_{(Q_{st})} \ .$$

3.7.6. Theorem. Any equilibrium distribution P of Cox type is induced by a unique stable J-invariant distribution Q (which will be called **intensity distribution** of P).

A glimpse at Theorems 3.4.8. to 3.4.10. yields:

3.7.7. Theorem. For $P = V_{(Q)}$, $Q \in V^{inv}$ of first order, there holds

$$\Lambda_P = (\Lambda_Q)_{st} = \Lambda_{Q_{st}} \ .$$

3.7.8. Theorem. A J-invariant distribution of first order is stable iff its intensity measure has this property.

3.7.9. Theorem. For all stable measures $\nu \in N_\kappa{}^{st}$, $Q \to V_{(Q)}$, is a 1-1 mapping from $V^{inv,\nu}$ into $P^{inv,\nu}$.

Theorem 3.5.1. together with Theorem 3.4.5. tells that the following is true:

3.7.10. Theorem. An equilibrium distribution P is of Cox type iff its associated family equilibrium distribution $\mathcal{G}_P$ is a Cox distribution $\mathfrak{Z}(S)$, where the distribution S on $\mathcal{N}$ has the property

(*) $$S(\Lambda_W \in N) = 1 \ .$$

In this case, $S = \underline{Q}(\Pi^\kappa{}_{(\gamma_i)} \in (.))$, where Q is the intensity distribution of P.

Let us mention that, even if an equilibrium distribution P is not of Cox type, its associated family equilibrium distribution may be a Cox distribution. Indeed, in the setting of Example 2.7.26., $\mathcal{G}_P = \mathfrak{Z}(S)$, where $S = \delta_{\tilde{P}}$, thus $S(\Lambda_W \in N) = 0$, and, in virtue of Theorem 3.7.10., P cannot be of Cox type.

If, however, P is of first order, then Theorem 3.5.2. as well as Theorem 3.7.10. boil down to

3.7.11. Theorem. An equilibrium distribution P of first order is of Cox type iff its associated family equilibrium distribution is a Cox distribution.

We now give a formula for the second factorial moment measure of an equilibrium distribution of Cox type.

3.7.12. Theorem. For any equilibrium distribution P of Cox type with intensity distribution Q of first order there holds

$$\Gamma^{(2)}{}_P = \Lambda^{(2)}{}_Q + \iint \sum_{i\geq 0} J^{[i]}(x,(.))\otimes J^{[i]}(y,(.))\, \Gamma^{(2)}{}_{\kappa_{(z)}}(d[x,y])\Lambda_Q(dz) \ .$$

Proof. By definition Q is stable, and hence

$$P \ = V_{(Q)} \ = \int \Pi^{\kappa}{}_{(\gamma_i)}(.)\underline{Q}(d(\gamma_i)) \ .$$

An application of Proposition 3.6.7. now yields

$$\Gamma^{(2)}{}_{V_{(Q)}} = \int (\gamma_0 \otimes \gamma_0)(.)\underline{Q}(d(\gamma_i))$$

$$+ \sum_{i\geq 0} \iiint J^{[i]}(x,(.))\otimes J^{[i]}(y,(.))\, \Gamma^{(2)}{}_{\kappa_{(z)}}(d[x,y])\gamma_{-(i+1)}(dz)\underline{Q}(d(\gamma_i))$$

$$= \Lambda^{(2)}{}_Q + \sum_{i\geq 0} \iint J^{[i]}(x,(.))\otimes J^{[i]}(y,(.))\, \Gamma^{(2)}{}_{\kappa_{(z)}}(d[x,y])\Lambda_Q(dz) \ . \blacksquare$$

A distribution on Q on $\mathcal{N}$ is certainly J-invariant if it is concentrated on N_J; in this case one has

$$V_{(Q)} = \int V_{(\gamma)}Q(d\gamma) \quad \text{and} \quad Q_{st} = Q(\gamma_{st}\in(.)) \ ,$$

hence Q is stable iff $Q(N_\kappa{}^{st}) = 1$.

If the intensity measure of a distribution Q, which is concentrated on N_J, is extremal in N_J, then Q is concentrated on the multiples of Λ_Q:

3.7.13. Any distribution Q on $\mathcal{N}$ with the properties $Q(N_J) = 1$ and $\Lambda_Q \in N_J{}^{ex}$ obeys

$$Q(\{r\cdot\Lambda_Q\}_{r\in \mathbf{R}_+}) = 1 \ .$$

Proof. Let us put $\nu := \Lambda_Q$, assuming without loss of generality that $\nu \neq o$. For any X_1, $X_2 \in \mathcal{B}$, define the set

$$\Gamma_{X_1,X_2} := \{\gamma\in N_J : \gamma(X_1)\nu(X_2) > \gamma(X_2)\nu(X_1)\}.$$

If $Q(\Gamma_{X_1,X_2})$ were not zero, then, in contradiction to $\nu\in N_J{}^{ex}$, $\gamma_0 := \int_{\Gamma_{X_1,X_2}} \gamma(.)Q(d\gamma)$ would define an element $\gamma_0 \in N_J$ which obeys $\gamma_0 \leq \nu$ and is not of the form $r\cdot\nu$, $0\leq r\leq 1$, (due to $\gamma_0(X_1)\, \nu(X_2) > \gamma_0(X_2)\, \nu(X_1)$). Hence follows, for all X_1, $X_2 \in \mathcal{B}$

$$\gamma(X_1)\, \nu(X_2) = \gamma(X_2)\, \nu(X_1) \ \text{ for Q-almost all } \gamma \ .$$

Fixing an $X_2 \in \mathcal{B}$ with $\nu(X_2) > 0$ and using the fact that $\mathcal{A}$ is generated by a countable subring of $\mathcal{B}$, we obtain

$$\gamma = \nu(X_2)^{-1}\, \gamma(X_2)\cdot\nu \ \text{ for Q-almost all } \gamma,$$

which shows that Q is concentrated on $\{r\cdot\nu\}_{r\in \mathbf{R}_+}$. $\blacksquare$

We give a simple example of a J-invariant distribution Q satisfying $\Lambda_Q \in N_J{}^{ex}$ and not being concentrated on N_J:

3.7.14. Example. Let $[A, \rho_A]$, κ, μ and σ be as in Example 2.2.4., i. e. $J(i,(.)) = (1/3)(\delta_{i-1}+2\cdot\delta_{i+1})$, $i\in\mathbf{Z}$. Let σ_1 and σ_2 be the restriction of σ to the even and to the odd integers, respectively.

Then $\sigma_1 * J = \sigma_2$ and $\sigma_2 * J = \sigma_1$, hence $Q := 2^{-1}(\delta_{\sigma_1} + \delta_{\sigma_2})$ is a J-invariant distribution which is not concentrated on N_J and satisfies $\Lambda_Q = \sigma/2 \in N_J^{ex}$. ■

The following section contains a sufficient "first-order criterion" (3.8.2.) for $Q\in V^{inv}$ being concentrated on N_J.

3.8. J-aperiodic Measures

A measure $\nu\in N_J$ will be called **J-aperiodic** if

$$\int |J^{[n]}(a,X) - J^{[n+1]}(a,X)|\, \nu(da) \longrightarrow 0 \qquad (X\in\mathcal{B}) .$$

The set of all J-aperiodic measures will be denoted by N_J^{ap}.

3.8.1. Example. Put $A := \mathbf{Z}$, $\rho_A(i,j) := |i-j| \quad (i,j\in\mathbf{Z})$, and, for all $i\in\mathbf{Z}$,

$$\kappa_{(i)} := c_i\, \delta_{\delta_i} + (1-c_i)\, \delta_{\delta_{i+1}} \qquad \text{for some } c_i\in(0,1) .$$

Then there holds

$$N_J = \{r\mu\}_{r\in\mathbf{R}_+} = N_J^{ex} ,$$

where $\mu(\{i\}) = (1-c_i)^{-1}$, $i\in\mathbf{Z}$.

Kerstan and Fichtner (1979) showed that μ belongs to N_J^{ap} if and only if $\sum_{i<0} c_i$ is finite. ■

3.8.2. Proposition. Any J-invariant distribution Q on $\mathcal{N}$ with $\Lambda_Q\in N_J^{ap}$ is concentrated on N_J.

Proof. For all $X\in\mathcal{B}$ and $n\in\mathbf{Z}_+$ we get

$$\begin{aligned} \int |\gamma(X) - \gamma * J(X)|\, Q(d\gamma) &= \int |\gamma * J^{[n]}(X) - \gamma * J^{[n+1]}(X)|\, Q(d\gamma) \\ &\le \int \langle \gamma, |J^{[n]}((.),X) - J^{[n+1]}((.),X)|\rangle\, Q(d\gamma) \\ &= \langle \Lambda_Q, |J^{[n]}((.),X) - J^{[n+1]}((.),X)|\rangle \underset{n\to\infty}{\longrightarrow} 0 . \end{aligned}$$

Since $\mathcal{A}$ is generated by a countable subring of $\mathcal{B}$, we obtain

$$Q\,((\gamma * J)(X) = \gamma(X) \quad \text{for all } X\in\mathcal{A}) = 1 \quad . ■$$

3.8.3. Example. Let $[A, \rho_A]$, κ and Q be as in Example 3.7.14. Then Proposition 3.8.2. implies that $\sigma = 2\Lambda_Q \notin N_J^{ap}$.

3.8.4. Proposition. For all $\nu \in N_J$ the following assertions are equivalent:

a) $\nu \in N_J^{ap}$.

b) $\langle \nu, |J^{[n]}*f - J^{[n+1]}*f| \rangle \xrightarrow[n\to\infty]{} 0$ for all $f \in F_\nu$.

c) Let D be a dual stochastic kernel of J with respect to ν. Then

$$\|\sigma * D^{[n]} - \sigma * D^{[n+1]}\| \longrightarrow 0$$

for all distributions σ on $\mathcal{A}$ which are absolutely continuous with respect to ν.

Proof. Due to the fact that $f \to J*f := \int f(b)J(a,db)$ defines a linear non-expansive mapping from $L_1(\nu)$ into itself, the set

$$E := \{f \in L_1(\nu) : J^{[n]}*(f - J*f) \xrightarrow{L_1(\nu)} 0\}$$

is a closed linear subspace of $L_1(\nu)$. Now, in view of the equality $J((.),X) = J*1_X$, assertion a) may be written as

a') $1_X \in E$,

whereas b) may be written as

b') $E = L_1(\nu)$.

But as the set of finite linear combinations of 1_X, $X \in \mathcal{B}$, is dense in $L_1(\nu)$, conditions a') and b') are clearly equivalent.

2. As to the equivalence of b) and c) note that for any dual stochastic kernel D of J with respect to ν, there holds by 1.5.13. and 1.5.15. for $n \in \mathbb{Z}_+$

$$\|\sigma * D^{[n]} - \sigma * D^{[n+1]}\| = \langle \nu, |d(\sigma * D^{[n]})/d\nu - d(\sigma * D^{[n+1]})/d\nu| \rangle$$

$$= \langle \nu, |J^{[n]}*(d\sigma/d\nu) - J^{[n+1]}*(d\sigma/d\nu)| \rangle \quad . \blacksquare$$

A simple sufficient criterion for J-aperiodicity of a measure $\nu \in N_J$ is given by

3.8.5. Proposition. If $\inf_{a \in A} J(a,\{a\}) > 0$, then $N_J = N_J^{ap}$.

Proof. Our assumption guarantees that

$$J = (1-p)I + pG ,$$

where $p \in (0,1)$, $I(a,(.)) := \delta_a$, $a \in A$, and G is some kernel from $[A, \mathcal{A}]$ to $[A, \mathcal{A}]$. Obviously there holds

$$\nu * G = \nu \qquad (\nu \in N_J) .$$

Due to

$$J^{[n]} = \sum_{0 \le i \le n} b_{n,p}(\{i\}) G^{[i]} ,$$

where $b_{n,p}$ is the binominal distribution with parameters n and p, we get for any $\nu \in N_J$ and $X \in \mathcal{B}$:

$$\langle \nu, |J^{[n]}((.),X) - J^{[n+1]}((.),X)| \rangle \le \sum_{0 \le i} |b_{n,p}(\{i\}) - b_{n+1,p}(\{i\})| \int G^{[i]}(a,X)\nu(da)$$

$$= \nu(X) \| b_{n,p} - b_{n+1,p} \| .$$

Since the sequence $(b_{n,p})_{n=1,2,...}$ is asymptotically equidistributed (cf.[Heyer (1977), Theorem 2.6.14.]), the last expression tends towards zero for $n \to \infty$, implying that $\nu \in N_J^{ap}$. ■

The following theorem characterizes J-aperiodicity combined with extremality of J-invariant measures via a certain "asymptotic balance equation" and also via dual stochastic kernels:

3.8.6. Theorem. For all $\nu \in N_J$ the following assertions are equivalent:

a) $\nu \in N_J^{ex} \cap N_J^{ap}$.

b) For all $X_1, X_2 \in \mathcal{B}$ there holds

$$\langle \nu, |\nu(X_1) \cdot J^{[n]}((.),X_2) - \nu(X_2) \cdot J^{[n]}((.),X_1)| \rangle \to 0 \quad .$$

c) For all $f, g \in F_\nu$ such that $\langle \nu, f \rangle = \langle \nu, g \rangle$ there holds

$$\langle \nu, |J^{[n]}*f - J^{[n]}*g| \rangle \to 0 \quad .$$

d) Let D be a dual stochastic kernel of J with respect to ν. Then there holds

$$\|\sigma_1 * D^{[n]} - \sigma_2 * D^{[n+1]}\| \longrightarrow 0$$

for all distributions σ_1, σ_2 on $\mathcal{A}$ which are absolutely continuous with respect to ν.

Proof. 1. Consider the sets

$$E_1 := \{f \in L_1(\nu) : \langle \nu, f \rangle = 0\} \ ,$$
$$E_2 := \{f \in E_1 : J^{[n]}*f \xrightarrow{L_1(\nu)} 0\} \ ,$$

both of which are closed linear subspaces of $L_1(\nu)$. Now b) may be written as

b') $$\nu(X_1) 1_{X_2} - \nu(X_2) 1_{X_1} \in E_2 \qquad (X_1, X_2 \in \mathcal{B}) \ ,$$

whereas c) is clearly equivalent to

c') $$E_1 = E_2 \quad .$$

The set of all finite linear combinations of functions g of the form

$$\nu(X_1) 1_{X_2} - \nu(X_2) 1_{X_1} \in E_2 \qquad (X_1, X_2 \in \mathcal{B})$$

is, however, dense in E_1. Indeed, by the Hahn-Banach-theorem, it suffices to check that any continuous linear functional ℓ defined on $L_1(\nu)$, which vanishes for all functions g of the above-mentioned type, vanishes on the whole of E_1. Now take such a functional ℓ, having a representation

$$\ell(g) = \int g(a) h(a) \nu(da) \qquad (g \in L_1(\nu))$$

for some $h \in L_\infty(\nu)$. Then

$$\nu(X_1) = \int_{X_2} h(a) \nu(da) = \nu(X_2) = \int_{X_1} h(a) \nu(da) \qquad (X_1, X_2 \in \mathcal{B}) \ ,$$

hence h is constant ν-almost everywhere, implying that ℓ vanishes on E_1. Thus b') and c') are equivalent.

2. We are now going to show that a) implies c). Take any continuous linear functional ℓ defined on $L_1(\nu)$. ℓ has a representation

$$\ell(g) = \int g(a)h(a)\nu(da) \qquad (g \in L^1(\nu))$$

for some $h \in L_\infty(\nu)$. By Proposition 3.8.4. our assumption $\nu \in N_J^{equ}$ implies that $f - J*f$ belongs to E_2 for any $f \in L_1(\nu)$; hence $\ell(f) = \ell(J*f)$ for any $f \in L_1(\nu)$. Putting

$$\sigma := \int_{(.)} h(a)\nu(da),$$

we observe that

$$\langle \sigma, g\rangle = \langle \sigma, J*g\rangle = \langle \sigma*J, g\rangle \qquad (g \in F_c) \quad ,$$

i. e. σ is J-invariant. But as σ is majorized by a multiple of $\nu \in N_J^{ex}$, σ itself must be a multiple of ν, which means that g is constant almost ν-almost everywhere, implying that ℓ vanishes on E_1. One more application of the Hahn-Banach-theorem now yields that $E_1 = E_2$.

3. Conversely, we show that c') implies a). In virtue of $E_1 = E_2$, any $f \in L_1(\nu)$ obeys $f - J*f \in E_2$, i. e.

$$(J^{[n]}*f - J^{[n+1]}*f) \xrightarrow{L_1(\nu)} 0.$$

Hence, by Proposition 3.8.4., the measure ν belongs to N_J^{ap}. Let further σ be some arbitrary minorant of ν in N_J. σ is of the form

$$\sigma = \int_{(.)} h(a)\nu(da)$$

for some $h \in L_\infty(\nu)$. For all $f \in E_1$ (= E_2) we now obtain

$$\int f(a)h(a)\nu(da) = \int (J^{[n]}*f)(a)h(a)\, \nu(da) \xrightarrow[n\to\infty]{} 0.$$

Therefore, h is constant ν-almost everywhere, i. e. σ is a multiple of ν. Consequently, $\nu \in N_J^{ex}$.

4. Like in the last step of the proof of Proposition 3.3.4. we conclude, for all distributions σ_1, σ_2 being absolutely continuous with respect to ν and all $n \in Z_+$, the equality

$$\|\sigma_1 * D^{[n]} - \sigma_2 * D^{[n]}\| \;=\; \langle \nu, |J^{[n]}*(d\sigma_1/d\nu) - J^{[n]}*(d\sigma_1/d\nu)| \rangle \ ,$$

which shows directly the equivalence of c) and d). ■

3.8.7. Example. Let A be a nonvoid, at most countable set, and let the metric ρ_A be such that it generates the discrete topology and makes only the finite sets ρ_A-bounded. Further let κ be a clustering field whose intensity kernel J is stochastic and generates an irreducible homogeneous recurrent Markov chain. Subject to these assumptions there exists (see, e. g. [Chung (1967), p. 53]) a (non-zero) measure μ in N_J, and there holds

$$N_J = \{r\cdot\mu\}_{r \in R_+} = N_J^{ex} \quad .$$

Bearing in mind that also the dual stochastic kernel D of J with respect to ν generates a recurrent irreducible Markov chain with the same period, we recognize by means of Theorem 3.3.6. that μ belongs to N_J^{ap} iff J is aperiodic, because only in this case 3.8.6. d) holds true . ■

If the kernel J is stochastic, then a distribution ν on $\mathcal{A}$ obviously belongs to N_J^{ex} iff it is an irreducible stationary distribution for the Markov chain with transition kernel J.

The following result characterizes, for a stochastic intensity kernel, the distributions $\nu \in N_J^{ex} \cap N_J^{ap}$:

3.8.8. Proposition. Let ν be a distribution in N_J, and assume that $J(a,A) = 1$ for ν-almost all a. Then the following assertions are equivalent:

a) $\nu \in N_J^{ex} \cap N_J^{ap}$.

b) For all $X \in \mathcal{A}$ there holds

$$\langle \nu, |J^{[n]}((.), X) - \nu(X)| \rangle \xrightarrow[n\to\infty]{} 0 \quad .$$

c) For all $f \in F_\nu$ there holds

$$\langle \nu, |(J^{[n]}*f) - \langle \nu,f \rangle| \rangle \xrightarrow[n\to\infty]{} 0 \quad .$$

d) Let D be a dual stochastic kernel of J with respect to ν. Then there holds

$$\|\sigma * D^{[n]} - \nu\| \longrightarrow 0$$

for any distribution σ on $\mathcal{A}$ which is absolutely continuous with respect to ν.

Proof. 1. The implication c) $\Rightarrow$b) is obvious: simply put $f := 1_X$.

2. Assume that b) holds. Then there results for all $X_1, X_2 \in \mathcal{B}$

$$\langle \nu, |\nu(X_1)J^{[n]}((.),X_2) - \nu(X_2)J^{[n]}((.),X_1)| \rangle$$
$$\leq \nu(X_1)\langle \nu, |J^{[n]}((.),X_2) - \nu(X_2)| \rangle + \nu(X_2)\langle \nu, |J^{[n]}((.),X_1) - \nu(X_1)| \rangle \xrightarrow[n\to\infty]{} 0 \quad ,$$

which is condition b) of Theorem 3.8.6. and thus implies that $\nu \in N_J^{ex} \cap N_J^{ap}$, i. e. a).

3. The implication a) $\Rightarrow$d) follows immediately from Theorem 3.8.6. in putting $\sigma_1 := \sigma$, $\sigma_2 := \nu$ and bearing in mind that $\nu * J^{[n]} = \nu$, $n \in \mathbf{Z}_+$.

4. As to the implication d) $\Rightarrow$ a), choose two distributions σ_1, σ_2 on $\mathcal{A}$ which are absolutely continuous with respect to ν and observe that

$$\| \sigma_1 * D^{[n]} - \sigma_2 * D^{[n]} \| \leq \| \sigma_1 * D^{[n]} - \nu \| + \| \sigma_2 * D^{[n]} - \nu \| \xrightarrow[n\to\infty]{} 0 \quad ,$$

which implies that $\nu \in N_J^{ex} \cap N_J^{ap}$ due to Theorem 3.8.6.

5. To check that a) entails c), apply the implication a) $\Rightarrow$ c) of Theorem 3.8.6., in putting $g := \langle \nu,f \rangle$. ■

We are now going to continue the investigations of section 2.10., in proving the following "local structural result":

3.8.9. Theorem. Let ν be a measure in N_κ^{ff} which is different from zero, $S \in \mathcal{A}_{(\Pi_\nu^\kappa)^{stoch}}$, and $\gamma := (\nu(S))^{-1}{}_S\nu$. Further, let $J_{[S]}$ denote the intensity kernel of the clustering field $\kappa_{[S]}$. Then there holds

$$\gamma * J_{[S]} = \gamma \quad ,$$

and $\mathcal{Q}_\gamma$ is an equilibrium distribution with respect to $\kappa_{[S]}$. Let P_γ denote the equilibrium distribution in $\mathbf{P}^{inv}$ with the property $_S(P_\gamma) = \mathcal{Q}_\gamma$ (note that P_γ exists due to

Proposition 2.10.14.).

If we assume in addition that $\gamma \in N_{J_{[S]}}{}^{ex} \cap N_{J_{[S]}}{}^{ap}$, then the mapping

$$p \to \sum_{n \in Z_+} p(\{n\})(P_\gamma)^n$$

provides a 1-1 correspondence between the set of all distributions p on $\mathcal{P}(Z_+)$ with the property $\sum_{n \in Z_+} p(\{n\}) = \nu(S)$ and the set $P^{inv,\nu}$. Its inverse mapping is given by

$$p \to P(\Phi(S) \in (.)) \quad .$$

Proof. According to Proposition 2.10.14. it only remains to show that

$$p \to \sum_{n \in Z_+} p(\{n\})(\mathcal{Q}_\gamma)^n$$

is a 1-1 mapping from the set of all distributions p on $\mathcal{P}(Z_+)$ with expectation value $\nu(S)$ onto the set of all equilibrium distributions with respect to $\omega := \kappa_{[S]}$ having intensity measure ${}_S\nu$. Obviously this mapping is 1-1 and *into* the set under consideration ; what remains to show is that it is onto. Let L be an equilibrium distribution with respect to ω having intensity measure ${}_S\nu$. Then L can be written as

$$L = \sum_{n \in Z_+, L(\Phi(A)=n)>0} L(\Phi(A) = n) \cdot L(\Phi \in (.) | \Phi(A) = n) \quad .$$

For any $n \in Z_+$ with $L(\Phi(A) = n) > 0$, the summand $H := L(\Phi \in (.) | \Phi(A) = n)$ remains to be an invariant distribution with respect to ω (because ω is, on the set S, "almost" a stochastic translation.) We are going to show that $H = (\mathcal{Q}_\gamma)^n$.

Let Φ be some counting measure with total mass n, i. e. of the form

$$\Phi = \sum_{1 \le i \le n} \Phi_{a_i} \quad .$$

Then there results for all natural numbers m and all $f \in F_c$

$$| \mathcal{L}_{\omega^{[m]}(\Phi)}(f) - \mathcal{L}_{(\mathcal{Q}_\gamma)^n}(f) |$$

$$= | \prod_{1 \le i \le n} \mathcal{L}_{\omega^{[m]}(a_i)}(f) - (\mathcal{L}_{\mathcal{Q}_\gamma}(f))^n | \quad \le \sum_{1 \le i \le n} | \mathcal{L}_{\omega^{[m]}(a_i)}(f) - \mathcal{L}_{\mathcal{Q}_\gamma}(f) |$$

$$= \sum_{1 \le i \le n} | \int e^{-f(b)} J_{[S]}{}^{[m]}(a_i, db) - \int e^{-f(b)} \gamma(db) |$$

$$= \langle \Phi, | \int e^{-f(b)} J_{[S]}{}^{[m]}((.), db) - \int e^{-f(b)} \gamma(db) | \rangle \quad .$$

This yields

$$| \mathcal{L}_H(f) - \mathcal{L}_{(\mathcal{Q}_\gamma)^n}(f) | = \limsup_{m \to \infty} | \mathcal{L}_{H_\omega{}^{[m]}}(f) - \mathcal{L}_{(\mathcal{Q}_\gamma)^n}(f) |$$

$$\le \limsup_{m \to \infty} \int | \mathcal{L}_{\omega^{[m]}(\Phi)}(f) - \mathcal{L}_{(\mathcal{Q}_\gamma)^n}(f) | \, H(d\Phi)$$

$$\le \limsup_{m \to \infty} \langle \Lambda_H, | \int e^{-f(b)} J_{[S]}{}^{[n]}((.), db) - \int e^{-f(b)} \gamma(db) | \rangle$$

$$\le (L(\Phi(A) = n))^{-1} \, \nu(S) \limsup_{m \to \infty} \langle \gamma, | \int e^{-f(b)} J_{[S]}{}^{[m]}((.), db) - \langle \gamma, e^{-f} \rangle | \rangle \quad .$$

The last expression, however, equals zero due to Proposition 3.8.8 and the condition that $\gamma \in N_{J_{[S]}}{}^{ex} \cap N_{J_{[S]}}{}^{ap}$. ■

The just mentioned condition on γ is equivalent to the condition $\nu \in N_J^{ex} \cap N_J^{ap}$. More precisely, there holds

3.8.10. Proposition. Let ν, S, γ and $J_{[S]}$ be as in Theorem 3.8.9.

Then there holds

(i) $\nu \in N_J^{ex}$ iff $\gamma \in N_{J_{[S]}}^{ex}$.

(ii) $\nu \in N_J^{ap}$ iff $\gamma \in N_{J_{[S]}}^{ap}$.

Proof. 1. Assertion (i) can easily be derived from Proposition 2.10.14.

2. The direction from the left to the right in (ii) is a quick consequence of the definitions.

3. Finally we assume that $\gamma \in N_{J_{[S]}}^{ap}$. For any $X \in \mathcal{B}$ we obtain

$$\limsup_{n\to\infty} \int | J^{[n]}(a,X) - J^{[n+1]}(a,X) | \nu(da)$$

$$\leq \nu(S) \limsup_{n\to\infty} \int | J^{[n]}(a,X) - J^{[n+1]}(a,X) | \gamma(da)$$

$$+ 2 \limsup_{n\to\infty} ({}_{A\setminus S}\nu * J^{[n]})(X) .$$

By assumption, the second term on the right hand side of this estimate vanishes. For all $m \in \mathbf{Z}_+$, the first term on the r.h.s. equals

$$\nu(S) \limsup_{n\to\infty} \int | \int J^{[m]}(b,X)\, J^{[n]}(a,db) - \int J^{[m]}(b,X)\, J^{[n+1]}(a,db) | \gamma(da) ,$$

which is bounded from above by

$$\nu(S) \limsup_{n\to\infty} \int | \int J^{[m]}(b,X) 1_S(b) J^{[n]}(a,db) - \int J^{[m]}(b,X)\, 1_S(b) J^{[n+1]}(a,db) | \gamma(da)$$

$$+ 2 \limsup_{n\to\infty} \iint J^{[m]}(b,X) 1_{A\setminus S}(b) J^{[n]}(a,db) \nu(da)$$

$$= \nu(S) \limsup_{n\to\infty} \int | \int J^{[m]}(b,X) J_{[S]}^{[n]}(a,db) - \int J^{[m]}(b,X)\, J_{[S]}^{[n+1]}(a,db) | \gamma(da)$$

$$+ 2 ({}_{A\setminus S}\nu * J^{[m]})(X) .$$

The first term on the r.h.s. vanishes due to Proposition 3.8.4. (note that the mapping f defined by $f(b) := J^{[m]}(b,X)$ belongs to F_γ), whereas the second term is arbitrarily small for m large enough.

Altogether, this shows that ν is a J-aperiodic measure. ■

4. CHARACTERISTIC J-INVARIANT DISTRIBUTIONS

While the preceding two chapters were focussing on the construction of classes of equilibrium distributions, namely those of Poisson and Cox type, the present chapter is of a more analytic nature. A main result (Thm. 4.7.25.) says that, for a given stable measure ν, any equilibrium distribution with intensity measure ν is of Cox type, provided that ν is "regular", i. e. (cf. Prop. 4.7.18.) satisfies the deconcentration property

$$\int J^{[n]}(a,X)\min(1, J^{[n]}(a,X))\, \nu(da) \longrightarrow 0 \qquad (X \in \mathcal{B}) \ .$$

Any equilibrium distribution P corresponds in a natural way to a "stationary κ-process", i. e. a time-shift-invariant distribution $\underline{P}$ on $\mathcal{M}^{\otimes Z}$ which is Markovian with transition probability $\Phi \rightarrow \kappa_{(\Phi)}$. $\underline{P}$ is extremal in the convex set of all (not necessarily stationary) κ-processes iff it is trival on the σ-algebra $\mathcal{F}_{-\infty}$ of infinite past. Proposition 4.7.18. says that for a Poisson type equilibrium distribution P this is the case iff its intensity measure is regular in the above mentioned sense. According to Proposition 4.1.14., an equilibrium distribution P is trivial on the "spatial tail σ-algebra" $_{\infty}\mathcal{M}$ (i. e. has "short range correlations," cf. Prop. 4.1.10.) if $\underline{P}$ is $\mathcal{F}_{-\infty}$ -trivial and $\kappa_{(a)}$ (χ is finite) = 1 holds for all $a \in A$.

If two distributions L, Q on $\mathcal{N}$ obey the convergence relation $L(\gamma * J^{[n]} \in (.)) \Rightarrow Q$, then we put $\mathcal{W}_L := Q$. If, in addition, Λ_L has a majorant in N_J, then $\mathcal{W}_P$ is a J-invariant distribution (Thm. 4.7.1.). According to 4.7.3., $\mathcal{W}_P$ exists for all equilibrium distributions P of first order. The mapping $P \rightarrow \mathcal{W}_P$ from the set of all equilibrium distributions P, for which $\mathcal{W}_P$ exists, into the set V^{inv}, is one-to-one (Thm. 4.7.4.); this motivates to call $\mathcal{W}_P$ the "characteristic J-invariant distribution" of P.

For any stable measure ν the injective mapping $P \rightarrow \mathcal{W}_P$ from $P^{inv,\nu}$ into $V^{inv,\nu}$ is onto iff ν is regular (Thm. 4.7.26.) ; in this case its inverse mapping is given by $Q \rightarrow V_{(Q)}$ (Thm. 4.7.25.)

By means of Propositions 2.5.1. and 4.7.12. and Theorem 4.7.13., the second factorial moment measure of a second order equilibrium distribution P may be expressed in terms of J, Λ_P and the second factorial moment measures of $\mathcal{W}_P$ and κ_a , $a \in A$. Another simple inference from $\mathcal{W}_P$ to properties of P is provided by Theorem 4.7.16., which states that, for any first order equilibrium distribution P, its corresponding stationary κ-process is $\mathcal{F}_{-\infty}$-trivial iff $\mathcal{W}_P = \delta_{\Lambda_P}$. For all stable measures ν , regularity is thus equivalent to the property $\mathcal{W}_{\Pi^{\kappa}_{\nu}} = \delta_{\nu}$. This, in turn , is equivalent to $\mathcal{W}_{\Pi_{\nu}} = \delta_{\nu}$, because $\mathcal{W}_{\Pi^{\kappa}_{\nu}}$ coincides with $\mathcal{W}_{\Pi_{\nu}}$ for all stable measures ν (Thm. 4.7.15.).

For any $P \in \mathbf{P}^{inv}$ the distribution $\mathcal{W}_P$ exists iff $\underline{P}$ can be represented as a mixture of (not necessarily stationary) κ-processes $\underline{H}$, for which all "one dimensional distributions" $\underline{H}(\Phi_i \in (.))$, $i \in \mathbf{Z}$, are of first order. In particular, this is the case for all equilibrium distributions of Cox type. If the branching dynamics is "globally deconcentrating", i. e.

$$\sup_{a \in A} \kappa^{[n]}{}_{(a)}(\chi(X)>0) \longrightarrow 0 \qquad (X \in \mathcal{B}) \quad ,$$

then the equilibrium distributions of Cox type are exactly those for which $\mathcal{W}_P$ exists (Thm. 4.7.31.).

Like in the preceding chapter, most concepts and results are extended to the case of (not necessarily stationary) κ-processes (instead of equilibrium distributions only), which provides a somewhat more general and even more natural context.

4.1. Extremal κ-processes

In order to investigate the structure of the convex set $\underline{\mathbf{P}}^{\kappa}$ of all κ-processes, let us introduce some further notations. A κ-process $\underline{P}$ is called **extremal** if any representation

$$\underline{P} = c_1\underline{P}_1 + c_2\underline{P}_2 \quad , \quad c_1, c_2 \in (0,1) \quad ; \quad \underline{P}_1, \underline{P}_2 \in \underline{\mathbf{P}}^{\kappa}$$

is of the form $\underline{P}_1 = \underline{P}_2 = \underline{P}$.

Let, for $n \in \mathbf{Z}$, $\mathcal{F}_n$ be the σ-algebra on $M^{\mathbf{Z}}$ which is generated by the mappings

$$(\Phi_i) \rightarrow \Phi_j(X) \qquad (j \leq n, X \in \mathcal{B}).$$

The **σ-algebra $\mathcal{F}_{-\infty}$ of infinite past** is defined as

$$\mathcal{F}_{-\infty} := \bigcap_{n \in \mathbf{Z}} \mathcal{F}_n \quad .$$

A distribution on $\mathcal{F}$ is called **$\mathcal{F}_{-\infty}$-trivial** if it takes only values 0, 1 on $\mathcal{F}_{-\infty}$. For κ-processes, this property is equivalent to extremality:

4.1.1. a) A κ-process is extremal iff it is $\mathcal{F}_{-\infty}$ - trivial.

b) There exists a kernel $\underline{K}$ from $[M^{\mathbf{Z}}, \mathcal{F}_{-\infty}]$ to $[M^{\mathbf{Z}}, \mathcal{F}]$ with the following properties

(i) For all $(\Phi_i) \in M^{\mathbf{Z}}$, $\underline{K}((\Phi_i),(.))$ is an extremal κ-process.

(ii) For all $\underline{P} \in \underline{\mathbf{P}}^{\kappa}$ there holds

$$\underline{P}((.)| \mathcal{F}_{-\infty})((\Phi_i)) = \underline{K}((\Phi_i),(.)) \qquad \underline{P}\text{-a.all } (\Phi_i) \quad .$$

(iii) For all extremal κ-processes P there holds

$$\underline{K}((\Phi_i),(.)) = \underline{P} \qquad \underline{P}\text{-a.all } (\Phi_i) \quad .$$

Proof. Any κ-process $\underline{P}$ may be viewed as distribution of a Markovian random sequence $(\Phi_n)_{n \in \mathbf{Z}}$ with state space $[M^{\kappa}, \mathcal{M} \cap M^{\kappa}]$ (where $M^{\kappa} := \bigcap_{n \in \mathbf{Z}_+} \kappa^{[n]}M$), and transition probability $\Phi \rightarrow \kappa_{\Phi}(.)$, $\Phi \in M^{\kappa}$ - and vice versa. (Note that indeed any κ-process is concentrated on $(M^{\kappa})^{\mathbf{Z}}$). Hence our assertions follow from the general theory of ergodic decompositions (see e. g. [Dynkin (1978), Theorems 3.1., 3.2. and 9.1.]). ■

In the following, a kernel $\underline{K}$ with the properties stated in 4.1.1.b) will be kept fixed.

4.1.2. For any κ-process $\underline{P}$, $m \in \mathbf{Z}$, and all bounded $\mathcal{M}$-measurable real valued functions f defined on M there holds for $\underline{P}$-almost all (Φ_i)

$$\int f(\chi)\kappa^{[n]}(\Phi_{m-n})(d\chi) \xrightarrow[n\to\infty]{} \int f(\chi_j)\underline{K}((\Phi_i), d(\chi_i)) \quad .$$

Proof. The left hand side equals $E_{\underline{P}}(f(\chi_m)|\mathcal{F}_{m-n})((\Phi_i))$ $\underline{P}$-a.s., which, for $n\to\infty$, tends towards $E_{\underline{P}}(f(\chi_m)\mathcal{F}_{-\infty})((\Phi_i))$ $\underline{P}$-a.s. by the martingale convergence theorem. But the latter expression equals the right hand side $\underline{P}$-a.s. by 4.1.1.b) ii). ■

For any natural number j we endow M^j with the product topology arising from τ_M, thus obtaining a Polish space whose σ-algebra of Borel sets coincides with $\mathcal{M}^{\otimes j}$. In this way, we have also defined weak convergence of distributions on $\mathcal{M}^{\otimes j}$.

4.1.3. For any κ-process $\underline{P}$ and any $m \in \mathbf{Z}$ there holds for $\underline{P}$-almost all (Φ_i)

$$\overrightarrow{(\delta_{\Phi_{-n-m}}})([\chi_n,\dots,\chi_{n+2m}]\in(.)) \underset{n\to\infty}{\Rightarrow} \underline{K}((\Phi_i), [\chi_{-m},\dots,\chi_m]\in(.)) \quad .$$

Proof. For all $\mathcal{M}^{\otimes(2m+1)}$-measurable bounded real valued mappings h defined on M^{2m+1} and $\underline{P}$-almost all $(\Phi_i)\in M^{\mathbf{Z}}$ we obtain from 4.1.2.:

$$\int h(\chi_n,\dots,\chi_{n+2m})\,\overrightarrow{(\delta_{\Phi_{-n-m}}})(d(\chi_i))$$

$$= \iint h(\Psi_0,\dots,\Psi_{2m})\,\overrightarrow{\delta_\chi}(d(\Psi_k))\,\kappa^{[n]}(\Phi_{-n-m})(d\chi)$$

$$\xrightarrow[n\to\infty]{} \iint h(\Psi_0,\dots,\Psi_{2m})\,\overrightarrow{\delta_{\chi_{-m}}}(d(\Psi_k))\,\underline{K}((\Phi_i), d(\chi_j))$$

But the latter expression equals $\int h(\chi_{-m},\dots,\chi_m)\,K((\Phi_i), d(\chi_j))$, because $K((\Phi_i),(.))$ is a κ-process (see 4.1.1. b)i)). Now the assertion results from the fact that weak convergence of distributions on $\mathcal{M}^{\otimes(2m+1)}$ is implied by the convergence of at most countably many suitably chosen "expectation sequences". ■

4.1.4. For any κ-process $\underline{P}$ there holds for $\underline{P}$-almost all (Φ_i)

$$\underline{K}((\Phi_i), (\chi_j)\in(.)) = \underline{K}((\Phi_{i+1}), (\chi_{j-1})\in(.)) \quad .$$

Proof. For any $m \in \mathbf{Z}$, 4.1.3. applied to the κ-process $\theta\underline{P}$ yields the convergence

$$\overrightarrow{(\delta_{\Phi_{-n-m-1}}})([\chi_n,\dots,\chi_{n+2m}]\in(.)) \underset{n\to\infty}{\Rightarrow} \underline{K}((\Phi_i), [\chi_{-m-1},\dots,\chi_{m-1}]\in(.))$$

for $\theta\underline{P}$- almost all (Φ_i). But this is equivalent to say that the convergence

$$\overrightarrow{(\delta_{\Phi_{-n-m}}})([\chi_n,\dots,\chi_{n+2m}]\in(.)) \underset{n\to\infty}{\Rightarrow} \underline{K}((\Phi_{i+1}), [\chi_{-m-1},\dots,\chi_{m-1}]\in(.))$$

holds true for $\underline{P}$-almost all (Φ_i). Together with the convergence asserted in 4.1.3. this gives, for $\underline{P}$-a.all (Φ_i), the equality

$$\underline{K}((\Phi_{i+1}), [\chi_{-m-1},\dots,\chi_{m-1}]\in(.)) = \underline{K}((\Phi_i), [\chi_{-m},\dots,\chi_m]\in(.)) \quad ,$$

which, as m was arbitrary, proves our assertion. ■

4.1.5. For any two κ-processes $\underline{P}$ and $\underline{P}'$ there holds for $\underline{P}\otimes\underline{P}'$-almost all $[(\Phi_i),(\Phi'_i)]$

$$\underline{K}((\Phi_i+\Phi'_i),(.)) = \underline{K}((\Phi_i),(.)) * \underline{K}((\Phi'_i),(.)) \quad .$$

Proof. From 4.1.3. we obtain $\underline{P}\otimes\underline{P}'$-a.s. for $m\in \mathbf{Z}_+$:

$$\overrightarrow{(\delta_{\Phi_{-n-m}+\Phi'_{-n-m}}})([\chi_n,\ldots,\chi_{n+2m}]\in(.))$$

$$= \overrightarrow{(\delta_{\Phi_{-n-m}}})([\chi_n,\ldots,\chi_{n+2m}]\in(.)) * \overrightarrow{(\delta_{\Phi'_{-n-m}}})([\chi_n,\ldots,\chi_{n+2m}]\in(.))$$

$$\underset{n\to\infty}{\Rightarrow} \underline{K}((\Phi_i),[\chi_{-m},\ldots,\chi_m]\in(.)) * \underline{K}((\Phi'_i),[\chi_{-m},\ldots,\chi_m]\in(.)) \quad .$$

In the same way there results $\underline{P}\otimes\underline{P}'$- a.s. for $m\in \mathbf{Z}_+$

$$\overrightarrow{(\delta_{\Psi_{-n-m}}})([\chi_n,\ldots,\chi_{n+2m}]\in(.)) \underset{n\to\infty}{\Rightarrow} \underline{K}((\Psi_i),[\chi_{-m},\ldots,\chi_m]\in(.)).$$

Hence follows for $\underline{P}\otimes\underline{P}'$-almost all $[(\Phi_i),(\Phi'_i)]$ and all natural numbers m the equality

$$\underline{K}((\Phi_i+\Phi'_i),[\chi_{-m},\ldots,\chi_m]\in(.)) = (\underline{K}((\Phi_i),(.)) * \underline{K}((\Phi'_i),(.)))([\chi_{-m},\ldots,\chi_m]\in(.)) \quad . \blacksquare$$

4.1.6. Let $\underline{P}$ be a κ-process and h be an $\mathcal{M}$-measurable mapping from M into $[0,+\infty]$. Then there holds for all $m\in\mathbf{Z}$

$$\text{(o)} \quad \int\exp\left(-\int h(\chi)\kappa^{[n]}_{(\Phi)}(d\chi)\right)P_{m-n}(d\Phi) \underset{n\to\infty}{\downarrow} \int\exp\left(-\int h(\chi_m)\underline{K}((\Phi_i),d(\chi_j))\right)\underline{P}(d(\Phi_i)) \quad .$$

Proof. 1. The convergence

$$(*) \qquad E_{\underline{P}}(h(\Phi_m)|\mathcal{F}_{m-n}) \underset{n\to\infty}{\longrightarrow} E_{\underline{P}}(h(\Phi_m)|\mathcal{F}_{-\infty})$$

clearly holds $\underline{P}$-a.s. if $E_{\underline{P}}(h(\Phi_m))$ is finite (see 4.1.2.). To show (*) for all nonnegative, $\mathcal{M}$-measurable h, let us first observe the convergence

$$E_{\underline{P}}(\min(j,h(\Phi_m))|\mathcal{F}_{m-n}) \underset{n\to\infty}{\longrightarrow} E_{\underline{P}}(\min(j,h(\Phi_m))|\mathcal{F}_{-\infty}) \quad ,$$

which holds $\underline{P}$-a.s. for all natural numbers j. Hence follows $\underline{P}$-a.s.

$$\liminf_{n\to\infty} E_{\underline{P}}(h(\Phi_m)|\mathcal{F}_{m-n}) \geq \sup_{j=1,2,\ldots} E_{\underline{P}}(\min(j,h(\Phi_m))|\mathcal{F}_{-\infty})$$
$$= E_{\underline{P}}(h(\Phi_m)|\mathcal{F}_{-\infty}).$$

Assertion (*) is thus valid $\underline{P}$-almost surely on the event $E_{\underline{P}}(h(\Phi_m)|\mathcal{F}_{-\infty}) = +\infty$.

For all natural numbers j, let Y_j be the $\mathcal{F}_{-\infty}$-measurable event $E_{\underline{P}}(h(\Phi_m)|\mathcal{F}_{-\infty}) \leq j$. We obtain

$$E_{\underline{P}}(1_{Y_j}h(\Phi_m)) = E_{\underline{P}}(E_{\underline{P}}(1_{Y_j}h(\Phi_m)|\mathcal{F}_{-\infty})) \leq j$$

and hence $\underline{P}$-a.s.

$$1_{Y_j}E_{\underline{P}}(h(\Phi_m)|\mathcal{F}_{m-n}) = E_{\underline{P}}(1_{Y_j}h(\Phi_m)|\mathcal{F}_{m-n})$$

$$\underset{n\to\infty}{\longrightarrow} E_{\underline{P}}(1_{Y_j}h(\Phi_m)|\mathcal{F}_{-\infty}) = 1_{Y_j}E_{\underline{P}}(h(\Phi_m)|\mathcal{F}_{-\infty}) \quad .$$

Assertion (*) therefore holds $\underline{P}$-almost everywhere.

2. For all $n \in \mathbf{Z}_+$ we obtain by Jensen's inequality:

$$\int \exp(-\int h(\chi)\kappa^{[n+1]}{}_{(\Phi)}(d\chi))P_{m-(n+1)}(d\Phi)$$
$$= E_{\underline{P}}\Big(\exp(-E_{\underline{P}}(h(\Phi_m)|\mathcal{F}_{m-(n+1)}))\Big)$$
$$= E_{\underline{P}}\Big(\exp(-E_{\underline{P}}(E_{\underline{P}}(h(\Phi_m)|\mathcal{F}_{m-n})|\mathcal{F}_{m-(n+1)}))\Big)$$
$$\leq E_{\underline{P}}\Big(E_{\underline{P}}(\exp(-E_{\underline{P}}(h(\Phi_m)|\mathcal{F}_{m-n}))|\mathcal{F}_{m-(n+1)})\Big)$$
$$= E_{\underline{P}}(\exp(-E_{\underline{P}}(h(\Phi_m)|\mathcal{F}_{m-n})))$$
$$= \int \exp(-\int h(\chi)\kappa^{[n]}{}_{(\Phi)}(d\chi))P_{m-n}(d\Phi) \quad ,$$

which shows that the sequence in the left hand side of (o) is decreasing. Its convergence towards the right hand side follows from (*) by dominated convergence. ■

The following proposition characterizes $\mathcal{F}_{-\infty}$-triviality - and hence extremality - of κ-processes.

4.1.7. Proposition. For any κ-process $\underline{P}$ the following assertions are equivalent:

a) $\underline{P}$ is $\mathcal{F}_{-\infty}$-trivial.

b) $\int |\kappa^{[n]}{}_{(\Phi)}(Y) - P_m(Y)| P_{m-n}(d\Phi) \xrightarrow[n\to\infty]{} 0 \qquad (m \in \mathbf{Z}, Y \in \mathcal{M})$.

c) $\int |\mathcal{L}_{\kappa^{[n]}{}_{(\Phi)}}(f) - \mathcal{L}_{P_m}(f)| P_{m-n}(d\Phi) \xrightarrow[n\to\infty]{} 0 \qquad (m \in \mathbf{Z}, f \in F_c)$.

d) For all bounded $\mathcal{M}$-measurable real valued functions h on M and all $m \in \mathbf{Z}$ there holds

(o) $$\int | \int h(\chi)\kappa^{[n]}{}_{(\Phi)}(d\chi) - \int h(\chi)P_m(d\chi) | P_{m-n}(d\Phi) \xrightarrow[n\to\infty]{} 0 \ .$$

Proof. 1. First we show the implication a) $\Rightarrow$ d). For any bounded $\mathcal{M}$-measurable real valued function h on M and all $m \in \mathbf{Z}$ there holds

$$\int | \int h(\chi)\kappa^{[n]}{}_{(\Phi)}(d\chi) - \int h(\chi)P_m(d\chi) | P_{m-n}(d\Phi)$$
$$= \int | \int h(\chi)\kappa^{[n]}{}_{(\Phi_{m-n})}(d\chi) - \int h(\chi)P_m(d\chi) | \underline{P}(d(\Phi_j))$$
$$= E_{\underline{P}} |E_{\underline{P}}(h(\Phi_m) | \mathcal{F}_{m-n}) - E_{\underline{P}}(h(\Phi_m))| \xrightarrow[n\to\infty]{} E_{\underline{P}} |E_{\underline{P}}(h(\Phi_m) | \mathcal{F}_{-\infty}) - E_{\underline{P}}(h(\Phi_m))| \ ,$$

which shows the implication a) $\Rightarrow$ d).

2. Conversely, we assume the validity of d). Let m be a natural number and g be a bounded $\mathcal{M}^{\otimes(2m+1)}$-measurable real function on M^{2m+1}. For all $\Phi \in M^\kappa$ we put

$$h(\Phi) := \int g(\chi_0, \ldots \chi_{2m}) \, \vec{\delta}_\Phi \, (d(\chi_j)_{j\geq 0}) \ .$$

Then we get by the Markov property of $\underline{P}$:

$$E_{\underline{P}} | E_{\underline{P}}(g(\Phi_{-m}, \ldots \Phi_m) | \mathcal{F}_{-\infty}) - E_{\underline{P}}(g(\Phi_{-m}, \ldots \Phi_m)) |$$
$$= \lim_{n\to\infty} E_{\underline{P}} | E_{\underline{P}}(g(\Phi_{-m}, \ldots \Phi_m) | \mathcal{F}_{-n-m}) - E_{\underline{P}}(g(\Phi_{-m}, \ldots \Phi_m)) |$$
$$= \lim_{n\to\infty} E_{\underline{P}} | \int h(\chi)\kappa^{[n]}{}_{(\Phi_{-n-m})}(d\chi) - E_{\underline{P}}(h(\Phi_{-m})) | = 0$$

in virtue of the assumed condition d).This shows that a) follows from d).

3. The implications d) $\Rightarrow$ b) and d) $\Rightarrow$ c) are clear by putting $h(\Phi) := 1_Y(\Phi)$ and $h(\Phi):= \exp(\langle\Phi,f\rangle)$, respectively.

4. Next we show that b) implies d). By linearity, it is clear that, for all fixed $m \in \mathbf{Z}$, (o) holds for the linear span E of the functions 1_Y, $Y \in M$. Now any bounded, $\mathcal{M}$-measurable real-valued function h on M is in the $L_1(P_m)$-closure of E; hence there exists, for any $\varepsilon>0$, an $h' \in E$ such that $\int |h'(\Phi) - h(\Phi)| \, P_m(d\Phi) < \varepsilon$. Then there follows

$$\limsup_{n\to\infty} \int |\int h(\chi)\, \kappa^{[n]}{}_{(\Phi)}(d\chi) - \int h(\chi) P_m(d\chi)|\, P_{-n+m}(d\Phi)$$

$$\leq \limsup_{n\to\infty} \int | \int h(\chi)\, \kappa^{[n]}{}_{(\Phi)}(d\chi) - \int h'(\chi)\, \kappa^{[n]}{}_{(\Phi)}(d\chi) |\, P_{-n+m}(d\Phi)$$

$$+ \limsup_{n\to\infty} \int | \int h'(\chi)\, \kappa^{[n]}{}_{(\Phi)}(d\chi) - \int h'(\chi) P_m(d\chi) |\, P_{-n+m}(d\Phi)$$

$$+ | \int h'(\chi) P_m(d\chi) - \int h(\chi) P_m(d\chi) |$$

$$\leq \iint |h(\Phi) - h'(\Phi)|\, P_m(d\Phi) + 0 + \varepsilon \leq 2\varepsilon \ .$$

As $\varepsilon>0$ was arbitrary, we arrive at d).

5. Concerning the implication c)$\Rightarrow$ d), we parallel the arguments in 4., now defining E to be the linear span of all mappings $\Phi \to \exp(-\langle\Phi,f\rangle)$, $f \in F_c$. It remains to show that, for any fixed $m \in \mathbf{Z}$, any bounded, $\mathcal{M}$-measurable real-valued function h on M is in the $L_1(P_m)$-closure of E. To this end, let $\mathcal{H}$ denote the space of all bounded $\mathcal{M}$-measurable real-valued functions having this property. $\mathcal{H}$ is a vector space which contains the constant 1, is closed under uniform convergence as well as under convergence of nonnegative, uniformly bounded increasing sequences of functions and contains E (which is closed under multiplication) as a subset, hence contains (cf. [Meyer (1966), Thm. 20] all the bounded functions measurable with respect to the σ-field generated by E (which is $\mathcal{M}$). The same argument as in 4. now shows the validity of d). ■

We are going to show that $\mathcal{F}_{-\infty}$-triviality of distributions on $\mathcal{F}$ (and, a fortiori, of κ-processes) is preserved under convolution. For this purpose we need the following result:

4.1.8. Proposition. A distribution $\underline{H}$ on $\mathcal{F}$ is $\mathcal{F}_{-\infty}$-trivial iff for all $Y \in \mathcal{F}$ with the property $\underline{H}(Y) > 0$ there holds

$$\text{(o)} \qquad \| H((\Phi_i)_{i\leq n} \in (.) | Y) - H((\Phi_i)_{i\leq n} \in (.)) \| \xrightarrow[n\to-\infty]{} 0 \ .$$

Proof. For all $n \in \mathbf{Z}$, let p_n and q_n denote the restrictions of $\underline{H}((.)|Y)$ and $\underline{H}$ to $\mathcal{F}_n$, respectively. Further, let f_n be a version of dp_n/dq_n. The sequence $(f_n)_{n\in\mathbf{Z}}$ is an $(\mathcal{F}_n)_{n\in\mathbf{Z}}$-martingale, hence there exists an $\mathcal{F}_{-\infty}$-measurable nonnegative real function $f_{-\infty}$ on $M^{\mathbf{Z}}$ such that the convergence

$$f_n \xrightarrow[n\to-\infty]{L_1(\underline{H})} f_{-\infty}$$

holds true.

2. Assume that $\underline{H}$ is $\mathcal{F}_{-\infty}$-trivial; then $f_{-\infty}$ is $\underline{H}$-a.s. equal to the expectation value of $f_0((\Phi_i))$, which equals one, and we obtain

$$\| \underline{H}((\Phi_i)_{i\leq n}\in(.)|Y) - \underline{H}((\Phi_i)_{i\leq n}\in(.)) \| = \|q_n - p_n\| = E_{\underline{H}}|f_n - 1| \xrightarrow[n\to-\infty]{} 0 \quad ,$$

i. e. the validity of (o).

3. Conversely, (o) implies for all $Y\in \mathcal{F}_{-\infty}$ with $\underline{H}(Y) > 0$, in virtue of

$$1 - \underline{H}(Y) \leq \| H((\Phi_i)_{i\leq n}\in(.)|Y) - \underline{H}((\Phi_i)_{i\leq n}\in(.)) \| \qquad (n\in \mathbf{Z}) \quad ,$$

that $\underline{H}(Y) = 1$. ■

4.1.9. Proposition. Let $(\underline{P}_s)_{s\in S}$ be an at most countable family of $\mathcal{F}_{-\infty}$-trivial distributions on $\mathcal{F}$ such that $\underline{P} := \mathop{*}\limits_{s\in S} \underline{P}_s$ exists. Then also $\underline{P}$ is $\mathcal{F}_{-\infty}$- trivial.

Proof. 1. For any two $\mathcal{F}_{-\infty}$-trivial distributions $\underline{U}, \underline{U}'$ on $\mathcal{F}$ and any $Y\in \mathcal{F}_{-\infty}$ there holds

$$(\underline{U}*\underline{U}')(Y) = \int \underline{U}((\Phi_i + \Phi'_i)_{i\in \mathbf{Z}}\in Y)\, \underline{U}'(d((\Phi'_i)) \quad .$$

For any fixed sequence $(\Phi'_i)\in M^{\mathbf{Z}}$, the set $\{(\Phi_i) : (\Phi_i)\in M^{\mathbf{Z}}, (\Phi_i + \Phi'_i)\in Y\}$ is $\mathcal{F}_{-\infty}$-measurable, hence the $\mathcal{F}$-measurable mapping

$$f((\Phi'_i)) := \underline{U}((\Phi_i + \Phi'_i)_{i\in \mathbf{Z}}\in Y) \qquad ((\Phi'_i)\in M^{\mathbf{Z}})$$

takes only the values 0,1. Bearing in mind that $f((\Phi'_i))$ depends, for all $n\in \mathbf{Z}$, only on $(\Phi'_i)_{i\leq n}$, one recognizes that f is even $\mathcal{F}_{-\infty}$-measurable, and hence is constant $\underline{U}'$-a.e. Thus there holds $(\underline{U}*\underline{U}')(Y)\in \{0,1\}$.

2. In view of 1., the assertion holds true for finite S.

3. Now let S be countable, and choose an arbitrary natural number m and an arbitrary set $X\in \mathcal{B}$. Further, let Y be a set in the subring of $\mathcal{F}$ which is generated by the mappings $\Phi\to\Phi_i(X')$, $-m \leq i \leq m$, $X'\in \mathcal{B}\cap X$. Writing, for abbreviation, $\underline{P}_S := \bigotimes\limits_{s\in S} \underline{P}_s$, we obtain for all finite subsets $L\subseteq S$, all $n\in \mathbf{Z}$ and $C\in \mathcal{M}^{\otimes\{i\in \mathbf{Z}: i\leq n\}}$:

$$| \underline{P}((\Phi_i)_{i\leq n}\in C, (\Phi_i)\in Y) - \underline{P}((\Phi_i)_{i\leq n}\in C)\, \underline{P}((\Phi_i)\in Y) |$$

$$\leq |\underline{P}_S ((\sum_{s\in S}\Phi_{i,s})_{i\leq n}\in C, (\sum_{s\in S}\Phi_{i,s})_{i\leq n}\in Y) - \underline{P}_S ((\sum_{s\in S}\Phi_{i,s})_{i\leq n}\in C, (\sum_{s\in S}\Phi_{i,s})_{i\leq n}\in Y)|$$

$$+4\, \underline{P}_S (\sum_{\substack{-m\leq i\leq m \\ s\in S\setminus L}} \Phi_{i,s}(X) > 0)$$

Now take, for any given $\varepsilon>0$, the set L large enough such that the last summand is smaller than ε. Thus we get as an upper estimate:

$$\int |(\mathop{*}\limits_{s\in L} \underline{P}_s)((\Phi_i + \Phi'_i)_{i\leq n}\in C, (\Phi_i)\in Y) - (\mathop{*}\limits_{s\in L} \underline{P}_s)((\Phi_i + \Phi'_i)_{i\leq n}\in C)(\mathop{*}\limits_{s\in S} \underline{P}_s)((\Phi_i)\in Y)|$$

$$(\mathop{*}\limits_{s\in S\setminus L} \underline{P}_s)(d(\Phi'_i)) + \varepsilon \quad .$$

Assuming in addition that $\underline{P}(Y) > 0$ we obtain from this estimate, using 2., Proposition 4.1.7. and the fact that $\varepsilon >0$ was arbitrary:

$$\limsup_{n\to-\infty} \|\underline{P}((\Phi_i)_{i\leq n}\in(.)|Y) - \underline{P}((\Phi_i)\in(.))\| = 0 \quad .$$

A monotone class argument shows that this latter relation holds true even for all $Y\in \mathcal{F}$ with $\underline{P}(Y) > 0$; hence one more application of Proposition 4.1.1. yields the asserted $\mathcal{F}_{-\infty}$ triviality of $\underline{P}$. ■

For all $X \in \mathcal{A}$, let ${}_X\mathcal{M}$ denote the sub-σ-algebra of $\mathcal{M}$ generated by the mappings $\Phi \to \Phi(X')$, $X' \in \mathcal{B} \cap X$. The σ-algebra

$$_\infty\mathcal{M} := \bigcap_{X \in \mathcal{B}} {}_{A \setminus X}\mathcal{M}$$

is called the **spatial tail σ-algebra**. A distribution P on $\mathcal{M}$ is called **$_\infty\mathcal{M}$-trivial** if it takes only values 0,1 on $_\infty\mathcal{M}$. Obviously all distributions having independent increments, and in particular the distributions Π_ν, $\nu \in N$, have this property. Similar to Proposition 4.1.8. there follows

4.1.10. Proposition. A distribution $P \in \mathbf{P}$ is $_\infty\mathcal{M}$-trivial iff for any $Y \in \mathcal{M}$ with the property $P(Y) > 0$ there holds

$$\inf_{X \in \mathcal{B}} \| P({}_{A\setminus X}\Phi \in (.)|Y) - P({}_{A\setminus X}\Phi \in (.)) \| = 0.$$

The proof of Proposition 4.1.9. carries over as well and leads to

4.1.11. Proposition. Let $(P_s)_{s \in S}$ be an at most countable family of $_\infty\mathcal{M}$-trivial distributions on $\mathcal{M}$ such that $P := \mathop{*}_{s \in S} P_s$ exists. Then also P is $_\infty\mathcal{M}$-trivial.

Conversely, a convolution product may be $_\infty\mathcal{M}$-trivial even if one of its factors is not:

4.1.12. Example. Let $[A, \rho_A]$ be the real axis and μ be Lebesgue measure. Further we put

$$\nu := \int_{(.)\cap \mathbf{R}_+} (1+x)^{-1}\mu(dx) \;\; ; \;\; P := \Pi_\mu * \mathcal{E}_{\Pi_\nu} \; .$$

As ν is an infinite measure, there exists a sequence $0 = a_0 < a_1 < a_2 < \ldots$ with the property $\nu([a_i, a_{i+1})) = 1$, $i \in \mathbf{Z}_+$. By the law of large numbers there follows for j= 1,2,...

$$\Pi_{j\nu}(n^{-1}\Phi([0, a_n)) \xrightarrow[n\to\infty]{} j) = 1 \quad .$$

For

$$\{n^{-1}\Phi([0, a_n)) \xrightarrow[n\to\infty]{} 1\} := Y \in {}_\infty\mathcal{M}$$

there holds

$$\mathcal{E}_{\Pi_\nu}(Y) = e^{-1} \sum_{j\geq 0} (j!)^{-1}\, \Pi_{j\nu}(Y) = e^{-1} \quad .$$

Hence - in contrast to Π_μ - the distribution $\mathcal{E}_{\Pi_\nu}$ is not $_\infty\mathcal{M}$-trivial. Nevertheless, P is $_\infty\mathcal{M}$-trivial! Indeed, for all natural numbers n we obtain

$$_{\mathbf{R}\setminus(-n,n)}\| \Pi_\mu - P\| = {}_{\mathbf{R}\setminus(-n,n)}\|\Pi_\mu - e^{-1}\sum_{j\geq 0} (j!)^{-1}\, \Pi_\mu * \Pi_{j\nu}\|$$

$$\leq e^{-1} \sum_{j\geq 1} (j!)^{-1}\; {}_{\mathbf{R}\setminus(-n,n)}\| \Pi_\mu - \Pi_{\mu+j\nu}\| \quad .$$

Bearing in mind that $\mu + j\nu << \mu$ with density

$$(d(\mu+j\nu)/d\mu)(x) = 1 + (j/(1+x)) \cdot 1_{[0, +\infty)}(x) \qquad (x \in \mathbf{R}),$$

we may, in virtue of inequality 1.12.2. in [MKM], proceed in our estimate by

$$\leq (8(1 - \exp(-(1/2)\int_n^\infty (1 - \sqrt{(1 + j/(1+x))})^2\mu(dx))))^{1/2}$$

$$\leq (8(1-\exp(-(1/2)\int_{n}^{\infty}(j/(1+x))^2\mu(dx))))^{1/2}$$

$$\leq (8(1-\exp(-j^2/2)\int_{n+1}^{\infty} x^{-2}\mu(dx))))^{1/2}$$

$$= (8(1-\exp(-(j^2/(2n+2)))))^{1/2} \underset{n\to\infty}{\longrightarrow} 0 \ .$$

Thus one gets for $Y\in {}_{\infty}\mathcal{M}$ and n =1,2,...

$$|\Pi_\mu(Y)-P(Y)|$$

$$= |\Pi_\mu({}_{\mathbf{R}\setminus(-n,n)}\Phi\in Y)-P({}_{\mathbf{R}\setminus(-n,n)}\Phi\in Y)|$$

$$\leq {}_{\mathbf{R}\setminus(-n,n)}\|\Pi_\mu - P\| \underset{n\to\infty}{\longrightarrow} 0$$

and hence $P(Y) = \Pi_\mu(Y)\in\{0,1\}$. ■

The following result says that clustering by a field with a.s. finite offsprings preserves ${}_{\infty}\mathcal{M}$-triviality of distributions.

4.1.13. Proposition. Let $[A', \rho_{A'}]$ be a complete, separable metric space, ω be a clustering field on $[A, \rho_A]$ with phase space $[A', \rho_{A'}]$ and P be an ${}_{\infty}\mathcal{M}$-trivial distribution on $\mathcal{M}$ such that P_ω exists and there holds

$$\omega_{(a)}(\chi \text{ is finite}) = 1 \qquad \Lambda_P\text{-a. all } a\in A.$$

Then P_ω is again ${}_{\infty}\mathcal{M}'$-trivial.

Proof. 1. Obviously we may assume without loss of generality that $\omega_{(a)}(\chi$ is finite$)=1$ holds for all $a\in A$. Now let Y be some set in ${}_{\infty}\mathcal{M}'$. For all $\Phi\in{}_{\omega}M$ and all $X\in\mathcal{B}$ there holds

$$\omega_{(\Phi)}(Y) = (\omega_{(_X\Phi)}\otimes\omega_{(_{A\setminus X}\Phi)})((\chi_1+\chi_2)\in Y) = (\omega_{(_X\Phi)}\otimes\omega_{(_{A\setminus X}\Phi)})(\chi_2\in Y) = \omega_{(_{A\setminus X}\Phi)}(Y)\ .$$

Hence the real valued mapping $\Phi\to\omega_{(\Phi)}(Y)$, which is defined on ${}_{\omega}M$, is measurable with respect to ${}_{\infty}\mathcal{M}$.

2. For all $Y\in{}_{\infty}\mathcal{M}'$ and $\Phi\in{}_{\omega}M$ there holds

$$\omega_{(\Phi)}(Y)\in\{0,1\} \ .$$

Indeed, if Φ is finite, we obtain according to 1.

$$\omega_{(\Phi)}(Y) = \begin{cases} 0 & \text{if } o\notin Y \\ 1 & \text{if } o\in Y. \end{cases}$$

If, on the other hand, $\Phi = \sum_{j\in\mathbf{Z}_+}\delta_{a_j}$ is some infinite counting measure in ${}_{\omega}M$, then there results similar to 1. for all $n\in\mathbf{Z}$

$$\omega_{(\Phi)}(Y) = (\bigotimes_{j\in\mathbf{Z}_+}\omega_{(a_j)})(\sum_{j\geq 0}\chi_j\in Y) = (\bigotimes_{j\in\mathbf{Z}_+}\omega_{(a_j)})(\sum_{j\geq n}\chi_j\in Y)$$

By Kolmogorov's zero-one law this implies

$$\omega_{(\Phi)}(Y)\in\{0,1\}.$$

3. The assumed ${}_{\infty}\mathcal{M}$-triviality of P leads, in virtue of 1. and 2., for all $Y\in{}_{\infty}\mathcal{M}'$ to

$$P_\omega(Y) = \int\omega_{(\Phi)}(Y)P(d\Phi) \in \{0,1\} \ .■$$

The following proposition says that the successive distributions of an extremal κ-process are ${}_{\infty}\mathcal{M}$-trivial, provided the offsprings of single individuals are a.s. finite.

4.1.14. Proposition. Let $\underline{P}$ be an extremal κ-process, and assume that

$$\kappa_{(a)}(\chi \text{ is finite}) = 1 \qquad \text{for all } a\in A \quad .$$

Then P_m is an ${}_{\infty}\mathcal{M}$-trivial distribution for all $m\in \mathbf{Z}$.

Proof. 1. An induction argument shows that

$$\kappa^{[n]}{}_{(a)}(\chi \text{ is finite}) = 1 \qquad (n\in \mathbf{Z}_+ ;\ a\in A) \quad .$$

2. Like in the second step of the proof of Proposition 4.1.13. we obtain for all $Y\in {}_{\infty}\mathcal{M}$

$$\kappa^{[n]}{}_{(\Phi)}(Y) \in \{0,1\} \qquad (n=1,2,\dots ;\ \Phi\in {}_{\kappa^{[n]}}M) \quad .$$

3. According to Proposition 4.1.7. there holds for all $Y\in {}_{\infty}\mathcal{M}$

$$\int| \kappa^{[n]}{}_{(\Phi)}(Y) - P_m(Y)| P_{-n+m}(d\Phi) \underset{n\to\infty}{\longrightarrow} 0 \qquad (m\in \mathbf{Z}) \quad ,$$

which, in view of 2., leads to $P_m(Y)\in \{0,1\}$. ■

Finally, we mention that extremality of a κ-processes is equivalent to extremality of its associated family process:

4.1.15. Proposition. A κ-process $\underline{P}$ is extremal iff the κ-process $\underline{\mathcal{G}}^{\underline{P}}$ has this property.

Proof. 1. Assume that $\underline{P}$ is extremal and that $\underline{\mathcal{G}}^{\underline{P}}$ is a convex combination

$$\underline{\mathcal{G}}^{\underline{P}} = c\, \underline{\mathcal{G}}_1 + (1-c)\, \underline{\mathcal{G}}_2$$

where $0 < c < 1$ and $\underline{\mathcal{G}}_1, \underline{\mathcal{G}}_2$ are κ-processes. Theorem 3.1.8. now implies the existence of κ-processes $\underline{P}_1, \underline{P}_2$ such that

$$\underline{\mathcal{G}}^{\underline{P}_i} = \underline{\mathcal{G}}_i \qquad (i=1,2) \quad ,$$

which leads to

$$\underline{P} = c\, \underline{P}_1 + (1-c)\, \underline{P}_2 \quad .$$

Extremality of $\underline{P}$ guarantees that $\underline{P}_1 = \underline{P}_2$ and hence $\underline{\mathcal{G}}_1 = \underline{\mathcal{G}}_2$, which shows that $\underline{\mathcal{G}}^{\underline{P}}$ is extremal.

2. Conversely, $\underline{\mathcal{G}}^{\underline{P}}$ cannot be extremal unless $\underline{P}$ has this property, because any convex combination

$$\underline{P} = c\, \underline{P}_1 + (1-c)\, \underline{P}_2 \qquad (0 < c < 1 \quad ;\quad \underline{P}_1, \underline{P}_2 \in \underline{P}^{\kappa})$$

induces a convex combination

$$\underline{\mathcal{G}}^{\underline{P}} = c\, \underline{\mathcal{G}}^{\underline{P}_1} + (1-c)\, \underline{\mathcal{G}}^{\underline{P}_2} \quad . ■$$

4.2. κ-processes with First Order Representation and Their Characteristic J-processes

Let us recall that 4.1.1. provides, for any κ-process P, via

$$\underline{P} = \int \underline{K}((\Phi_i), (.))\underline{P}(d(\Phi_i))$$

an integral representation of $\underline{P}$ over the set of extremal κ-processes. $\underline{P}$ is said to have a **first order representation** if $\underline{K}((\Phi_i),(.))$ is of first order for $\underline{P}$ -a. all (Φ_i). Clearly, any κ-process of first order has a first order representation.

For any $(\Phi_i) \in M^{\mathbf{Z}}$, let $(S_n^{(\Phi_i)})_{n \in \mathbf{Z}}$ denote the intensity sequence of the κ-process $\underline{K}((\Phi_i),(.))$. By 3.1.3. there holds

4.2.1. For all $(\Phi_i) \in M^{\mathbf{Z}}$, $(S_n^{(\Phi_i)})$ is a stable J-sequence iff $\underline{K}((\Phi_i),(.))$ is of first order.

Now let $\underline{P}$ be a κ-process having a first order representation. The distribution of the random J-sequence $(S_n^{(\Phi_i)})$ under $\underline{P}$ will be called **characteristic J-process** of $\underline{P}$, and will be denoted by $\underline{\mathcal{W}}^{\underline{P}}$. For the corresponding J-entrance law, we will write $(\mathcal{W}_m^{\underline{P}})$. We are now going to state some basic properties of the distribution $\underline{\mathcal{W}}^{\underline{P}}$.

4.2.2. For any κ-process $\underline{P}$ having a first order representation, $\underline{\mathcal{W}}^{\underline{P}}$ is a stable J-process having the same intensity sequence as P and obeying

$$\theta(\underline{\mathcal{W}}^{\underline{P}}) = \underline{\mathcal{W}}^{(\theta \underline{P})} \quad .$$

Proof. 4.2.1. implies that $\underline{\mathcal{W}}^{\underline{P}}$ is a stable J-process.
From 4.1.1. b) ii) we get for all $m \in \mathbf{Z}$

$$\begin{aligned} \Lambda_{\mathcal{W}_m^{\underline{P}}}(.) &= \int \chi_m(.)\, \underline{K}((\Phi_i),d(\chi_j))\underline{P}(d(\Phi_i)) \\ &= \int \Phi_m(.)\, \underline{P}(d(\Phi_i)) = \Lambda_{P_m}(.) \quad . \end{aligned}$$

Finally we note, using 4.1.4., that

$$\begin{aligned} \theta(\underline{\mathcal{W}}^{\underline{P}}) &= \underline{P}((\Lambda_{\underline{K}((\Phi_i),\chi_{j+1} \in (.))})_{j \in \mathbf{Z}} \in (.)) \\ &= \underline{P}((\Lambda_{\underline{K}((\Phi_{i+1}),\chi_j \in (.))})_{j \in \mathbf{Z}} \in (.)) = \underline{\mathcal{W}}^{(\theta \underline{P})} \quad . \blacksquare \end{aligned}$$

4.2.3. For any two κ-processes $\underline{P}, \underline{P}'$ having a first order representation there hold

$$\underline{\mathcal{W}}^{c\underline{P}+(1-c)\underline{P}'} = c\,\underline{\mathcal{W}}^{\underline{P}} + (1-c)\, \underline{\mathcal{W}}^{\underline{P}'} \qquad (0 \le c \le 1),$$

$$\underline{\mathcal{W}}^{\underline{P}*\underline{P}'} = \underline{\mathcal{W}}^{\underline{P}} * \underline{\mathcal{W}}^{\underline{P}'} \quad .$$

Proof. The first statement is immediate; the second follows from 4.1.5. ■

4.2.4. Proposition. For any κ-process $\underline{P}$ having a first order representation and all $m \in \mathbf{Z}$ there holds:

$$(*) \qquad P_{m-n}((\Phi * J^{[n]}) \in N) = 1 \quad \text{for all } n \in \mathbf{Z}_+ \ .$$

For all $f \in F$, the convergence

$$\mathcal{L}_{P_{m-n}((\Phi * J^{[n]}) \in (.))}(f) = \mathcal{L}_{P_{m-n}}(J^{[n]} * f) \downarrow \mathcal{L}_{\mathcal{W}_m \underline{P}}(f)$$

and, a fortiori, the convergence

$$P_{m-n}((\Phi * J^{[n]}) \in (.)) \underset{n \to \infty}{\Rightarrow} \mathcal{W}_m \underline{P}$$

holds true.

Proof. 1. We show first that (*) holds under the additional assumption that $\underline{P}$ is of first order. Indeed, for any $X \in \mathcal{B}$ we have

$$\int (\Phi * J^{[n]})(X)\, P_{m-n}(d\Phi) = (\Lambda_{P_{m-n}} * J^{[n]})(X) = \Lambda_{P_m}(X) < +\infty \ .$$

2. In the general case we obtain, using 1. together with the assumption that $\underline{K}((\Phi_i),(.))$ is of first order $\underline{P}$ - almost surely:

$$P_{m-n}((\Phi * J^{[n]}) \in N) = \int \underline{K}((\Phi_i),(\chi_{m-n} * J^{[n]}) \in N)\, P(d(\Phi_i)) = 1$$

3. To show the asserted convergence, we get, using 4.1.6.

$$\begin{aligned}
&\mathcal{L}_{P_{m-n}((\Phi * J^{[n]}) \in (.))}(f) \\
&= \int \exp(-\langle \Phi * J^{[n]}, f\rangle) P_{m-n}(d\Phi) \\
&= \int \exp(-\langle \Phi, J^{[n]} * f\rangle) P_{m-n}(d\Phi) = \mathcal{L}_{P_{m-n}}(J^{[n]} * f) \\
&= \int \exp\left(-\int \langle \chi, f\rangle \kappa^{[n]}_{(\Phi)}(d\chi)\right) P_{m-n}(d\Phi) \\
&\underset{n \to \infty}{\downarrow} \int \exp\left(-\int \langle \chi_m, f\rangle \underline{K}((\Phi_i), d(\chi_j))\right) \underline{P}(d(\Phi_i)) \\
&= \mathcal{L}_{\mathcal{W}_m \underline{P}}(f) \ . \blacksquare
\end{aligned}$$

4.2.5. Example. Let us put, like in Example 3.1.1., for some $c \in (0,1)$, $\kappa := \vartheta_c$. As stated in 3.1.1., any κ-process $\underline{P}$ is of the form

$$(*) \qquad \underline{P} = \int \underline{P}_\gamma(.) Q(d\gamma) \ ,$$

where $\underline{P}_\gamma$ is the (extremal) κ-process corresponding to the entrance law $(\Pi_{c^i \gamma})_{i \in Z}$, and Q is an arbitrary distribution on $\mathcal{N}$. For all $\gamma \in N$, $X \in \mathcal{B}$, $\varepsilon > 0$ and $n \in \mathbf{Z}_+$ we obtain by means of the Chebyshev inequality

$$\begin{aligned}
&\underline{P}_\gamma(|c^n \Phi_{-n}(X) - \gamma(X)| > \varepsilon) \\
&= \Pi_{c^{-n}\gamma}(|c^n \Phi(X) - \gamma(X)| > \varepsilon) \le c^n \gamma(X) \varepsilon^{-2} \ .
\end{aligned}$$

By the Borel-Cantelli Lemma, one infers that

$$c^n \Phi_{-n} \xrightarrow[n \to \infty]{N} \gamma \qquad \underline{P}_\gamma\text{-a .all } (\Phi_i).$$

We may thus put

$$\underline{K}((\Phi_i),(.)) := \begin{cases} \underline{P}_\gamma & \text{if } c^n\Phi_{-n} \xrightarrow[n\to\infty]{N} \gamma \\ \delta_{(o)} & \text{otherwise} \end{cases} ,$$

where (o) stands for the sequence with constant value $o \in M$. In particular, we obtain for any κ-process $\underline{P}$ of the form (∗):

$$\underline{\mathcal{W}}^{\underline{P}} = Q((c^i\gamma) \in (.)) . \blacksquare$$

In the following we need the approximation lemma

4.2.6. Let $L_0, L_1, \ldots$, be a sequence of distributions on $\mathcal{M}$ such that for a J-sequence (λ_i) the inequalities

$$\Lambda_{L_n} \leq \lambda_{-n} \qquad (n \in \mathbf{Z}_+)$$

are valid. Subject to these assumptions there holds for all $f \in F_b$

$$c_n := \sup_{m=0,1,\ldots} \int | \mathcal{L}_{\kappa^{[m+n]}_{(\Phi)}}(f) - \mathcal{L}_{\delta_\Phi}\left(J^{[m]} * (-\ln \mathcal{L}_{\kappa^{[n]}_{(.)}}(f))\right) | L_{m+n}(d\Phi) \xrightarrow[n\to\infty]{} 0.$$

Proof. 1. In virtue of 1.4.7. and 1.5.9. we obtain for all $m, n \in \mathbf{Z}_+$ and all $a \in A$:

$$-\ln \mathcal{L}_{\kappa^{[m+n]}_{(a)}}(f) = -\ln \mathcal{L}_{\kappa^{[m]}_{(a)}}(-\ln \mathcal{L}_{\kappa^{[n]}_{(.)}}(f))$$

$$\leq J^{[m]} * (-\ln \mathcal{L}_{\kappa^{[n]}_{(.)}}(f))\,(a) \quad .$$

2. Using 1. and the estimate

$$|\exp(x) - \exp(y)| \leq |x-y| \qquad (x,y \leq 0)$$

we obtain

$$\int | \mathcal{L}_{\kappa^{[m+n]}_{(\Phi)}}(f) - \mathcal{L}_{\delta_\Phi}(-J^{[m]} * \ln \mathcal{L}_{\kappa^{[n]}_{(.)}}(f)) | L_{m+n}(d\Phi)$$

$$= \int |\exp(\int \ln \mathcal{L}_{\kappa^{[m+n]}_{(a)}}(f)\Phi(da)) - \exp(\int J^{[m]} * \ln \mathcal{L}_{\kappa^{[n]}_{(.)}}(f)(a)\Phi(da))| L_{m+n}(d\Phi)$$

$$\leq \iint | \ln \mathcal{L}_{\kappa^{[m+n]}_{(a)}}(f) - J^{[m]} * \ln \mathcal{L}_{\kappa^{[n]}_{(.)}}(f)(a) | \,\Phi(da) L_{m+n}(d\Phi)$$

$$\leq \int \left(J^{[m]} * (-\ln \mathcal{L}_{\kappa^{[n]}_{(.)}}(f))(a) - (-(\ln \mathcal{L}_{\kappa^{[m+n]}_{(a)}}(f))\right) \lambda_{-m-n}(da)$$

$$= \langle \lambda_{-n}, -\ln \mathcal{L}_{\kappa^{[n]}_{(.)}}(f) \rangle - \langle \lambda_{-m-n}, -\ln \mathcal{L}_{\kappa^{[m+n]}_{(.)}}(f) \rangle$$

3. From 2. we get immediately that the sequence

$$s_n := \langle \lambda_{-n}, -\ln \mathcal{L}_{\kappa^{[n]}_{(.)}}(f) \rangle$$

is nonincreasing. Moreover,

$$s_0 = \langle \lambda_0, -\ln \mathcal{L}_{\kappa^{[0]}_{(.)}}(f) \rangle = \langle \lambda_0, f \rangle < +\infty ,$$

hence (s_n) converges in $\mathbf{R}_+$. This together with 2. yields

$$c_n \leq \sup_{m=0,1,\ldots} (s_n - s_{n+m}) \xrightarrow[n\to\infty]{} 0 \quad . \blacksquare$$

Combining 4.2.6. and Proposition 4.2.4., we are in the position to prove the following **inversion formula**:

4.2.7. Theorem. For any κ-process $\underline{P}$ having a first order representation, and all j∈**Z** there holds

$$\mathcal{L}_{\mathcal{W}_{j-n}^{\underline{P}}}(-\ln \mathcal{L}_{\kappa^{[n]}_{(.)}}(f)) \xrightarrow[n\to\infty]{} \mathcal{L}_{P_j}(f) \qquad (f\in F_b) \ .$$

Proof. In view of 4.2.2. it suffices to consider the case j=0. For any $n\in \mathbf{Z}_+$ we obtain from Proposition 4.2.4.

$$|\mathcal{L}_{\mathcal{W}_{-n}^{\underline{P}}}(-\ln \mathcal{L}_{\kappa^{[n]}_{(.)}}(f)) - \mathcal{L}_{P_0}(f)|$$

$$= |\lim_{m\to\infty} \mathcal{L}_{P_{-m-n}}(-J^{[m]}*\ln \mathcal{L}_{\kappa^{[n]}_{(.)}}(f)) - \mathcal{L}_{P_0}(f)|$$

$$\le \limsup_{m\to\infty} \int |\mathcal{L}_{\delta_\Phi}(-J^{[m]}*\ln \mathcal{L}_{\kappa^{[n]}_{(.)}}(f)) - \mathcal{L}_{\kappa^{[m+n]}_{(\Phi)}}(f)| P_{-m-n}(d\Phi)$$

$$= \limsup_{m\to\infty} \iint |\mathcal{L}_{\delta_\Phi}(-J^{[m]}*\ln \mathcal{L}_{\kappa^{[n]}_{(.)}}(f)) - \mathcal{L}_{\kappa^{[m+n]}_{(\Phi)}}(f)| \, \underline{K}((\Phi_i),(.))_{-m-n}(d\Phi)) \, \underline{P}(d(\Phi_i)) \ .$$

This expression, however, tends to zero for n→∞ by 4.2.6. and Lebesgue's dominated convergence theorem. ■

Theorem 4.2.7. tells in particular that the mapping $\underline{P}\to \mathcal{W}^{\underline{P}}$ from the set of κ-processes having a first order representation into the set of J-processes is 1-1.

The next proposition says that the mapping $\underline{P}\to \mathcal{W}^{\underline{P}}$ preserves absolute continuity:

4.2.8. Proposition. Let $\underline{H}$ and $\underline{P}$ be κ-processes. Assume that $\underline{H}$ has a first order representation and that $\underline{P}<<\underline{H}$. Then also $\underline{P}$ has a first order representation, and there holds $\mathcal{W}^{\underline{P}} << \mathcal{W}^{\underline{H}}$.

Proof. Due to the assumed absolute continuity, $\underline{K}((\Phi_i),(.))$ is of first order not only for $\underline{H}$-almost all (Φ_i) but also for $\underline{P}$ -almost all (Φ_i). Moreover, then also $\mathcal{W}^{\underline{P}} = \underline{P}((S_n^{(\Phi_i)})\in (.))$ is absolutely continuous with respect to $\mathcal{W}^{\underline{H}} = H((S_n^{(\Phi_i)})\in (.))$. ■

Together with Theorem 4.2.7. and Proposition 4.2.8., the following result reveals that the mapping $\underline{P}\to \mathcal{W}^{\underline{P}}$ establishes, for each κ-process $\underline{H}$ having a first order representation, a 1-1 correspondence between the set of all κ-processes $\underline{P}$ which are absolutely continuous with respect to $\underline{H}$ and the set of all J-processes which are absolutely continuous with respect to $\mathcal{W}^{\underline{H}}$.

4.2.9. Proposition. Let $\underline{H}$ be a κ-process having a first order representation. Further, let $\underline{Q}$ be a J-process which is absolutely continuous with respect to $\mathcal{W}^{\underline{H}}$. Then there exists a κ-process $\underline{P}$ which obeys

$$\underline{P} << \underline{H} \quad \text{and} \quad \mathcal{W}^{\underline{P}} = \underline{Q}.$$

Proof. Let h be a version of the Radon - Nikodym derivative $d\underline{Q}/d\mathcal{W}^{\underline{H}}$. We define

$$f((\Phi_i)) := h((S_n^{(\Phi_i)})),$$

thus obtaining a real-valued function f on $M^{\mathbf{Z}}$ having expectation one under $\underline{H}$.
The distribution

$$\underline{P} := \int_{(.)} f((\Phi_i))\underline{H}(d(\Phi_i))$$

is a κ-process due to the $\mathcal{F}_{-\infty}$-measurability of f. Moreover, there holds for all $Y \in \mathcal{N}_J^{entr}$:

$$\mathcal{W}^{\underline{P}}(Y) = \int 1_Y((S_n^{(\Phi_i)}))\, \underline{P}(d(\Phi_i))$$

$$= \int (1_Y \cdot h)((S_n^{(\Phi_i)}))\, \underline{H}(d(\Phi_i)) = \int (1_Y \cdot h)((\gamma_n))\, \mathcal{W}^{\underline{H}}(d(\gamma_n)) = \underline{Q}(Y)\ . \blacksquare$$

It is now easy to deduce that the set of characteristic J-processes is a "face" in the convex set of all J-processes:

4.2.10. Proposition. Let $\underline{P}$ be a κ-process having a first order representation and assume that

$$\mathcal{W}^{\underline{P}} = c\underline{Q}_1 + (1-c)\underline{Q}_2$$

for some $c \in (0,1)$ and some J-processes $\underline{Q}_1, \underline{Q}_2$. Then there exist κ-processes $\underline{P}_1, \underline{P}_2$ having a first order representation and such that

$$\mathcal{W}^{\underline{P}_i} = Q_i \qquad (i=1,2)\ ,$$

$$\underline{P} = c\underline{P}_1 + (1-c)\underline{P}_2\ .$$

Proof. The first part of the assertion follows directly from Proposition 4.2.9., whereas the second part is a consequence of 4.2.3. and Theorem 4.2.7. ■

The preceding result allows to characterize the extremal κ-process of first order in terms of their characteristic J-processes.

4.2.11. Theorem. Let $\underline{P}$ be a κ-process having a first order representation. Then $\underline{P}$ is an extremal κ-process iff there holds

(*) $$\mathcal{W}^{\underline{P}} = \delta_{(\Lambda_{\underline{P}_j})}$$

(i. e. its characteristic J-process is deterministic).

Proof. 1. Assume that $\underline{P}$ is not extremal, i. e. there exist κ-processes $\underline{P}_1, \underline{P}_2$ (which are different from $\underline{P}$) and a real number $c \in (0,1)$ such that

$$\underline{P} = c\underline{P}_1 + (1-c)\underline{P}_2\ .$$

By Proposition 4.2.8., $\underline{P}_1$ as well as $\underline{P}_2$ have a first order representation, and according to 4.2.3. there holds

$$\mathcal{W}^{\underline{P}} = c\mathcal{W}^{\underline{P}_1} + (1-c)\mathcal{W}^{\underline{P}_2}\ .$$

According to Theorem 4.2.7. , $\mathcal{W}^{\underline{P}_1}$ is different from $\mathcal{W}^{\underline{P}}$, which shows that $\mathcal{W}^{\underline{P}}$ cannot be of the form (*).

2. Assume, conversely that $\mathcal{W}^{\underline{P}}$ is not of the form (*). Then there exists a set $Y \in \mathcal{N}_J^{entr}$ with the property

$$0 < \mathcal{W}^{\underline{P}}(Y) < 1\ ,$$

which shows that $\mathcal{W}^{\underline{P}}$ is a genuine convex combination of the two J-processes $Q_1 := \mathcal{W}^{\underline{P}}((.)|Y)$ and $Q_2 := \mathcal{W}^{\underline{P}}((.)|N_J^{entr} \setminus Y)$. Proposition 4.2.10. together with Theorem 4.2.7. now reveal that $\underline{P}$ is not an extremal κ–process. ■

4.3. Characteristic J-processes of Poisson-type κ-processes

The first result of this section is a converse to 4.2.4.:

4.3.1. Proposition. If $\underline{P}$ is a κ-process with the properties

$$P_{m-n}(\Phi * J^{[m]} \in N) = 1 \qquad (m \in \mathbf{Z},\ n \in \mathbf{Z}_+)$$

and

$$P_{m-n}(\Phi * J^{[m]} \in (.)) \underset{n\to\infty}{\Rightarrow} Q_m \qquad (m \in \mathbf{Z})$$

for some sequence (Q_m) of distributions on $\mathcal{N}$, then $\underline{P}$ has a first order representation and there holds

$$\mathcal{W}_m^{\underline{P}} = Q_m \qquad (m \in \mathbf{Z}) \ .$$

Proof. For all $f \in F_c$ we get, using the arguments in the first step of the proof of 4.1.6., for $\underline{P}$ -a.all (Φ_i)

$$\langle \Phi_{m-n} * J^{[n]}, f\rangle = \int \langle \chi, f\rangle \kappa^{[n]}(\Phi_{m-n})(d\chi)$$

$$= E_{\underline{P}}(\langle \Phi_m, f\rangle | \mathcal{F}_{m-n})$$

$$\underset{n\to\infty}{\longrightarrow} E_{\underline{P}}(\langle \Phi_m, f\rangle | \mathcal{F}_{-\infty}) = \int \langle \chi_m, f\rangle \, \underline{K}((\Phi_i), d(\chi_j)) \ .$$

This latter random variable is almost surely finite due to the asssumed convergence in distribution of $\langle \Phi_{-n+m} * J^{[n]}, f\rangle$ under $\underline{P}$, which implies that

$$\underline{P}(\underline{K}((\Phi_i),(.)) \text{ is of first order}) = 1 \quad ,$$

i. e. $\underline{P}$ has a first order representation. The rest of the assertion follows from Proposition 4.2.4. ■

The characterization of $\mathcal{W}^{\underline{P}}$ relying on Proposition 4.3.1. suggests the following **definition:**

If $(L_i)_{i\in \mathbf{Z}}$ is a sequence of distributions on $\mathcal{N}$ with the properties

$$L_{m-n}(\gamma * J^{[n]} \in N) = 1 \qquad (m \in \mathbf{Z},\ n \in \mathbf{Z}_+)$$

and

$$L_{m-n}(\gamma * J^{[n]} \in (.)) \underset{n\to\infty}{\Rightarrow} Q_m \qquad (m \in \mathbf{Z})$$

for some sequence (Q_m) of distributions on $\mathcal{N}$, then we put

$$\mathcal{W}_m^{(L_i)} := Q_m \qquad (m \in \mathbf{Z}) \ .$$

Note that for any κ-processes P having a first order representation there holds by Proposition 4.2.4

$$\mathcal{W}_m^{\underline{P}} = \mathcal{W}_m^{(P_i)} \qquad (m \in \mathbf{Z}) \ .$$

If $(L_i)_{i\in \mathbf{Z}}$ is a J-entrance law, one clearly has

$$\mathcal{W}_m^{(L_i)} = L_m \qquad (m \in \mathbf{Z}) \ .$$

Next, we give a condition on (L_i) under which the sequence $(\mathcal{W}_m^{(L_i)})_{m\in \mathbf{Z}}$ is a J-entrance law.

4.3.2. Proposition. Let (L_i) be a sequence of distributions on $\mathcal{N}$ for which $\mathcal{W}_m^{(L_i)}$, $m\in \mathbf{Z}$, exists, and assume that there holds

$$\Lambda_{L_i} \le \lambda_i \qquad (i\in \mathbf{Z})$$

for some sequence $(\lambda_i)\in N_J^{entr}$. Then the sequence $(\mathcal{W}_m^{(L_i)})_{m\in \mathbf{Z}}$ is a J-entrance law, obeying, for all $m\in \mathbf{Z}$,

$$\Lambda_{\mathcal{W}_m^{(L_i)}} \le \lambda_m$$

and

$$\mathcal{L}_{L_{m-n}}(J^{[n]}*f) \xrightarrow[n\to\infty]{} \mathcal{L}_{\mathcal{W}_m^{(L_i)}}(f) \qquad (f\in F_{\lambda_m}) \ .$$

Proof. 1. For all $m\in \mathbf{Z}$ and $n\in \mathbf{Z}_+$ there holds

$$\Lambda_{L_{m-n}(\gamma * J^{[n]}\in(.))} = \Lambda_{L_{m-n}} * J^{[n]} \le \lambda_{m-n} * J^{[n]} = \lambda_m \ ,$$

leading to

$$\Lambda_{\mathcal{W}_m^{(L_i)}} \le \lambda_m \ .$$

2. Because of $\mathcal{L}_{L_{m-n}}(J^{[n]}*f) = \mathcal{L}_{L_{m-n}(\gamma * J^{[n]}\in(.))}(f)$, the asserted convergence follows from 1. together with 1.3.7.

3. For all $f\in F_c$ and $m\in \mathbf{Z}$ one has

$$\langle \lambda_m, J*f\rangle = \langle \lambda_{m+1}, f\rangle < +\infty \quad ;$$

thus we obtain from 2.

$$\mathcal{L}_{L_{m+1-n}}(J^{[n]}*f) = \mathcal{L}_{L_{m-(n-1)}}(J^{[n-1]}*(J*f))$$

$$\xrightarrow[n\to\infty]{} \mathcal{L}_{\mathcal{W}_m^{(L_i)}}(J*f) = \mathcal{L}_{\mathcal{W}_m^{(L_i)}(\gamma * J\in(.))}(f) \ .$$

On the other hand there holds

$$\mathcal{L}_{L_{m+1-n}}(J^{[n]}*f) \xrightarrow[n\to\infty]{} \mathcal{L}_{\mathcal{W}_{m+1}^{(L_i)}}(f),$$

which reveals that

$$\mathcal{W}_m^{(L_i)}(\gamma * J\in(.)) = \mathcal{W}_{m+1}^{(L_i)} \qquad (m\in \mathbf{Z}) \ ,$$

i. e. $(\mathcal{W}_m^{(L_i)})$ is a J-entrance law. ■

Let (L_i) be a sequence of distributions on $\mathcal{N}$ which satisfies the assumptions of Proposition 4.3.2. The J-process corresponding to the J-entrance law $(\mathcal{W}_m^{(L_i)})$ will then be denoted by $\underline{\mathcal{W}}^{(L_i)}$.

4.3.3. Theorem. Let (L_i) be a sequence of distributions on $\mathcal{M}$ which obeys the assumptions of Proposition 4.3.2. If $\underline{\mathcal{W}}^{(L_i)}$ equals the characteristic J-process $\underline{\mathcal{W}}^{\underline{P}}$ of some κ-processes $\underline{P}$ of first order, then there holds

$$(L_{j-n})_{\kappa^{[n]}} \underset{n\to\infty}{\Rightarrow} P_j \quad (j\in \mathbf{Z}) \ .$$

Proof. As our assumptions remain valid "under time shift", we may restrict ourselves to the case j=0.

For all $f \in F_c$ there results

$$\limsup_{n\to\infty} |\mathcal{L}_{(L_{-n})_{\kappa^{[n]}}}(f) - \mathcal{L}_{P_0}(f)|$$

$$= \limsup_{n\to\infty} \limsup_{m\to\infty} |\mathcal{L}_{(L_{-m-n})_{\kappa^{[m+n]}}}(f) - \mathcal{L}_{P_0}(f)|$$

$$\leq \limsup_{n\to\infty} \sup_{m=0,1,2,\dots} \int |\mathcal{L}_{\kappa^{[m+n]}_{(\Phi)}}(f) - \mathcal{L}_{\delta_{(\Phi)}}(-J^{[m]} * \ln \mathcal{L}_{\kappa^{[n]}_{(.)}}(f))| L_{-m-n}(d\Phi)$$

$$+ \limsup_{n\to\infty} \limsup_{m\to\infty} |\mathcal{L}_{L_{-m-n}}(-J^{[m]} * \ln \mathcal{L}_{\kappa^{[n]}_{(.)}}(f)) - \mathcal{L}_{\mathcal{W}_{-n}^{(L_i)}}(-\ln \mathcal{L}_{\kappa^{[n]}_{(.)}}(f))|$$

$$+ \limsup_{n\to\infty} |\mathcal{L}_{\mathcal{W}_{-n}^{(L_i)}}(-\ln \mathcal{L}_{\kappa^{[n]}_{(.)}}(f)) - \mathcal{L}_{P_0}(f)|$$

The first and third term vanish due to 4.2.6. and Theorem 4.2.7., respectively. To make sure that also the second term vanishes, recall that, due to 1.5.9.,

$$-\ln \mathcal{L}_{\kappa^{[n]}_{(.)}}(f) \in F_{\lambda_{-n}} \qquad (n \in \mathbf{Z}_+)$$

and apply Proposition 4.3.2. ■

The convergence relation

$$(L_{j-n})_{\kappa^{[n]}} \underset{n\to\infty}{\Rightarrow} P_j \qquad (j \in \mathbf{Z})$$

for some κ-process $\underline{P}$ of first order, together with the assumption

$$\Lambda_{L_i} \leq \lambda_i$$

for some sequence $(\lambda_i) \in N_J^{entr}$ does not yet guarentee the existence of $(\mathcal{W}_m^{(L_i)})$, as the following example shows:

4.3.4. Example. Let $[A, \rho_A]$ and κ be as in Example 2.2.4., put $\Phi := \sum_{n\in\mathbf{Z}} \delta_{2n}$ and $L_i := \delta_\Phi$, $i \in \mathbf{Z}$. Note that

$$\Lambda_{L_i} = \Phi \leq \mu := \sum_{n\in\mathbf{Z}} \delta_n$$

and recall that, due to 2.6.5., μ belongs to N_κ^{nst}. Hence Proposition 2.6.6. yields that

$$(L_{j-n})_{\kappa^{[n]}} \underset{n\in\mathbf{Z}}{\Rightarrow} \delta_o \qquad (j \in \mathbf{Z}) .$$

On the other hand there holds

$$\Phi * J = \sum_{n\in\mathbf{Z}} \delta_{2n+1} \quad ; \quad \Phi * J^{[2]} = \Phi \quad ,$$

revealing that $\mathcal{W}_0^{(L_i)}$ does not exist. ■

The main result of this section is

4.3.5. Theorem. Let $\underline{P}$ be a κ-process of Poisson type with intensity sequence (ν_i), and put

$$L_i := \Pi_{\nu_i} \qquad (i \in \mathbf{Z}) .$$

Then there holds

$$\underline{\mathcal{W}}^{(L_i)} = \underline{\mathcal{W}}^{\underline{P}} .$$

In the proof of Theorem 4.3.5., and also later on, we use the following lemma, whose proof is elementary and will be omitted:

4.3.6. The real function g defined by

$$g(0) := 1 \quad , \quad g(x) := x^{-1}(1-e^{-x})$$

is strictly decreasing, convex, differentiable, and has the properties

$$(\alpha)\ \ 0 < g \le 1 \qquad (x \in \mathbf{R}_+)$$

$$(\beta)\ \ 0 \le g(x) - g(x+y) \le 1 - g(y) \qquad (x,y \in \mathbf{R}_+)$$

$$(\gamma)\ \ g(x) - g(x+y) \le \min(1,y) \qquad (x,y \in \mathbf{R}_+)$$

$$(\delta)\ \ g(x) - g(x+y) \ge y\,|g'(x+y)| \qquad (x,y \in \mathbf{R}_+)$$

$$(\varepsilon)\ \ g(x) - g(x+y) \le y \cdot g(x) \cdot g(y) \qquad (x,y \in \mathbf{R}_+)\,.$$

Proof of Theorem 4.3.5. Obviously it suffices to show that

$$\mathcal{W}_0^{(L_i)} = \mathcal{W}_0^{\underline{P}}\ .$$

Using the function g defined in 4.3.6. we obtain for all $f \in F_c$ and all natural numbers n

$$|\,\mathcal{L}_{P_{-n}}(J^{[n]}*f) - \mathcal{L}_{L_{-n}}(J^{[n]}*f)\,|$$

$$= |\exp(-\int(1-\exp(-\langle\Psi,J^{[n]}*f\rangle)\,\widetilde{P}_{-n}(d\Psi))$$
$$- \exp(-\int(1-\exp((J^{[n]}*f)(a))\,\nu_{-n}(da))\,|$$
$$\le |\int(1-\exp(-\langle\Psi,J^{[n]}*f\rangle))\,\widetilde{P}_{-n}(d\Psi) - \int(1-\exp((J^{[n]}*f)(a)))\,\nu_{-n}(da)\,|$$
$$= |\int g(\langle\Psi,J^{[n]}*f\rangle)\,\langle\Psi,J^{[n]}*f\rangle\,\widetilde{P}_{-n}(d\Psi) - \int g((J^{[n]}*f)(a))\,(J^{[n]}*f)(a)\,\nu_{-n}(da)\,|$$
$$= |\iint g(\langle\Psi+\delta_a,\ J^{[n]}*f\,\rangle)(J^{[n]}*f)(a)\,(\widetilde{P_{-n}})^!_a(d\Psi)\nu_{-n}(da)$$
$$- \iint g(J^{[n]}*f)(a))(J^{[n]}*f)(a)\,(\widetilde{P_{-n}})^!_a(d\Psi)\,\nu_{-n}(da)\,|\quad .$$

Setting $\|f\| := \sup_{a\in A} f(a)$; $X := \operatorname{supp} f$, and using the monotonicity as well as property 4.3.6.(γ) of the mapping g, we continue our estimation for arbitrary constants c,d>0 by

$$|\mathcal{L}_{P_{-n}}(J^{[n]}*f) - \mathcal{L}_{L_{-n}}(J^{[n]}*f)|$$

$$\le \|f\| \iint\Big(g((J^{[n]}*f)(a)) - g((J^{[n]}*f)(a)+\|f\|((\Psi*J^{[n]})(X)))\Big)J^{[n]}(a,X)(\widetilde{P_{-n}})^!_a(d\Psi)\nu_{-n}(da)$$

$$\le \|f\| \iint J^{[n]}(a,X)\min(1,\|f\|(\Psi*J^{[n]})(X))(\widetilde{P_{-n}})^!_a(d\Psi)\nu_{-n}(da)$$

$$= \|f\| \iint J^{[n]}(a,X)\min(1,\ \|f\|\cdot\int\chi(X)\kappa^{[n]}_{(\Psi)}(d\chi))(\widetilde{P_{-n}})^!_a(d\Psi)\nu_{-n}(da)$$

$$\le \|f\| \iint J^{[n]}(a,X)\min\Big(1,c\|f\|\kappa^{[n]}_{(\Psi)}(\chi(X)>0)+\|f\| \int_{\{\chi:\chi(X)>c\}} \chi(X)\kappa^{[n]}_{(\Psi)}(d\chi)\Big)(\widetilde{P_{-n}})^!_a(d\Psi)\nu_{-n}(da)$$

$$\leq \|f\| \iint J^{[n]}(a,X)\min(1,c\|f\|\kappa^{[n]}_{(\Psi)}(\chi(X)>0))\, (\widetilde{P_{-n}})^!_a(d\Psi)\, \nu_{-n}(da)$$

$$+ \|f\| \iint J^{[n]}(a,X)\min(1,\|f\| \int_{\{\chi:\chi(X)>c\}} \chi(X)\kappa^{[n]}_{(\Psi)}(d\chi))\, (\widetilde{P_{-n}})^!_a(d\Psi)\, \nu_{-n}(da)$$

$$\leq \|f\|^2 c \int J^{[n]}(a,X)((\widetilde{P_{-n}})^!_a)_{\kappa^{[n]}}(\Phi(X)>0)\, \nu_{-n}(da)$$

$$+ (1+\|f\|)^2 \iint J^{[n]}(a,X) \min(1, \int_{\{\chi:\chi(X)>c\}} \chi(X)\kappa^{[n]}_{(\Psi)}(d\chi))\, (\widetilde{P_{-n}})^!_a(d\Psi)\, \nu_{-n}(da)$$

$$\leq \|f\|^2 c \int J^{[n]}(a,X)((\widetilde{P_{-n}})^!_a)_{\kappa^{[n]}}(\Phi(X)>0)\, \nu_{-n}(da)$$

$$+ (1+ \|f\|)^2 \iint (\Psi * J^{[n]})(X) \min(1, \int_{\{\chi:\chi(X)>c\}} \chi(X)\kappa^{[n]}_{(\Psi)}(d\chi))\, (\widetilde{P_{-n}})(d\Psi)$$

$$\leq \|f\|^2 c \int J^{[n]}(a,X)((\widetilde{P_{-n}})^!_a)_{\kappa^{[n]}}(\Phi(X)>0)\, \nu_{-n}(da)$$

$$+ (1+ \|f\|)^2 \int_{\{\Psi:(\Psi * J^{[n]})(X)>d\}} (\Psi * J^{[n]})(X)\, (\widetilde{P_{-n}})(d\Psi)$$

$$+ (1+\|f\|)^2 d \int_{\{\Phi:\Phi(X)>c\}} \Phi(X)\, (\widetilde{P_{-n}})_{\kappa^{[n]}}(d\Phi) \quad .$$

Hence we obtain, in virtue of Proposition 4.2.4. and Theorem 3.3.7.

$$r_f := \limsup_{n\to\infty} |\mathcal{L}_{\mathcal{W}_0^P}(f) - \mathcal{L}_{L_{-n}}(J^{[n]} * f)|$$

$$= \limsup_{n\to\infty} |\mathcal{L}_{P_{-n}}(J^{[n]} * f) - \mathcal{L}_{L_{-n}}(J^{[n]} * f)|$$

$$\leq 0 + (1+ \|f\|)^2 \limsup_{n\to\infty} \int_{\{\Psi:(\Psi * J^{[n]})(X)>d\}} (\Psi * J^{[n]})(X)\widetilde{P}_{-n}(d\Psi)$$

$$+ (1+\|f\|)^2 d \int_{\{\Phi:\Phi(X)>c\}} \Phi(X)(\widetilde{P_0})(d\Phi) \quad .$$

For $c\to\infty$, the last term on the r.h.s. vanishes, and we obtain for all $d>0$, using 3.6.5. in the second estimate:

$$r_f \leq (1+\|f\|)^2 \limsup_{n\to\infty} \int_{\{\Psi:(\Psi * J^{[n]})(X)>d\}} (\Psi * J^{[n]})(X)\, \widetilde{P}_{-n}(d\Psi)$$

$$= (1+\|f\|)^2 \limsup_{n\to\infty} \int_{\{\Psi:(\Psi * J^{[n]})(X)>d\}} \int (J^{[n]}(a,X)\, \Psi(da)\, \widetilde{P}_{-n}(d\Psi)$$

$$= (1+\|f\|)^2 \limsup_{n\to\infty} \int J^{[n]}(a,X)(\widetilde{P_{-n}})_a((\Psi * J^{[n]})(X)>d))\nu_{-n}(da)$$

$$\leq (1+\|f\|)^2 \limsup_{n\to\infty} \int J^{[n]}(a,X)(P_{-n})_a((\Psi * J^{[n]})(X)>d)\nu_{-n}(da)$$

$$= (1+\|f\|)^2 \limsup_{n\to\infty} \int_{\{\Phi:(\Phi * J^{[n]})(X)>d\}} (\Phi * J^{[n]})(X)\, P_{-n}(d\Phi) \quad .$$

Thus there holds for $f \in F_c$

$$r_f \leq (1+\|f\|)^2 \inf_{d>0} \sup_{n=0,1,\ldots} \int_{\{\Phi:(\Phi * J^{[n]})(X)>d\}} (\Phi * J^{[n]})(X)\, P_{-n}(d\Phi) \quad .$$

In view of the equality

$$(\Phi_n * J^{[n]})(X) = E_{\underline{P}}(\Phi_0(X) | \mathcal{F}_{-n}) \quad ,$$

which holds for all $n \in \mathbf{Z}_+$ and $\underline{P}$-almost all (Φ_i), the sequence of random variables $(J^{[n]} * \Phi_{-n})(X)$, $n \in \mathbf{Z}_+$, is uniformly integrable with respect to $\underline{P}$, which allows to conclude that $r_f = 0$. ■

We are now in the position to prove a converse of Theorem 4.3.3.:

4.3.7. Theorem. Let $(L_i)_{i \in \mathbf{Z}}$ be a sequence of distributions on $\mathcal{M}$ such that there exists a sequence $(\lambda_i) \in N_\kappa^{entr,st}$ and a κ-process $\underline{P}$ with the properties

$$\Lambda_{L_i} \leq \lambda_i \quad i \in \mathbf{Z} \ ; \ (L_{j-n})_{\kappa^{[n]}} \underset{n\to\infty}{\Rightarrow} P_j \quad , \quad j \in \mathbf{Z} \quad .$$

Then $\underline{P}$ is of first order, $\underline{\mathcal{W}}^{(L_i)}$ exists and equals $\underline{\mathcal{W}}^{\underline{P}}$.

Proof 1. Because of

$$\Lambda_{(L_{j-n})_{\kappa^{[n]}}} = \Lambda_{L_{j-n}} * J^{[n]} \leq \lambda_j \qquad (j \in \mathbf{Z})$$

our assumptions guarantee that $\Lambda_{P_j} \leq \lambda_j$, $j \in \mathbf{Z}$; hence $\underline{P}$ is of first order.

2. In virtue of 1.5.9. there holds for all $h \in F$, $m \in \mathbf{Z}_+$ and $a \in A$:

$$\ln \mathcal{L}_{\kappa^{[m]}_{(a)}}(h) \geq -(J^{[m]} * h)(a) \quad .$$

3. Together with the elementary inequality

$$\prod_{i\in I} \beta_i - \prod_{i \in I} \alpha_i \leq \sum_{i \in I} (\beta_i - \alpha_i) \quad ,$$

which is true for all at most countable families $(\alpha_i)_{i \in \mathbf{Z}}$ and $(\beta_i)_{i \in \mathbf{Z}}$ of real numbers obeying $0 \leq \alpha_i \leq \beta_i \leq 1$, $i \in I$, step 2. leads, for all $\Phi \in M$ and $m, n \in \mathbf{Z}_+$, to the estimate

$$0 \leq \exp\langle \Phi, \ln \mathcal{L}_{\kappa^{[m]}_{(.)}}(J^{[n]} * f)\rangle - \exp\langle \Phi, -J^{[m+n]} * f\rangle$$

$$\leq \langle \Phi, \mathcal{L}_{\kappa^{[m]}_{(.)}}(J^{[n]} * f) - \exp(-J^{[m+n]} * f)\rangle \quad .$$

4. Using 3., we obtain for all $f \in F_c$

$$\limsup_{n\to\infty} \limsup_{m\to\infty} |\mathcal{L}_{L_{-m-n}}(J^{[m+n]} * f) - \mathcal{L}_{(L_{-m-n})_{\kappa^{[m]}}}(J^{[n]} * f)|$$

$$= \limsup_{n\to\infty} \limsup_{m\to\infty} |\mathcal{L}_{L_{-m-n}}(J^{[m+n]} * f) - \mathcal{L}_{L_{-m-n}}(-\ln \mathcal{L}_{\kappa^{[m]}_{(.)}}(J^{[n]} * f))|$$

$$\leq \limsup_{n\to\infty} \limsup_{m\to\infty} \int \langle \Phi, \mathcal{L}_{\kappa^{[m]}_{(.)}}(J^{[n]}*f) - \exp(-J^{[m+n]}*f)\rangle L_{-m-n}(d\Phi)$$

$$\leq \limsup_{n\to\infty} \limsup_{m\to\infty} (\langle \lambda_{-m-n}, 1-\exp(-J^{[m+n]}*f)\rangle - \langle \lambda_{-m-n}, 1-\mathcal{L}_{\kappa^{[m]}_{(.)}}(J^{[n]}*f)\rangle)$$

$$= \limsup_{n\to\infty} \limsup_{m\to\infty} (-\ln \mathcal{L}_{\Pi_{\lambda_{-m-n}}}(J^{[m+n]}*f) - (-\ln \mathcal{L}_{(\Pi_{\lambda_{-m-n}})\kappa^{[m]}}(J^{[n]}*f))$$

For the moment, let $\underline{H}$ denote the κ-process of Poisson type with intensity sequence (λ_j). Theorem 4.3.5. and 1.3.7., respectively, yield that

$$\mathcal{L}_{\Pi_{\lambda_{-m-n}}}(J^{[m+n]}*f) \xrightarrow[m\to\infty]{} \mathcal{L}_{\mathcal{W}_0\underline{H}}(f)$$

and

$$\mathcal{L}_{(\Pi_{\lambda_{-m-n}})\kappa^{[m]}}(J^{[n]}*f) \xrightarrow[m\to\infty]{} \mathcal{L}_{H_{-n}}(J^{[n]}*f) \quad .$$

The latter expression also tends, for $n\to\infty$, towards $\mathcal{L}_{\mathcal{W}_0\underline{H}}(f)$, which shows that the last term in the above chain of equalities vanishes.

5. Finally, we verify the two conditions

(i) $$L_{-n+j}(\Phi*J^{[n]}\in N) = 1 \qquad (j\in \mathbf{Z},\ n\in \mathbf{Z}_+)$$

and

(ii) $$L_{-n+j}(\Phi*J^{[n]}\in (.)) \Rightarrow P_j \qquad (j\in \mathbf{Z}) \quad .$$

As to (i), note that for all $j\in \mathbf{Z}$, $n\in \mathbf{Z}_+$ and $X\in \mathcal{B}$ there even holds

$$\int (\Phi*J^{[n]})(X)\, L_{-n+j}(d\Phi) \leq \lambda_j(X) < +\infty \quad .$$

Concerning (ii), one obtains

$$\limsup_{n\to\infty} | \mathcal{L}_{L_{-n}}(J^{[n]}*f) - \mathcal{L}_{\mathcal{W}_0\underline{P}}(f) |$$

$$\leq \limsup_{n\to\infty} \limsup_{m\to\infty} | \mathcal{L}_{L_{-m-n}}(J^{[n+m]}*f) - \mathcal{L}_{(L_{-m-n})\kappa^{[m]}}(J^{[n]}*f) |$$

$$+ \limsup_{n\to\infty} \limsup_{m\to\infty} | \mathcal{L}_{(L_{-m-n})\kappa^{[m]}}(J^{[n]}*f) - \mathcal{L}_{P_{-n}}(J^{[n]}*f) |$$

$$+ \limsup_{n\to\infty} | \mathcal{L}_{P_{-n}}(J^{[n]}*f) - \mathcal{L}_{\mathcal{W}_0\underline{P}}(f) | \quad .$$

The first term on the right side vanishes in virtue of 4. Due to 1.3.7. and our assumptions, the second term is zero, whereas the third term vanishes according to Proposition 4.2.4. Hence we have proved (ii) in the special case j=0; the general case follows by time-shifting. ■

4.4. Regular J-sequences and Extremal κ-processes of Poisson Type

A stable sequence $(\nu_i)_{i\in \mathbf{Z}}$ of measures in N is called **regular** if the κ-process of Poisson type with intensity sequence (ν_i) is extremal. The set of all regular stable sequences is denoted by $N_\kappa^{entr,reg}$. An immediate consequence of 3.3.4. and Proposition 4.1.9. is

4.4.1. Proposition. If the (component-wise) sum of at most countably many regular stable sequences is a sequence of measures in N, then it is a regular stable sequence.

Theorems 4.2.11. and 4.3.5. yield immediately:.

4.4.2. For any stable sequence (ν_i) of measures in N, the following assertions are equivalent:

a) (ν_i) is regular.

b) $\underline{\mathcal{W}}^{\Pi^\kappa(\nu_i)} = \delta_{(\nu_i)}$.

c) $\underline{\mathcal{W}}^{(\Pi_{\nu_i})} = \delta_{(\nu_i)}$.

This admits the following extension of the above definition:

A J-sequence $(\nu_i)_{i\in \mathbf{Z}}$ is called **regular** if

$$\underline{\mathcal{W}}^{(\Pi_{\nu_i})} = \delta_{(\nu_i)} .$$

The set of all regular J-sequences is denoted by $N_J^{entr,reg}$.

4.4.3. Proposition. A J-sequence (ν_i) is regular iff it has the following **deconcentration property**

$$\int J^{[n]}(a,X)\min(1, J^{[n]}(a,X))\nu_{m-n}(da) \xrightarrow[n\to\infty]{} 0 \qquad (m\in \mathbf{Z},\ X\in \mathcal{B}) .$$

Proof. For any $m\in \mathbf{Z}_+$ and $X\in \mathcal{B}_{\nu_m}$, the following assertions obviously are equivalent:

$$\Pi_{\nu_{m-n}}((\Phi * J^{[n]})(X)\in (.)) \underset{n\to\infty}{\Rightarrow} \delta_{\nu_m(X)}$$

$$\mathcal{L}_{\Pi_{\nu_{m-n}}((\Phi * J^{[n]})(X)\in (.))}(s) \xrightarrow[n\to\infty]{} \exp(-s\nu_m(X)) \qquad (s\in \mathbf{R}_+)$$

$$\exp\Big(-\int(1-\exp(-s\, J^{[n]}(a,X))\, \nu_{m-n}(da)\Big) \xrightarrow[n\to\infty]{} \exp(-s\nu_m(X)) \qquad (s\in \mathbf{R}_+)$$

$$\int(1-\exp(-s\, J^{[n]}(a,X)))\nu_{m-n}(da) \xrightarrow[n\to\infty]{} s\nu_m(X) \qquad (s\in \mathbf{R}_+)$$

$$\int(\exp(-s\,J^{[n]}(a,X))-1+s\cdot J^{[n]}(a,X))\,\nu_{m-n}(da) \xrightarrow[n\to\infty]{} 0 \qquad (s\in\mathbf{R}_+) \;.\;,$$

In view of the inequality

$$e^{-1}x\min(1,x) \le e^{-x}-1+x \le x\min(1,x) \qquad (x\in\mathbf{R}_+)\;,$$

the latter assertion is equivalent to

$$\int s\,J^{[n]}(a,X)\min(1,s\,J^{[n]}(a,X))\,\nu_{m-n}(da) \xrightarrow[n\to\infty]{} 0 \quad (s\in\mathbf{R}_+)\;,$$

which, because of the inequality

$$(\min(1,s))^2x(\min 1,x) \le sx\min(1,sx) \le (\max(1,s))^2x\min(1,x) \qquad (s,x\in\mathbf{R}_+)$$

is equivalent to

$$\text{(o)} \qquad \int J^{[n]}(a,X)\min(1,J^{[n]}(a,X))\,\nu_{m-n}(da) \xrightarrow[n\to\infty]{} 0\;.$$

Hence (ν_i) is regular iff (o) holds for all $X\in\mathcal{B}_{\nu_m}$, $m\in\mathbf{Z}$. But since any $X\in\mathcal{B}$ is contained in some $X'\in\mathcal{B}_{\nu_m}$, $m\in\mathbf{Z}$, this is equivalent to the deconcentration property stated in the proposition. ■

All regular J-sequences (ν_i) have the **weak deconcentration property**

$$(*) \qquad \int(\kappa^{[n]}{}_{(a)}(\chi(X)>0))^2\nu_{m-n}(da) \xrightarrow[n\to\infty]{} 0 \quad (m\in\mathbf{Z}\;;\; X\in\mathcal{B})\;;$$

on the other hand (*) does not necessarily imply regularity of (ν_i):

4.4.4. Example. Put

$$\kappa_{(a)} := 2^{-1}(\delta_{2\delta_a}+\delta_o) \qquad (a\in A)\;.$$

In this case, any constant sequence $(\nu_i)=(\nu)$ (with $\nu\in N$) is a J-sequence and obeys

$$\int\kappa^{[n]}{}_{(a)}(\chi(X)>0)\nu_{m-n}(da) \xrightarrow[n\to\infty]{} 0 \quad (m\in\mathbf{Z}\,;\,X\in\mathcal{B})\;.$$

It is, however, not regular unless $\nu=o$. ■

For stable sequences (ν_i), the weak deconcentration property (*) is equivalent to regularity, as the following theorem tells:

4.4.5. Theorem. For all stable sequences (ν_i) of measures in N the following assertions are equivalent:

a) (ν_i) is regular.

b) $\int(\kappa^{[n]}{}_{(a)}(\chi(X)>0))^2\nu_{m-n}(da) \xrightarrow[n\to\infty]{} 0 \qquad (m\in\mathbf{Z};\,X\in\mathcal{B})\;.$

c) $\nu_{m-n}(\{a: a\in A,\, J^{[n]}(a,X)>\varepsilon\}) \xrightarrow[n\to\infty]{} 0 \qquad (m\in\mathbf{Z};\,X\in\mathcal{B};\,\varepsilon>0)\;.$

d) $\int|(\Phi_{m-n}*J^{[n]})(X)-\nu_m(X)|\,\amalg^{\kappa}{}_{(\nu_i)}(d(\Phi_i)) \xrightarrow[n\to\infty]{} 0 \quad (m\in\mathbf{Z};\,X\in\mathcal{B})\;.$

Proof. 1. As mentioned above, b) is a consequence of a).

2. Assuming the validity of b), we obtain for all $m\in\mathbf{Z}$ and $c, \varepsilon>0$:

$$\nu_{m-n}\Big(\{a: a\in A;\ (\sum_{0\le j\le c} j\ \kappa^{[n]}_{(a)}(\chi(X)=j)) >\varepsilon\}\Big) \xrightarrow[n\to\infty]{} 0 \ .$$

On the other hand, stability of (ν_i) guarantees by Theorem 3.3.1. that

$$\sup_{n=0,1,\dots} \int (\sum_{j>c} j\ \kappa^{[n]}_{(a)}(\chi(X)=j))\ \nu_{m-n}(da) \xrightarrow[c\to\infty]{} 0 \ .$$

Hence b) implies c).

3. The assumed stability of (ν_i) yields, for all $m\in\mathbf{Z}$ and $X\in\mathcal{B}$, the convergence relation

$$\sup_{n=0,1,\dots} \int_{\{a: J^{[n]}(a,X)>c\}} J^{[n]}(a,X)\ \nu_{m-n}(da) \xrightarrow[c\to\infty]{} 0 \ .$$

Indeed, if $\underline{P}$ is some κ-process with intensity sequence (ν_i), then, for any $m\in\mathbf{Z}$, the sequence of random variables

$$E_{\underline{P}}(\Phi_m(X)|\mathcal{F}_{m-n}) = (\Phi_{m-n} * J^{[n]})(X) \qquad (n\in\mathbf{Z}_+)$$

is uniformly integrable under P, and one obtains like in the proof of Proposition 2.2.9.:

$$\sup_{n=0,1,2,\dots} \int_{\{a: J^{[n]}(a,X)>c\}} J^{[n]}(a,X)\ \nu_{m-n}(da)$$

$$= \sup_{n=0,1,2,\dots} \int \int_{\{a: J^{[n]}(a,X)>c\}} J^{[n]}(a,X)\Phi_{m-n}(da)\ \underline{P}(d(\Phi_i))$$

$$\le \sup_{n=0,1,2,\dots} \int_{\{\Phi_{m-n}: (\Phi_{m-n} * J^{[n]})(X)>c\}} (\Phi_{m-n} * J^{[n]})(X)\ \underline{P}(d(\Phi_i)) \xrightarrow[n\to\infty]{} 0 \ .$$

4. If (ν_i) obeys condition c), then we obtain for all $m\in\mathbf{Z}$, $X\in\mathcal{B}$ and all ε,η such that $0<\varepsilon<1<\eta$:

$$\limsup_{n\to\infty} \int J^{[n]}(a,X)\min(1, J^{[n]}(a,X))\nu_{m-n}(da)$$

$$\le \limsup_{n\to\infty} \int_{\{a: J^{[n]}(a,X)\le\varepsilon\}} (J^{[n]}(a,X))^2\ \nu_{m-n}(da) + \limsup_{n\to\infty} \int_{\{a:\varepsilon< J^{[n]}(a,X)\le 1\}} (J^{[n]}(a,X))^2\ \nu_{m-n}(da)$$

$$+ \limsup_{n\to\infty} \int_{\{a: 1< J^{[n]}(a,X)\le\eta\}} J^{[n]}(a,X)\ \nu_{m-n}(da) + \limsup_{n\to\infty} \int_{\{a:\eta< J^{[n]}(a,X)\}} J^{[n]}(a,X)\ \nu_{m-n}(da)$$

$$\le \varepsilon \limsup_{n\to\infty} \int J^{[n]}(a,X)\ \nu_{m-n}(da) + \limsup_{n\to\infty} \nu_{m-n}(\{a: a\in A,\ J^{[n]}(a,X)>\varepsilon\})$$

$$+ \eta \limsup_{n\to\infty} \nu_{m-n}(\{a: a\in A,\ J^{[n]}(a,X)>1\}) + \sup_{n=1,2,\dots} \int_{\{a:\eta< J^{[n]}(a,X)\}} J^{[n]}(a,X)\ \nu_{m-n}(da)$$

$$= \varepsilon\nu_m(X) + \sup_{n=1,2,\dots} \int_{\{a:\eta< J^{[n]}(a,X)\}} J^{[n]}(a,X)\ \nu_{m-n}(da) \ .$$

It is now obvious from 3. and Proposition 4.4.3. that a) is a consequence of c).

5. Up to now we have established the equivalence of a), b), and c). Finally we are going to show the equivalence of a) and d).

Condition d) implies that

$$\mathcal{W}^{(\Pi_{\nu_i})} = \delta_{(\nu_i)} \ ,$$

which means regularity of (ν_i) by 4.4.2.

Conversely, assume that (ν_i) is regular and hence has the deconcentration property of Proposition 4.4.3. Put, for abbreviation,

$$\underline{P} := \Pi^{\kappa}_{(\nu_i)} \quad .$$

Then we obtain from Proposition 4.1.7. for all $X \in \mathcal{B}$, $c>0$ and $m \in \mathbf{Z}$:

$$\int | \sum_{0 \le j \le c} j(\kappa^{[n]}_{(\Phi)}(\chi(X)=j) - P_m(\chi(X)=j) \,|\, P_{m-n}(d\Phi) \underset{n\to\infty}{\longrightarrow} 0 \quad .$$

On the other hand there holds

$$\sup_{n=1,2,\ldots} \int \sum_{j>c} j\, \kappa^{[n]}_{(\Phi)}(\chi(X)=j)\, P_{m-n}(d\Phi)$$

$$= \sup_{n=1,2,\ldots} \int_{\{\Phi:\Phi(X)>c\}} \Phi(X) P_m(d\Phi) \underset{c\to\infty}{\longrightarrow} 0 \quad .$$

This may be combined to yield

$$\int | \sum_{j\ge 0} j(\kappa^{[n]}_{(\Phi)}(\chi(X)=j) - \nu_m(X) \,|\, P_{m-n}(d\Phi) \underset{n\to\infty}{\longrightarrow} 0 \quad (m \in \mathbf{Z},\, X \in \mathcal{B}) \quad ,$$

which is condition d). ■

The preceding theorem reveals, as a supplement to Proposition 4.4.1., the following properties of the set of regular stable sequences.

4.4.6. $N_\kappa^{\text{entr,reg}} \in \mathcal{N}_J^{\text{entr}}$.

4.4.7. $N_\kappa^{\text{entr,reg}}$ is a descending subset of $N_\kappa^{\text{entr,st}}$ with respect to the semi-order $\le$.

4.4.8. For all $c \in \mathbf{R}_+$, together with (ν_i) also $(c\cdot\nu_i)$ belongs to $N_\kappa^{\text{entr,reg}}$.

Let us say that κ is **globally deconcentrating** if

$$\sup_{a\in A} \kappa^{[n]}_{(a)}(\chi(X)>0) \underset{n\to\infty}{\longrightarrow} 0 \quad (X \in \mathcal{B}) \quad .$$

In this case, any J-sequence (ν_i) has the weak deconcentration property (*), because of the following estimate which is valid for all $X \in \mathcal{B}$, $m \in \mathbf{Z}$ and $n \in \mathbf{Z}_+$:

$$\int (\kappa^{[n]}_{(a)}(\chi(X)>0))^2\, \nu_{m-n}(da) \le (\sup_{a\in A} \kappa^{[n]}_{(a)}(\chi(X)>0)) \int J^{[n]}(b,X) \nu_{m-n}(db)$$

$$= \nu_m(X) \sup_{a\in A} \kappa^{[n]}_{(a)}(\chi(X)>0) \quad .$$

Hence Theorem 4.4.5. yields immediately:

4.4.9. Proposition. If κ is globally deconcentrating, then every stable sequence of measures in N is regular.

An example of a branching dynamics which is globally deconcentrating is the "thinning" ϑ_c with $c<1$ (cf. Example 3.1.1.).

In the following we need the inequaity

4.4.10. For any distribution $L \in T$ which is of first order there holds

$$\mathcal{L}_{\Pi_{\Lambda_L}}(f) \leq \mathcal{L}_L(f) \qquad (f \in F) \ .$$

Proof. According to 1.6.11. and 1.6.12. there holds

$$L = (\Pi_{\tilde{L}})_\alpha \cdot$$

On the other hand there follows for all $X \in \mathcal{A}$

$$(\tilde{L} * J_\alpha)(X) = \int \Psi(X) \tilde{L}(d\Psi) = \Lambda_L(X) \ .$$

The inequality in question thus turns out as a special case of 1.5.10. ■

4.4.11. Proposition. Any extremal κ-process $\underline{P}$ which is of first order and infinitely divisible, is of Poisson type.

Proof By Theorem 4.2.11., extremality of $\underline{P}$ comes up to

$$\underline{\mathcal{W}}^{\underline{P}} = \delta_{(v_i)} \ ,$$

where (v_i) denotes the intensity sequence of $\underline{P}$. Together with Jensen's inequality and 4.4.10., this yields for all $m \in \mathbf{Z}$, $f \in F_c$ and $n \in \mathbf{Z}_+$

$$e^{-\langle v_m, f\rangle} = \exp\left(-\int \langle \Phi, J^{[n]} * f\rangle \Pi_{v_{m-n}}(d\Phi)\right)$$

$$\leq \mathcal{L}_{\Pi_{v_{m-n}}}(J^{[n]} * f) \leq \mathcal{L}_{P_{m-n}}(J^{[n]} * f) \xrightarrow[n \to \infty]{} e^{-\langle v_m, f\rangle} \ .$$

Hence follows

$$\underline{\mathcal{W}}^{(\Pi_{v_i})} = \delta_{(v_i)},$$

and by Theorem 4.3.5. also

$$\underline{\mathcal{W}}^{\underline{\Pi}^\kappa_{(v_i)}} = \delta_{(v_i)} \cdot$$

which by means of the inversion formula 4.2.7. yields equality of $\underline{\Pi}^\kappa_{(v_i)}$ and $\underline{P}$. ■

The assumption of infinite divisibility of $\underline{P}$ in Proposition 4.4.11. may not be cancelled:

4.4.12. Example. Put $A := \{1\}$, $\kappa_{(1)} := \delta_{\delta_1}$. Then the κ-process

$$\underline{P} := \delta_{(\Phi_i)} \ , \quad \Phi_i = \delta_1 \qquad (i \in \mathbf{Z})$$

is extremal and not of Poisson type. ■

Proposition 4.1.15. combined with Theorem 3.3.11. yields immediately

4.4.13. Proposition. A stable sequence of measures $(v_m)_{m \in \mathbf{Z}}$ on $\mathcal{N}$ is regular iff the sequence $\left(\widetilde{\Pi^\kappa_{(v_{i+m})}}\right)_{m \in \mathbf{Z}}$ belongs to $N_\kappa^{entr,reg}$.

4.5. Which κ-processes are of Cox Type?

The first statement is an immediate consequence of Theorem 4.3.5.

4.5.1. Any κ-process $\underline{P}$ of Cox type has a first order representation, and its characteristic J-processes $\underline{\mathcal{W}}^{\underline{P}}$ obeys

$$\underline{\mathcal{W}}^{\underline{P}} = \int \underline{\mathcal{W}}^{\Pi^{\kappa}(\gamma_i)} \underline{Q}(d(\gamma_i)) = \int \underline{\mathcal{W}}^{(\Pi_{\gamma_i})} \underline{Q}(d(\gamma_i)) = \underline{\mathcal{W}}^{(\mathfrak{I}(Q_i))} \quad ,$$

where $\underline{Q}$ denotes the intensity sequence distribution of $\underline{P}$.

A J-process $\underline{Q}$ is called **regular** if $\underline{Q}(N_J^{entr,reg}) = 1$, i. e. if it is concentrated on the set of regular J-sequences. The following proposition tells that the intensity sequence distribution of a Cox type κ-process is regular if and only if it coincides with its characteristic J-process.

4.5.2. Proposition. For any stable J-process $\underline{Q}$, the following assertions are equivalent:

a) $\underline{Q}$ is regular .

b) $\underline{\mathcal{W}}^{\underline{V}\underline{Q}} = \underline{Q}$.

Proof. 1. For all $\gamma \in N$ and $h \in F$ we obtain

$$\mathcal{L}_{\Pi_\gamma}(h) = \exp(-\langle\gamma, 1-\exp(-h)\rangle) \geq \exp(\langle-\gamma, h\rangle) = \mathcal{L}_{\delta_\gamma}(h) \quad .$$

2. For all $(\gamma_i) \in N_\kappa^{entr,st}$ there follows from 1. for all $m \in \mathbf{Z}$, $n \in \mathbf{Z}_+$ and $f \in F_c$:

$$\mathcal{L}_{\Pi_{\gamma_{m-n}}}(J^{[n]} * f) \geq \exp(\langle-\gamma_{m-n}, J^{[n]} * f\rangle) = \exp(\langle-\gamma_m, f\rangle) = \mathcal{L}_{\delta_{\gamma_m}}(f) \quad ,$$

hence by Theorem 4.3.5.

$$\mathcal{L}_{\mathcal{W}_m \Pi^{\kappa}(\gamma_i)}(f) \geq \mathcal{L}_{\delta_{\gamma_m}}(f) \quad .$$

3. Step 2. together with 4.5.1. yields, for all $m \in \mathbf{Z}_+$ and $f \in F_c$

$$(*) \qquad \begin{aligned} \mathcal{L}_{\mathcal{W}_m \underline{V}\underline{Q}}(f) &= \int \mathcal{L}_{\mathcal{W}_m \Pi^{\kappa}(\gamma_i)}(f)\, \underline{Q}(d(\gamma_i)) \\ &\geq \int \mathcal{L}_{\delta_{\gamma_m}}(f) \underline{Q}(d(\gamma_i)) = \mathcal{L}_{Q_m}(f) \quad . \end{aligned}$$

4. Assertion b) is valid iff equality holds in (*) for all $m \in \mathbf{Z}$ and $f \in F_c$. In virtue of 2., this is equivalent to

$$(**) \qquad \mathcal{W}_m^{\Pi^{\kappa}(\gamma_i)} = \delta_{\gamma_m} \quad \underline{Q}\text{-a. all } (\gamma_i) \quad ,$$

since equality of two distributions follows from the equality of their Laplace transforms at a suitable at most countable subset of F_c. Assertion (**), however, is equivalent to regularity of $\underline{Q}$ by 4.4.2. ■

Part of the next theorem is the statement that every κ-process having a first order representation, whose characteristic J-process is regular, is of Cox type.

4.5.3. Theorem. Let $\underline{P}$ be a κ-process having a first order representation. Then the following assertions are equivalent:

a) $\underline{P} = \underline{V}^{\underline{\mathcal{W}}^{\underline{P}}}$ (i.e. $\underline{P}$ is a κ-process of Cox type induced by its characteristic J-process).

b) $\underline{\mathcal{W}}^{\underline{P}}$ is regular.

c) $\underline{K}((\Phi_i),(.))$ is of Poisson type for $\underline{P}$-a. all (Φ_i).

d) $\underline{K}((\Phi_i),(.))$ is infinitely divisible for $\underline{P}$-a. all (Φ_i).

Proof. Let us put $\underline{Q} := \underline{\mathcal{W}}^{\underline{P}}$.

1. By the inversion formula 4.2.7., a) is equivalent to

$$\underline{Q} = \underline{\mathcal{W}}^{\underline{V}^{\underline{Q}}} \quad ,$$

which in turn is equivalent to b) according to Proposition 4.5.2. Hence we have proved the equivalence of a) and b).

Now assume that a) (and due to 1. also b)) is valid. By a) $\underline{P}$ has the integral representation

(*) $$\underline{P} = \int \underline{\Pi}^{\kappa}{}_{(v_i)}(.)\underline{Q}(d(v_i)) \quad .$$

By b) and the definition of regularity, $\underline{\Pi}^{\kappa}{}_{(v_i)}$ is an extremal κ-process for $\underline{Q}$-almost all (v_i). Due to 4.1.1. b)iii) this entails, for $\underline{Q}$-almost all (v_i)

$$\underline{\Pi}^{\kappa}{}_{(v_i)} = \underline{K}((\Phi_i),(.)) \qquad \underline{\Pi}^{\kappa}{}_{(v_i)} \text{ a. all } (\Phi_i),$$

which proves c).

3. The implication c) ⇒ d) is obvious.

4. Finally let us assume d). Then Proposition 4.4.11. implies that, for $\underline{P}$-almost all (Φ_i), the κ-process $\underline{K}((\Phi_i),(.))$ coincides with the κ-process of Poisson type with intensity sequence $(S_n^{(\Phi_i)})$. Hence follows

$$\underline{P} = \int \underline{K}((\Phi_i),(.))\underline{P}(d(\Phi_i)) = \int \underline{\Pi}^{\kappa}{}_{(S_n^{(\Phi_i)})}(.)\underline{P}(d(\Phi_i)) = \int \underline{\Pi}^{\kappa}{}_{(v_i)}(.)\underline{Q}(d(v_i)) \quad ,$$

which proves a). ■

Proposition 4.4.9. yields immediately:

4.5.4. If κ is globally deconcentrating, then every stable J-process $\underline{Q}$ is regular.

The next theorem states, in particular, that every κ-process having a first order representation is of Cox type, provided that κ is globally deconcentrating.

4.5.5. Theorem. If κ is globally deconcentrating, then the mapping $\underline{Q} \to \underline{V}^{\underline{Q}}$ provides a 1-1 correspondence between the set of all stable J-processes $\underline{Q}$ and the set of all κ-processes $\underline{P}$ having a first order representation. Its inverse mapping is given by $\underline{P} \to \underline{\mathcal{W}}^{\underline{P}}$.

Proof. By 4.2.2. and 4.5.4., the characteristic J-process of any κ-process $\underline{P}$ having a first order representation is regular. Hence Theorem 4.5.3. yields

$$\underline{P} = \underline{V}^{\mathcal{W}^{\underline{P}}} .$$

Finally it follows from 4.5.1. and Theorem 3.4.6., respectively, that the mapping $\underline{Q} \to \underline{V}^{\underline{Q}}$ maps the set of all stable J-processes into the set of all κ-processes having a first order representation and is 1-1. ■

As a counterpart to Theorem 3.4.9. there holds

4.5.6. Proposition. A stable J-process $\underline{Q}$ of first order is regular iff its intensity sequence has this property.

Proof. The intensity sequence of $\underline{Q}$, which we denote by (ν_i), is stable due to Theorem 3.4.9. Further there holds for all $X \in \mathcal{B}$, $m \in \mathbf{Z}$ and $n \in \mathbf{Z}_+$

$$\int (\kappa^{[n]}{}_{(a)}(\chi(X)>0))^2 \nu_{m-n}(da)$$
$$= \int\int (\kappa^{[n]}{}_{(a)}(\chi(X)>0))^2 \gamma_{m-n}(da) \underline{Q}(d(\gamma_i)) \quad .$$

An application of Theorem 4.5.5. together with Lebesgue's dominated convergence theorem shows that regularity of (ν_i) is implied by regularity of $\underline{Q}$.

Conversely, assume that (ν_i) is regular. Then Proposition 4.4.3. together with Fatou's lemma implies, for all $m \in \mathbf{Z}$, $X \in \mathcal{B}$ and $\underline{Q}$-almost all (γ_i)

$$\liminf_{n \to \infty} \int J^{[n]}(a,X) \min(1, J^{[n]}(a,X)) \gamma_{m-n}(da) = 0$$

i. e. the existence of a strictly increasing sequence (n_k) of natural numbers such that

$$\lim_{k \to \infty} \int J^{[n_k]}(a,X) \min(1, J^{[n_k]}(a,X)) \gamma_{m-n_k}(da) = 0 \quad .$$

The same reasoning as in the proof of Proposition 4.4.3. now yields

$$(*) \qquad \Pi_{\gamma_{m-n_k}}((\Phi * J^{[n_k]}(X) \in (.)) \underset{k \to \infty}{\Rightarrow} \delta_{\gamma_m}(X) \quad .$$

Stability of (γ_i) and Theorem 4.3.5. ensure the existence of $\mathcal{W}^{(\Pi_{\gamma_i})}$, which by (*) is revealed to coincide with $\delta_{(\gamma_i)}$; i. e. (γ_i) - and in summary also $\underline{Q}$ - is regular. ■

4.5.7. Proposition. Any κ-process $\underline{P}$ of first order, whose intensity sequence is regular, is of Cox type, and its intensity sequence distribution coincides with its characteristic J-process.

Proof. By 4.2.2., the intensity sequence of $\mathcal{W}^{\underline{P}}$ equals that of $\underline{P}$ and thus is regular. Due to Proposition 4.5.6., also $\mathcal{W}^{\underline{P}}$ is regular; hence Theorem 4.5.3. implies that $\underline{P} = \underline{V}^{\mathcal{W}^{\underline{P}}}$. ■

We now obtain the following "local version" of Theorem 4.5.5.:

4.5.8. Theorem. For all regular stable sequences $(v_i)_{i\in Z}$ of measures on $\mathcal{N}$, the mapping $\underline{Q}\rightarrow \underline{V}^{\underline{Q}}$ provides a 1-1 mapping from the set $V_\kappa^{entr,(v_i)}$ of all J-processes with intensity sequence (v_i) onto the set $\underline{P}^{\kappa,(v_i)}$ of all κ-processes with intensity sequence (v_i); its inverse mapping is given by $\underline{P}\rightarrow \underline{\mathcal{W}}^{\underline{P}}$.

Proof. For any stable J-process $\underline{Q}$, the distributions $\underline{V}^{\underline{Q}}$ and $\underline{Q}$ have the same intensity sequence, i. e. the mapping $\underline{Q}\rightarrow \underline{V}^{\underline{Q}}$ maps $V_\kappa^{entr,(v_i)}$ into $\underline{P}^{\kappa,(v_i)}$. Injectivity of this mapping is guaranteed by Theorem 3.4.6., whereas Proposition 4.5.7. implies its surjectivity as well as the asserted form of its inverse mapping. ■

As a supplement to the preceding theorem we are going to prove:

4.5.9. Theorem. For all stable sequence $(v_i)_{i\in Z}$ of measures in N, the following assertions are equivalent:

a) (v_i) is regular.

b) Any J-process $\underline{Q}$ with intensity sequence (v_i) arises as characteristic J-process of some κ-process $\underline{P}\in\underline{P}^{\kappa,(v_i)}$.

c) For all $c\in \mathbf{R}_+$ there exists an extremal κ-process $\underline{P}_c$ with intensity sequence (cv_i).

Proof. 1. In virtue of Theorem 4.5.8., b) is a consequence of a).

2. Let us now assume the validity of c). Due to Theorem 4.2.8. and 4.2.3. there hold for all r, s> 0 :

$$\underline{\mathcal{W}}^{\underline{P}_{r+s}} = \delta_{((r+s)v_i)} \quad ,$$

$$\underline{\mathcal{W}}^{\underline{P}_r * \underline{P}_s} = \underline{\mathcal{W}}^{\underline{P}_r} * \underline{\mathcal{W}}^{\underline{P}_s} = \delta_{(rv_i)} * \delta_{(sv_i)} = \delta_{((r+s)v_i)} \quad ,$$

which yields by the inversion formula 4.3.7. that

$$\underline{P}_r * \underline{P}_s = \underline{P}_{r+s} \qquad (r, s > 0) \quad .$$

In virtue of [MKM, Prop. 2.4.1.] the κ-process $\underline{P}_1$ is thus infinitely divisible, and hence of Poisson type due to Proposition 4.4.11. But this means that (v_i) is regular, and shows that c) implies a).

3. Next we assume that b) holds true. To show the validity of c), let us fix an arbitrary number c>0, put $t := c^{-1}$, choose a number $\alpha\in(0,1)$ such that $\alpha c+(1-\alpha)t = 1$ and define

$$\underline{Q} := \alpha\delta_{(cv_i)} + (1-\alpha)\delta_{(tv_i)} \quad .$$

By assumption, there exists a κ-process $\underline{P}$ with intensity sequence (v_i), whose characteristic J-process equals $\underline{Q}$. The set

$$Y := \{(\Phi_i): (\Phi_i)\in M^{\mathbf{Z}} \quad , \quad (S_n^{(\Phi_i)}) = (cv_n)\}$$

belongs to $\mathcal{F}_{-\infty}$ and has positive $\underline{P}$-probability (namely 1 if c=1 and α if c≠1). Therefore $\underline{P}_c := \underline{P}((.)|Y)$ is a κ-process which has intensity sequence (cv_i) and obeys

$$\underline{\mathcal{W}}^{\underline{P}_c} = \delta_{(cv_i)} \quad .$$

Hence $\underline{P}_c$ is extremal by Theorem 4.2.11., and c) follows from b). ■

If κ is globally deconcentrating, then, according to Theorem 4.5.5., every κ-process having a first order representation is of Cox type. There may still exist, however, κ-processes which are not of Cox type, as the following example shows:

4.5.10. Example. Let $[A, \rho_A]$ be the phase space of Examples 2.1.6. and 2.7.25., and let us slightly modify their stochastic branching dynamics by putting

$$\kappa_{(j)} := \begin{cases} 2^{-1}(\delta_{\delta_{j-1}} + \delta_{\delta_{j+1}}) & \text{for } j=0, -1, -2, \ldots \\ 2^{-1}(\delta_{\delta_{j-1}} + \delta_{\delta_{j+1}}) * (j^{-2}\delta_{j\delta_\infty} + (1-j^{-2})\delta_o) & \text{for } j=1, 2, \ldots \end{cases},$$

$$\kappa_{(\infty)} := \delta_o \ .$$

Then κ is globally deconcentrating. For $\mu := \sum_{i\in\mathbb{Z}} \delta_i$ the distribution $P := (\Pi_\mu)_\kappa$ belongs to P^{inv}, and there holds

$$\mathcal{G}_P = \Pi_{\tilde{P}} = \Pi_{(\varrho_\mu)_\kappa}$$

In view of $\Lambda_{\tilde{P}} \notin N$ and Theorem 3.5.1., the equilibrium distribution P (together with its correponding stationary κ-process) is not of Cox type. ■

The next theorem tells that, for a globally deconcentrating branching dynamics κ, all family processes associated with some κ-process are Coxian.

4.5.11. Theorem. Let κ be globally deconcentrating and $\underline{P}$ be an arbitrary κ-process. Then the family process $\underline{\mathcal{G}}^{\underline{P}}$ associated with $\underline{P}$ is Coxian.

Proof 1. For all $n\in\mathbb{Z}_+$ we put

$$\tau^{(n)}{}_{(a)} := \kappa^{[n]}{}_{(\delta_a)} \qquad (a\in A) \ ,$$

thus obtaining a substochastic clustering field on $[A, \rho_A]$ with phase space $[\mathbb{A}, \rho_{\mathbb{A}}]$. Then there holds for all $m\in\mathbb{Z}$

$$(P_{m-n})_{\tau^{(n)}} = \mathcal{G}_{P_{m-n},n} \underset{n\to\infty}{\Rightarrow} \mathcal{G}_m{}^{\underline{P}} \quad .$$

2. We are now going to show that not only $(P_{m-n})_{\tau^{(n)}}$, but also the clustered Cox transforms $(\mathfrak{I}(P_{m-n}))_{\tau^{(n)}}$ converge, for $n\to\infty$, weakly towards $\mathcal{G}_m{}^{\underline{P}}$. The existence of the distributions $(\mathfrak{I}(P_{m-n}))_{\tau^{(n)}}$, $m\in\mathbb{Z}$, $n\in\mathbb{Z}_+$, follows from Lemma 4.5.12. below; thus a combination of 1.4.7. and 1.3.12. yields for all $f\in\mathbb{F}_c$

$$(*) \qquad \mathcal{L}_{(\mathfrak{I}(P_{m-n}))_{\tau^{(n)}}}(f) = \mathcal{L}_{P_{m-n}}(1 - \mathcal{L}_{(\tau^{(n)})_{(.)}}(f))$$

Now choose a set $X\in\mathcal{B}$ such that $\operatorname{supp} f \subseteq X$.

The following estimate is valid for all $n\in\mathbb{Z}_+$ and $a\in A$:

$$1 \geq \mathcal{L}_{(\tau^{[n]})_{(a)}}(f) = \int \exp(-\langle\Phi,f\rangle)(\tau^{(n)})_{(a)}(d\Phi)$$
$$\geq (\tau^{(n)})_{(a)}(\Phi(X) = 0) \quad \geq \ \kappa^{[n]}{}_{(a)}(\Phi(X) = 0) \quad .$$

This estimate together with the assumption that κ is uniformly deconcentrating shows that

$$g_n := -\ln \mathcal{L}_{(\tau^{[n]})_{(.)}}(f)$$

tends, for $n \to \infty$, uniformly towards zero. Hence there exists a sequence (ε_n) of positive real numbers tending towards zero, and such that

$$(1-\varepsilon_n)g_n \leq 1-e^{-g_n} \leq g_n \quad ,$$

which implies

$$\mathcal{L}_{(P_{m-n})}((1-\varepsilon_n)g_n) \geq \mathcal{L}_{(P_{m-n})}(1-\mathcal{L}_{(\tau^{[n]}_{(.)})}(f)) \geq \mathcal{L}_{(P_{m-n})}(g_n) \quad .$$

Due to 1., $\mathcal{L}_{(P_{m-n})}(g_n)$ tends, for $n \to \infty$, towards $\mathcal{L}_{\mathcal{G}_m P}(f)$, whereas the difference

$$\mathcal{L}_{(P_{m-n})}((1-\varepsilon_n)g_n) - \mathcal{L}_{P_{m-n}}(g_n)$$

tends, for $n \to \infty$, towards zero.

Indeed, this difference obeys, for any $C > 0$, and $n \in \mathbf{Z}_+$ such that $\varepsilon_n < 1/2$, the estimate

$$\int \big(\exp(-\langle \Phi, (1-\varepsilon_n)\, g_n\rangle) - \exp(-\langle \Phi, g_n\rangle)\big) P_{m-n}(d\Phi)$$

$$\leq \int_{\{\langle \Phi, g_n\rangle \geq C\}} \exp(-\langle \Phi, g_n/2\rangle) P_{m-n}(d\Phi)$$

$$+ \int_{\{\langle \Phi, g_n\rangle < C\}} \exp(-\langle \Phi, g_n\rangle)\,(\exp(\varepsilon_n \langle \Phi, g_n\rangle) - 1) P_{m-n}(d\Phi)$$

$$\leq 2 \cdot C^{-1} + \exp(\varepsilon_n \cdot C) - 1 \quad .$$

In summary, we have proved that also $\mathcal{L}_{(P_{m-n})}((1-\mathcal{L}_{(\tau^{(n)}_{(.)})})(f))$ converges, for $n \to \infty$, towards $\mathcal{L}_{\mathcal{G}_m P}(f)$.

Combined with (*) this latter result shows that $(\mathfrak{I}(P_{m-n}))_{\tau^{(n)}}$ converges, for $n \to \infty$, weakly towards $\mathcal{G}_m P$. But as, for all $m \in \mathbf{Z}$ and $n \in \mathbf{Z}_+$, $(\mathfrak{I}(P_{m-n}))_{\tau^{(n)}}$ is a Cox distribution due to the substochastic nature of $\tau^{(n)}$, also the weak limit $\mathcal{G}_m P$ is a Cox distribution (cf [MKM, Prop. 7.3.2.]). ■

The following lemma has been used in the preceding proof:

4.5.12. Let P be a distribution on $\mathcal{M}$ and τ be a clustering field on $[A, \rho_A]$ with a phase space $[A', \rho_{A'}]$. Then P_τ exists iff $(\mathfrak{I}(P))_\tau$ exists.

Proof. According to 1.4.1., the existence of P_τ is equivalent to

$$P\Big(\int \tau_{(a)}(\chi(X) > 0)\Phi(da) < +\infty \quad \text{for all } X \in \mathcal{B}'\Big) = 1 \quad .$$

In view of the "clustering theorem" 1.4.6., this is equivalent to

$$P((\Pi_\Phi)_\tau \text{ exists}) = 1 \quad ,$$

which, in turn, is equivalent to the existence of $(\mathfrak{I}(P))_\tau$. ■

Theorem 4.5.3. characterizes, among the κ-processes having a first order representation, those of Cox type. In the following we give an example of a κ-process having a first order representation, which is not Cox type, even though its associated family process in Poissonian.

4.5.13. Example. Put $A := \mathbf{Z} \cup \{\infty\}$, ρ_A like in Example 2.1.6., and

$$\kappa_{(j)} := \begin{cases} \delta_{\delta_j} * \delta_{|j|\,\delta_\infty} & \text{for } j \in \mathbf{Z} \\ \delta_o & \text{for } j = \infty \end{cases}$$

Then there holds

$$ {}_{\kappa}M = \{\Phi \in M : \Phi(Z) < +\infty\} \quad ; $$

$$ \kappa_{(\Phi)} = \delta_{Z\Phi} * \delta_{n(\Phi)\cdot\delta_\infty} \quad , \text{ where } n(\Phi) := \int |j| \, \Phi(dj) \qquad (\Phi \in {}_{\kappa}M) \ . $$

Hence any κ-process is concentrated on sequences (Φ_i) in ${}_{\kappa}M$ with the properties

$$ {}_Z\Phi_i = {}_Z\Phi_{i+1} \quad ; \quad \Phi_{i+1}(\{\infty\}) = n(\Phi_i) \qquad (i \in \mathbf{Z}) \ , $$

which is equivalent to the existence of some $\Phi \in {}_{\kappa}M$ such that

$$ \Phi_i = \Phi \quad , \quad i \in \mathbf{Z} \quad ; \quad \Phi(\{\infty\}) = n(\Phi) \ . $$

The extremal κ-processes are just the Dirac measures concentrated on a sequence (Φ_i) of that kind. As they are of first order, clearly any κ-process has a first order representation. Now put

$$ \lambda := \sum_{i \geq 1} i^{-2} \delta_i \quad ; \quad L := \Pi_\lambda \ . $$

Because of $\lambda(Z) < +\infty$ there holds $\Pi_V({}_{\kappa}M) = 1$. $P := (\Pi_V)_\kappa$ is an equilibrium distribution; its associated family equilibrium distribution is Poissonian, given by

$$ \mathcal{G}_P := \Pi_S \quad ; \quad S := (\mathcal{Q}_V)_\kappa \ . $$

Because of

$$ \Lambda_S(\{\infty\}) = \sum_{i \geq 1} i^{-2} \cdot i = +\infty $$

and Theorem 3.5.1., however, P is not of Cox type. ■

We conclude the section with

4.5.14. Proposition. Each of the two conditions

a) $\kappa_{(a)} \in \mathbf{T}$ for all $a \in A$,

b) $\kappa_{(a)}$ is Coxian for all $a \in A$,

implies

c) $N_\kappa^{entr,st} = N_\kappa^{entr,reg}$.

Proof. 1. Let us first assume the validity of a). By means of 1.4.6. and because of

$$ \kappa^{[m+1]}{}_{(a)} = (\kappa^{[m]}{}_{(a)})_\kappa \qquad (a \in A \quad ; \quad m \in \mathbf{Z}_+) $$

we conclude by induction that

$$ \kappa^{[m]}{}_{(a)} \in \mathbf{T} \quad (a \in A \quad ; \quad n=1,2,\ldots) \ . $$

Due to 1.4.5. there results

$$ (*) \qquad \kappa^{[n]}{}_{(\Phi)} \in \mathbf{T} \qquad (n=1,2,\ldots \quad ; \quad \Phi \in {}_{\kappa^{[n]}}M) \ . $$

If now $\underline{P}$ is some extremal κ-process and $m \in \mathbf{Z}_+$, then we get in virtue of 4.1.3.

$$ \kappa^{[n]}{}_{(\Phi_{m-n})} \underset{n\to\infty}{\Rightarrow} P_m \qquad \text{for } \underline{P}\text{-a.all } (\Phi_i) \ , $$

and hence because of (*) and 1.2.8. also $P_m \in \mathbf{T}$. Therefore, $\underline{P}$ is infinitely divisible.

By means of Theorem 4.5.3. and Proposition 4.5.6. one now infers immediately the validity of c).

2. Let us assume, from now on, that b) is fulfilled. For any $\Phi \in {}_{\kappa}M$, the distribution $\kappa^{[n]}{}_{(\Phi)}$ is Coxian due to 1.2.14. and 1.2.15. Hence, for any $L \in \mathbf{P}$, the clustered distribution L_{κ}, provided it exists, is Coxian as a mixture of Cox distributions.

3. According to 2., any κ-process $\underline{P}$ is Coxian; due to Theorem 3.5.1. also its associated family process $\underline{G}^{P}$ has this property.

4. To show the validity of c), take some stable sequence of measure (v_i) and consider a convex combination

$$\Pi^{\kappa}{}_{(v_i)} = c\underline{P}' + (1-c)\underline{P}'' \quad ,$$

where $c \in (0,1)$ and $\underline{P}', \underline{P}'' \in \underline{P}^{\kappa}$.

According to 3. and Theorem 3.5.2., $\underline{P}'$ and $\underline{P}''$ are of Cox type, i. e.

$$\underline{P}' = \underline{V}^{Q'} \quad , \quad \underline{P}'' = \underline{V}^{Q''}$$

for some J-processes $\underline{Q}', \underline{Q}''$, and we obtain by 3.2.9. and Theorem 3.4.6.

$$\delta_{(v_i)} = c\underline{Q}' + (1-c)\underline{Q}'' \quad .$$

This yields

$$\underline{Q}' = \delta_{(v_i)} = \underline{Q}'' \quad ,$$

i. e. $\underline{P}' = \Pi^{\kappa}{}_{(v_i)} = \underline{P}''$, which reveals that $(v_i) \in N_{\kappa}{}^{entr,reg}$.

4.6. Convergence Theorems

In Proposition 4.3.2. it has been proved that for any sequence $(L_i)_{i\in \mathbf{Z}}$ of distributions on $\mathcal{N}$, whose intensity sequence is dominated by some sequence $(\lambda_i) \in N_J{}^{entr}$ and for which $\mathcal{W}_m{}^{(L_i)}$ exists for all $m \in \mathbf{Z}$, the latter sequence is a J-entrance law. The corresponding J-process has been denoted by $\underline{\mathcal{W}}^{(L_i)}$.

The following theorem says that, in the just described situation, the "clustered Cox transforms of (L_i) converge towards the entrance law of Cox type induced by $\underline{\mathcal{W}}^{(L_i)}$ ".

4.6.1. Theorem. Let $(L_i)_{i\in \mathbf{Z}}$ be a sequence of distributions on $\mathcal{N}$ for which $\mathcal{W}_m{}^{(L_i)} := Q_m$, $m \in \mathbf{Z}$, exists, and assume that

$$\Lambda_{L_i} \leq \lambda_i \qquad (i \in \mathbf{Z})$$

for some sequence $\lambda_i \in N_J{}^{entr}$. Then there holds

$$(\mathfrak{I}(L_{j-n}))_{\kappa^{[n]}} \underset{n\to\infty}{\Rightarrow} V_j{}^{Q} \qquad (j \in \mathbf{Z}) \quad .$$

Proof. 1. Let f be a function in F_b.
For all $j \in \mathbf{Z}$ and all $m, n \in \mathbf{Z}_+$ we obtain, using the mapping g defined in 4.3.6. and applying the backward technique (1.9.2.) as well as property 4.3.6. ε) of the mapping g, the following chain of computations:

$$\int | \mathcal{L}_{(\Pi_\gamma)_{\kappa[m+n]}}(f) - \mathcal{L}_{(\Pi_{\gamma * J[m]})_{\kappa[n]}}(f) \,|\, L_{j-m-n}(d\gamma)$$

$$= \int | \exp\Big(-\int (1-e^{-\langle \Psi, f\rangle}) \widetilde{(\Pi_\gamma)_{\kappa[m+n]}}(d\Psi)\Big)$$

$$- \exp\Big(-\int (1-e^{-\langle \Psi, f\rangle}) \widetilde{(\Pi_{\gamma * J[m]})_{\kappa[n]}}(d\Psi)\Big) \,|\, L_{j-m-n}(d\gamma)$$

$$\leq \int | \int (1-e^{-\langle \Psi, f\rangle}) \, \widetilde{(\Pi_\gamma)_{\kappa[m+n]}}(d\Psi)$$

$$- \int (1-e^{-\langle \Psi, f\rangle}) \widetilde{(\Pi_{\gamma * J[m]})_{\kappa[n]}}(d\Psi) \,|\, L_{j-m-n}(d\gamma)$$

$$= \int | \int g(\langle \Psi, f\rangle) \langle \Psi, f\rangle \widetilde{(\Pi_\gamma)_{\kappa[m+n]}}(d\Psi)$$

$$- \int g(\langle \Psi, f\rangle) \langle \Psi, f\rangle \widetilde{(\Pi_{\gamma * J[m]})_{\kappa[n]}}(d\Psi) \,|\, L_{j-m-n}(d\gamma)$$

$$= \int | \int g(\langle \Psi, f\rangle) f(a) C_{\widetilde{(\Pi_\gamma)_{\kappa[m+n]}}}(d[a,\Psi])$$

$$- \int g(\langle \Psi, f\rangle) f(a) C_{\widetilde{(\Pi_{\gamma * J[m]})_{\kappa[n]}}}(d[a,\Psi]) \,|\, L_{j-m-n}(d\gamma)$$

$$= \int | \int g(\langle \Psi, f\rangle) f(a) C_{\widetilde{((\Pi_\gamma)_{\kappa[m]})_{\kappa[n]}}}(d[a,\Psi])$$

$$- \int g(\langle \Psi, f\rangle) f(a) C_{\widetilde{(\Pi_{\gamma * J[m]})_{\kappa[n]}}}(d[a,\Psi]) \,|\, L_{j-m-n}(d\gamma)$$

$$= \int | \iiint g(\langle \Psi, f\rangle) f(a) \, (\kappa^{[n]}{}_{(b)})_a(d\Psi) \, J^{[n]}(b,da) \, (\gamma * J^{[m]})(db)$$

$$- \iiint g(\langle \Psi, f\rangle) f(a) \Big(\Big(\big(\widetilde{(\Pi_\gamma)_{\kappa[m]}}\big)^{!}{}_{b} \Big)_{\kappa[n]} * (\kappa^{[n]}{}_{(b)})_a \Big)(d\Psi) \, J^{[n]}(b,da) \, (\gamma * J^{[m]})(db) \,|\, L_{j-m-n}(d\gamma)$$

$$= \iiiint\!\!\int (g(\langle \Psi_1, f\rangle) - g(\langle \Psi_1 + \Psi_2, f\rangle)) f(a) \, (\kappa^{[n]}{}_{(b)})_a(d\Psi_1)$$

$$\Big(\big(\widetilde{(\Pi_\gamma)_{\kappa[m]}}\big)^{!}{}_{b} \Big)_{\kappa[n]}(d\Psi_2) J^{[n]}(b,da)(\gamma * J^{[m]})(db) L_{j-m-n}(d\gamma)$$

$$\leq \iiiint\!\!\int \langle \Psi_2, f\rangle g(\langle \Psi_1, f\rangle) g(\langle \Psi_2, f\rangle) f(a) \, (\kappa^{[n]}{}_{(b)})_a(d\Psi_1)$$

$$\Big(\big(\widetilde{(\Pi_\gamma)_{\kappa[m]}}\big)^{!}{}_{b} \Big)_{\kappa[n]}(d\Psi_2) \, J^{[n]}(b,da) \, (\gamma * J^{[m]})(db) \, L_{j-m-n}(d\gamma)$$

$$= \iiiint \langle \Psi_2, f\rangle g(\langle \Psi_1, f\rangle) g(\langle \Psi_2, f\rangle) \langle \Psi_1, f\rangle \, \kappa^{[n]}{}_{(b)}(d\Psi_1)$$

$$\Big(\big(\widetilde{(\Pi_\gamma)_{\kappa[m]}}\big)^{!}{}_{b} \Big)_{\kappa[n]}(d\Psi_2) \, (\gamma * J^{[m]})(db) \, L_{j-m-n}(d\gamma)$$

$$= \iiiint (1-e^{-\langle \Psi_1, f\rangle})(1-e^{-\langle \Psi_2, f\rangle})\, \kappa^{[n]}_{(b)}(d\Psi_1)$$

$$\Big(\big(\widetilde{(\Pi_\gamma)_{\kappa^{[m]}}}\big)^!_b\Big)_{\kappa^{[n]}}(d\Psi_2)\,(\gamma * J^{[m]})(db)\, L_{j-m-n}(d\gamma)$$

$$= \iiint (1-\mathcal{L}_{\kappa^{[n]}_{(b)}}(f))(1-\mathcal{L}_{\kappa^{[n]}_{(\chi)}}(f))$$

$$\Big(\big(\widetilde{(\Pi_\gamma)_{\kappa^{[m]}}}\big)^!_b\Big)(d\chi)\,(\gamma * J^{[m]})(db)\, L_{j-m-n}(d\gamma)$$

$$= \iiint (1-\mathcal{L}_{\kappa^{[n]}_{(b)}}(f))(1-\mathcal{L}_{\kappa^{[n]}_{(\chi-\delta_b)}}(f))\,\chi(db)\,(\mathcal{Q}_\gamma)_{\kappa^{[m]}}(d\chi)\, L_{j-m-n}(d\gamma)$$

$$\leq \iint (1-\mathcal{L}_{\kappa^{[n]}_{(b)}}(f))(1-\mathcal{L}_{\kappa^{[n]}_{(\chi-\delta_b)}}(f))\chi(db)\,(\mathcal{Q}_{\lambda_{j-m-n}})_{\kappa^{[m]}}(d\chi)$$

$$\leq \iint (1-\mathcal{L}_{\kappa^{[n]}_{(b)}}(f))(1-\mathcal{L}_{\kappa^{[n]}_{(\chi-\delta_b)}}(f))\,\chi(db)\,(\mathcal{Q}_{\lambda^{st}_{j-m-n}})_{\kappa^{[m]}}(d\chi)$$

$$+ \int (1-\mathcal{L}_{\kappa^{[n]}_{(b)}}(f))(\lambda_{j-n}-\lambda^{st}_{j-n})(db)$$

$$= \int (1-\mathcal{L}_{\kappa^{[n]}_{(b)}}(f))(1-\mathcal{L}_{\kappa^{[n]}_{(\chi)}}(f))\,C^!_{\widetilde{(\Pi_{\lambda^{st}_{j-m-n}})_{\kappa^{[m]}}}}(d[b,\chi])$$

$$+ \int (1-\mathcal{L}_{\kappa^{[n]}_{(b)}}(f))(\lambda_{j-n}-\lambda^{st}_{j-n})(db)$$

$$\leq \int (1-\mathcal{L}_{\kappa^{[n]}_{(b)}}(f))(1-\mathcal{L}_{\kappa^{[n]}_{(\chi)}}(f))\,C^!_{\widetilde{\Pi^{\kappa}_{(\lambda^{st}_{j-n})}}}(d[b,\chi])$$
$$+ \int (1-\mathcal{L}_{\kappa^{[n]}_{(b)}}(f))(\lambda_{j-n}-\lambda^{st}_{j-n})(db)\ ,$$

where we used 3.6.6. in the last estimate.

We are now going to show that both summands on the right hand side of the preceding chain of inequalities tend to zero for $n\to\infty$. Rewriting the first summand, one gets by means of Theorem 3.3.7.

$$\iint (1-\mathcal{L}_{\kappa^{[n]}_{(b)}}(f))(1-\mathcal{L}_{\kappa^{[n]}_{(\chi)}}(f))\big(\widetilde{\Pi^{\kappa}_{(\lambda^{st}_{j-n})}}\big)^!_b(d\chi)\,\lambda^{st}_{j-n}(db)$$

$$= \int (1-\mathcal{L}_{\kappa^{[n]}_{(b)}}(f)(1-\mathcal{L}_{((\widetilde{\Pi^{\kappa}_{\lambda^{st}_{j-n}}})^!_b)_{\kappa^{[n]}}}(f))\,\lambda^{st}_{j-n}(db)$$

$$\leq \int J^{[n]}(b, \operatorname{supp} f)\Big(\big(\widetilde{\Pi^{\kappa}_{(\lambda^{st}_{j-n})}}\big)^!_b\Big)_{\kappa^{[n]}}(\chi(\operatorname{supp} f)>0)\,\lambda^{st}_{j-n}(db) \xrightarrow[n\to\infty]{} 0\ .$$

Concerning the second summand, we obtain because of

$$(\Pi_{\lambda_{j-n}-\lambda^{st}_{j-n}})_{\kappa^{[n]}} \underset{n\to\infty}{\Rightarrow} \delta_o$$

the relations

$$\int(1-\mathcal{L}_{\kappa^{[n]}_{(b)}}(f))(\lambda_{j-n}-\lambda^{st}_{j-n})(db) = -\ln \mathcal{L}_{(\Pi_{\lambda_{j-n}-\lambda^{st}_{j-n}})_{\kappa^{[n]}}}(f) \xrightarrow[n\to\infty]{} 0 .$$

In summary, we have proved, for all $j\in \mathbf{Z}$:

$$\sup_{m=0,1,\dots} \int | \mathcal{L}_{(\Pi_\gamma)_{\kappa^{[m+n]}}}(f) - \mathcal{L}_{(\Pi_{\gamma * J^{[m]}})_{\kappa^{[n]}}}(f) | L_{j-m-n}(d\gamma) \xrightarrow[n\to\infty]{} 0 .$$

2. For all $j\in \mathbf{Z}$ we obtain, writing $P_j := V_j^{\Omega}$ for abbreviation, the following estimate:

$$\limsup_{n\to\infty} | \mathcal{L}_{(\mathfrak{S}(L_{j-n}))_{\kappa^{[n]}}}(f) - \mathcal{L}_{P_j}(f) |$$

$$\leq \limsup_{n\to\infty} \limsup_{m\to\infty} \int | \mathcal{L}_{(\Pi_\gamma)_{\kappa^{[m+n]}}}(f) - \mathcal{L}_{(\Pi_{\gamma * J^{[m]}})_{\kappa^{[n]}}}(f) | L_{j-m-n}(d\gamma)$$

$$+ \limsup_{n\to\infty} \limsup_{m\to\infty} | \int \mathcal{L}_{(\Pi_{\gamma * J^{[m]}})_{\kappa^{[n]}}}(f) L_{j-m-n}(d\gamma) - \mathcal{L}_{(\mathfrak{S}(Q_{j-n}))_{\kappa^{[n]}}}(f) |$$

$$+ \limsup_{n\to\infty} | \mathcal{L}_{(\mathfrak{S}(Q_{j-n}))_{\kappa^{[n]}}}(f) - \mathcal{L}_{P_j}(f) | .$$

3. In virtue of step 1. and 3.2.8., respectively, the first and third term on the right hand side of the estimate in 2. vanish. Concerning the second term, we first observe, for all m, $n\in \mathbf{Z}_+$, the identity

$$\int \mathcal{L}_{(\Pi_{\gamma * J^{[m]}})_{\kappa^{[n]}}}(f) L_{j-m-n}(d\gamma) = \mathcal{L}_{(\mathfrak{S}(L_{j-m-n}(\gamma * J^{[m]} \in (.))))_{\kappa^{[n]}}}(f) .$$

By assumption, $L_{j-m-n}(\gamma * J^{[m]} \in (.))$ tends, for $m\to\infty$, weakly towards Q_{j-n}, which implies the convergence

$$\mathfrak{S}(L_{j-m-n}((\gamma * J^{[m]}) \in (.))) \underset{m\to\infty}{\Rightarrow} \mathfrak{S}(Q_{j-n}) .$$

In view of 1.5.11. this yields, for all $n\in \mathbf{Z}_+$, the convergence

$$\mathcal{L}_{(\mathfrak{S}(L_{j-m-n}((\gamma * J^{[m]}) \in (.)))_{\kappa^{[n]}}}(f) \xrightarrow[m\to\infty]{} \mathcal{L}_{(\mathfrak{S}(Q_{j-n}))_{\kappa^{[n]}}}(f) ,$$

which means that also the second term on the right hand side of the estimate in step 2. vanishes. ■

The preceding theorem helps to prove

4.6.2. Theorem. Let (L_i) be a sequence of distributions on $\mathcal{M}$ for which $\mathcal{W}_m^{(L_i)} := Q_m$ exists for all $m\in \mathbf{Z}$ and assume that

$$\Lambda_{L_i} \leq \lambda_i \qquad (i\in \mathbf{Z})$$

for some sequence $(\lambda_i)\in N_J^{entr}$ obeying the weak deconcentration property

$$\int (\kappa^{[n]}_{(a)}(\chi(X)>0))^2 \lambda_{m-n}(da) \xrightarrow[n\to\infty]{} 0 \qquad (m\in \mathbf{Z} ; X\in \mathcal{B}) .$$

Then there holds

$$(L_{j-n})_{\kappa^{[n]}} \underset{n\to\infty}{\Rightarrow} V_j^{\Omega} \qquad (j\in \mathbf{Z}) .$$

Proof. For all $j \in \mathbf{Z}$ and $f \in F_b$ we obtain by means of Theorem 4.6.1.

$$\limsup_{n\to\infty} |\mathcal{L}_{(L_{j-n})_{\kappa^{[n]}}}(f) - \mathcal{L}_{V_j\mathcal{Q}}(f)|$$

$$\leq \limsup_{n\to\infty} |\mathcal{L}_{(L_{j-n})_{\kappa^{[n]}}}(f) - \mathcal{L}_{(\mathfrak{I}(L_{j-n}))_{\kappa^{[n]}}}(f)| + 0$$

$$\leq \limsup_{n\to\infty} \int |\mathcal{L}_{\kappa^{[n]}_{(\Phi)}}(f) - \mathcal{L}_{(\Pi_\Phi)_{\kappa^{[n]}}}(f)| \, L_{j-n}(d\Phi)$$

$$= \limsup_{n\to\infty} \int \Big| \prod_{a\in \operatorname{supp}\Phi} (\mathcal{L}_{\kappa^{[n]}_{(a)}}(f))^{\Phi(\{a\})} - \prod_{a\in \operatorname{supp}\Phi} \big(\exp(-(1-\mathcal{L}_{\kappa^{[n]}_{(a)}}(f)))\big)^{\Phi(\{a\})}\Big| \, L_{j-n}(d\Phi)$$

$$\leq \limsup_{n\to\infty} \iint |\mathcal{L}_{\kappa^{[n]}_{(a)}}(f) - \exp(-(1-\mathcal{L}_{\kappa^{[n]}_{(a)}}(f)))| \, \Phi(da) \, L_{j-n}(d\Phi)$$

$$\leq \limsup_{n\to\infty} \iint (1-\mathcal{L}_{\kappa^{[n]}_{(a)}}(f))^2 \, \Phi(da) \, L_{j-n}(d\Phi)$$

$$\leq \limsup_{n\to\infty} \int (1-\mathcal{L}_{\kappa^{[n]}_{(a)}}(f))^2 \, \lambda_{j-n}(da)$$

$$\leq \limsup_{n\to\infty} \int (\kappa^{[n]}_{(a)}(\chi(\operatorname{supp} f > 0)))^2 \, \lambda_{j-n}(da) \xrightarrow[n\to\infty]{} 0 \ . \ \blacksquare$$

4.7. The Stationary Case: Characteristic J-invariant Distributions

Specializing the definition in Section 4.3., we arrive at the following: If a distribution L on $\mathcal{N}$ satisfies the conditions

$$L(\gamma * J^{[n]} \in N) = 1 \ , \quad n \in \mathbf{Z}_+ \quad , \qquad \text{and} \qquad L(\gamma * J^{[n]} \in (.)) \Rightarrow Q$$

for some distribution Q on $\mathcal{N}$, then we put

$$\mathcal{W}_L := Q \ .$$

Obviously there holds $\mathcal{W}_L = L$ for all J-invariant distributions L.

An immediate consequence of Proposition 4.3.2. and Theorem 4.6.1. is the following result on convergence towards Cox type equilibrium:

4.7.1. Theorem. Let L be a distribution on $\mathcal{N}$ with N_J-bounded intensity measure, and assume that $\mathcal{W}_L$ exists. Then $\mathcal{W}_L$ is a J-invariant distribution, and there holds

$$(\mathfrak{I}(L))_{\kappa^{[n]}} \Rightarrow V_{(\mathcal{W}_L)} \ .$$

In the sequel, the stationary κ-process corresponding to an equilibrium distribution P will always be denoted by $\underline{P}$.

Combining Propositions 4.2.4. and 4.3.1., one obtains

4.7.2. Proposition. For any $P \in \mathbf{P}^{inv}$, the distribution $\mathcal{W}_L$ exists iff $\underline{P}$ has a first order representation. In this case there hold

a) $\mathcal{W}_P = \underline{P}(S_n^{(\Phi_i)} \in (.)) \qquad (n \in \mathbf{Z}_+)$.

b) $\Lambda_{(\mathcal{W}_L)} = \Lambda_P$.

c) $\mathcal{W}_L \in \mathbf{V}^{inv}$.

A fortiori one gets

4.7.3. $\mathcal{W}_P$ exists for all equilibrium distributions P of first order.

Theorem 4.2.7. together with Proposition 4.7.2. yields immediately:

4.7.4. Theorem. The mapping $P \rightarrow \mathcal{W}_P$ from the set $\{P: P \in \mathbf{P}^{inv}, \underline{P}$ has a first order representation$\}$ into the set $\mathbf{V}^{inv}$ is 1-1.

This justifies to call $\mathcal{W}_P$ the **characteristic J-invariant distribution** of an equilibrium distribution P.

In view of 4.5.1. one obtains

4.7.5. For all stable J-invariant distributions Q, the distribution $\mathcal{W}_{V_{(Q)}}$ exists and equals $\mathcal{W}_{\mathfrak{Z}(Q)}$. (In other words: For any equilibrium distribution P of Cox type, $\mathcal{W}_P$ exists and equals $\mathcal{W}_{\mathfrak{Z}(Q)}$, where Q denotes the intensity distribution of P.)

Combining Theorems 4.3.3. and 4.3.7. we arrive at

4.7.6. Theorem. Let L be a distribution on $\mathcal{M}$ with N_κ^{st}-bounded intensity measure and P be an equilibrium distribution of first order. Then there holds

$$L_{\kappa^{[n]}} \Rightarrow P$$

iff $\mathcal{W}_L$ exists and equals $\mathcal{W}_P$.

We are now going to adapt the structural results of Section 4.2. to the stationary case:

4.7.7. Proposition. Let H and P be equilibrium distributions such that $\mathcal{W}_H$ exists, and assume that P<<H. Then also $\mathcal{W}_P$ exists and obeys $\mathcal{W}_P << \mathcal{W}_H$.

Proof. The absolute continuity P<<H implies $\underline{P} << \underline{H}$, where $(dP/dH)(\Phi_0)$ represents a version of $(dP/dH)((\Phi_i))$. Now Proposition 4.2.9. together with Proposition 4.7.2 implies our assertion. ■

4.7.8. Proposition. Let H be an equilibrium distribution such that $\mathcal{W}_H$ exists. Further, let Q be a distribution in $\mathbf{V}^{inv}$ which is absolutely continuous with respect to $\mathcal{W}_H$. Then there exists an equilibrium distribution P which obeys P<<H and whose characteristic J-invariant distribution $\mathcal{W}_P$ equals Q.

Proof. Let $\underline{Q}$ be the stationary J-process corresponding to the J-invariant distribution Q. The absolute continuity $Q << \mathcal{W}_P$ implies $\underline{Q} << \underline{\mathcal{W}}^H$, where $(dQ/d\mathcal{W}_H)(\gamma_0)$ represents a version of $(d\underline{Q}/d\underline{\mathcal{W}}^H)((\gamma_i))$. Now Proposition 4.2.9. ensures the existence of a κ-process

$\underline{P}$ which is absolutely continuous with respect to $\underline{H}$ and whose characteristic J-process $\mathcal{W}^{\underline{P}}$ equals $\underline{Q}$. By stationarity of $\underline{Q} = \mathcal{W}^{\underline{P}}$ and the inversion formula 4.2.7., $\underline{P}$ is even a stationary κ-process. Thus, the equilibrium distribution $P := \underline{P}(\Phi_0 \in (.))$ has all desired properties.■

As a "stationary pendant" to Proposition 4.2.10. one obtains

4.7.9. Proposition. Let P be an equilibrium distribution such that $\mathcal{W}_P$ exists, and assume that

$$\mathcal{W}_P = cQ_1 + (1-c)Q_2$$

for some $c \in (0,1)$ and some $Q_1, Q_2 \in V^{inv}$. Then there exist $P_1, P_2 \in P^{inv}$ with the properties

$$\mathcal{W}_{P_i} = Q_i \qquad (i=1,2) \quad ,$$
$$P = cP_1 + (1-c) P_2 \quad .$$

We turn to now the investigation of "second order quantities": A first result in this direction is

4.7.10. For all equilibrium distributions P of second order also $\mathcal{W}_P$ is of second order, and there holds for all $X_1, X_2 \in \mathcal{B}$

$$\int J^{[n]}(x, X_1) J^{[n]}(y, X_2) \Lambda^{(2)}{}_P(d[x,y]) \xrightarrow[n\to\infty]{} \Lambda^{(2)}{}_{\mathcal{W}_P}(X_1 \times X_2) \quad .$$

Proof. 1. Because of the estimate

$$Z_n := E_{\underline{P}}(\Phi_0(X_1)|\mathcal{F}_{-n})\, E_{\underline{P}}(\Phi_0(X_2)|\mathcal{F}_{-n})$$
$$\leq (E_{\underline{P}}(\Phi_0(X_1 \cup X_2)|\mathcal{F}_{-n}))^2 \leq E_{\underline{P}}((\Phi_0(X_1 \cup X_2))^2|\mathcal{F}_{-n}) \quad ,$$

the sequence of random variables Z_n, $n \geq 1$, is uniformly integrable under $\underline{P}$ and hence converges in $L_1(\underline{P})$ towards

$$Z_\infty := E_{\underline{P}}(\Phi_0(X_1)|\mathcal{F}_{-\infty}) E_{\underline{P}}(\Phi_0(X_2)|\mathcal{F}_{-\infty})$$

2. Due to 1. there follows

$$\int J^{[n]}(x, X_1) J^{[n]}(y, X_2)\, \Lambda^{(2)}{}_P(d[x,y])$$

$$= \iint J^{[n]}(x, X_1) J^{[n]}(y, X_2) (\Phi \otimes \Phi)(d[x,y])\, P(d\Phi)$$

$$= \int (\Phi * J^{[n]})(X_1)(\Phi * J^{[n]})(X_2)\, P(d\Phi) = E_{\underline{P}}(Z_n)$$

$$\longrightarrow E_{\underline{P}}(Z_\infty) = \Lambda^{(2)}{}_{\mathcal{W}_P}(X_1 \times X_2) < +\infty \quad .■$$

An immediate consequence of 1.2.18. and 1.9.5. is

4.7.11. For all equilibrium distributions P of first order there holds

$$\Gamma^{(2)}{}_P + \int J^{[n]}(a,(.)) \otimes J^{[n]}(a,(.)) \Lambda_P(da)$$
$$= \iint \sum_{0 \leq i < n} J^{[i]}(x,(.)) \otimes J^{[i]}(y,(.)) \Gamma^{(2)}{}_{\kappa(z)}(d[x,y])\, \Lambda_P(dz)$$
$$+ \int J^{[n]}(x,(.)) \otimes J^{[n]}(y,(.))\, \Lambda^{(2)}{}_P(d[x,y]) \quad .$$

Proposition 2.5.1. provides a formula for the second factorial moment measure $\Gamma^{(2)}{}_P$ of a Poisson type equilibrium distribution P. Combining this formula with the special choice of $P := \Pi^{\kappa}{}_{\nu}$ in 4.7.11. and 4.7.10., we arrive immediately at the following result on the "covariance measure" of the distribution $\mathcal{W}_{\Pi^{\kappa}{}_{\nu}}$:

4.7.12. Proposition. For all $\nu \in N_{\kappa}{}^{st(2)}$ there holds

$$\int J^{[n]}(a, X_1) J^{[n]}(a, X_2)\nu(da) \longrightarrow \Lambda^{(2)}{}_{\mathcal{W}_{\Pi^{\kappa}{}_{\nu}}}(X_1 \times X_2) - \nu(X_1)\nu(X_2) \qquad (X_1, X_2 \in \mathcal{B}).$$

4.7.10. and Theorem 2.5.4. tell that for all stable measures ν of second order the distributions $\mathcal{W}_P$, $\Pi^{\kappa}{}_{\nu}$ and $\mathcal{W}_{\Pi^{\kappa}{}_{\nu}}$ are of second order. Passage to the limit $n \to \infty$ in 4.7.11. yields in view of Proposition 2.5.1., 4.7.10. and 4.7.12.

4.7.13. Theorem. Let P be an equilibrium distribution of second order and put $\nu := \Lambda_P$. Then there holds

$$\Gamma^{(2)}{}_P + \Lambda^{(2)}{}_{\mathcal{W}_{\Pi^{\kappa}{}_{\nu}}} = \Gamma^{(2)}{}_{\Pi^{\kappa}{}_{\nu}} + \Lambda^{(2)}{}_{\mathcal{W}_P} \quad .$$

Now let Q be some J-invariant distribution, whose intensity measure ν is stable of second order. In virtue of Theorems 3.7.7. and 3.7.8., also $P := V_{(Q)}$ has intensity measure ν. Combining Theorem 3.7.12. and Proposition 2.5.1. we obtain

$$\Gamma^{(2)}{}_P = \Lambda^{(2)}{}_Q + \Gamma^{(2)}{}_{\Pi^{\kappa}{}_{\nu}} - \nu \otimes \nu \quad .$$

Thus also P is of second order, and from Theorem 4.7.13. one recognizes the validity of

4.7.14. Let Q be a J-invariant distribution whose intensity measure ν is stable of second order, and put $P := V_{(Q)}$. Then there holds

$$\Lambda^{(2)}{}_{\mathcal{W}_P} + \nu \otimes \nu = \Lambda^{(2)}{}_Q + \Lambda^{(2)}{}_{\mathcal{W}_{\Pi^{\kappa}{}_{\nu}}} \quad .$$

Immediate consequences of Theorem 4.3.5. and Theorem 4.2.11., respectively, are

4.7.15. Theorem. For all $\nu \in N_{\kappa}{}^{st}$ there holds

$$\mathcal{W}_{\Pi_{\nu}} = \mathcal{W}_{\Pi^{\kappa}{}_{\nu}} \, .$$

4.7.16. Theorem. Let P be an equilibrium distribution such that $\mathcal{W}_P$ exists. Then P is an extremal κ-process iff P is of first order and $\mathcal{W}_P = \delta_{\Lambda_P}$.

Let us say that a stable measure ν is **regular** if the constant sequence (ν) is regular, i.e. if the stationary κ-process of Poisson type with intensity sequence (ν) is extremal. The set of all regular stable measures will be denoted by $N_{\kappa}{}^{reg}$. In virtue of Theorems 4.7.15. and 4.7.16. there holds

4.7.17. For all stable measures ν, the following are equivalent

(a) ν is regular.
(b) $\mathcal{W}_{\Pi^{\kappa}{}_{\nu}} = \delta_{\nu}$.
(c) $\mathcal{W}_{\Pi_{\nu}} = \delta_{\nu}$.

A measure $\nu \in N_J$ is called **regular** if the constant sequence (ν) is regular, i. e. if $\mathcal{W}_{\Pi_\nu}$ exists and equals δ_ν. In view of 4.7.14., this is a straightforward extension of the definition of regularity given above. The set of regular J-invariant measure is denoted by N_J^{reg}. Proposition 4.4.3. boils down to

4.7.18. Proposition. A J-invariant measure ν is regular iff it has the following **deconcentration property** :

$$\int J^{[n]}(a,X)\min(1, J^{[n]}(a,X))\ \nu(da) \underset{n\to\infty}{\longrightarrow} 0 \qquad (X\in\mathcal{B}) .$$

The "stationary version" of Theorem 4.4.5. is

4.7.19. Theorem. For any stable measure ν the following assertions are equivalent:

a) ν is regular.

b) $\int(\kappa^{[n]}{}_{(a)}(\chi(X)>0))^2\nu(da) \longrightarrow 0 \qquad (X\in\mathcal{B})$.

c) $\nu(\{a{:}a\in A, J^{[n]}(a,X)>\varepsilon\}) \underset{n\to\infty}{\longrightarrow} 0 \qquad (X\in\mathcal{B}\,;\,\varepsilon>0)$.

d) $\int |(\Phi * J^{[n]})(X) - \nu(X)|\ \Pi^\kappa{}_\nu(d\Phi) \longrightarrow 0 \qquad (X\in\mathcal{B})$.

For measures which are stable of second order, the preceding theorem is supplemented by

4.7.20. Proposition. A measure $\nu\in N_\kappa^{st(2)}$ is regular iff there holds

(*) $$\int(J^{[n]}(a,X))^2\ \nu(da) \longrightarrow 0 \qquad (X\in\mathcal{B}) .$$

Proof. Due to Proposition 4.7.12. there holds

$$\int(J^{[n]}(a,X))^2\nu(da) \longrightarrow \int(\gamma(X) - \nu(X))^2\mathcal{W}_{\Pi^\kappa{}_\nu}(d\gamma) .$$

Hence (*) is valid iff $\mathcal{W}_{\Pi^\kappa{}_\nu} = \delta_\nu$, i. e. iff ν is regular . ■

As the following example shows, not every regular stable measure has the deconcentration property (*) of Proposition 4.7.20.:

4.7.21. Example. Put $A := \mathbf{R}\cup\{\infty\}$, $\rho_A(x,y) := |x-y|$, $x,y\in\mathbf{R}$; $\rho_A(x,\infty) := (x^2+1)^{1/2}$, $x\in\mathbf{R}$. Define

$$p := 2^{-2}(\delta_{-1}+2\delta_0+\delta_1),$$

and consider the mapping f given by

$$f(a) := 0 \quad (a\in A\setminus(0,1)) \quad ; \quad f(a) := 2a^{-1/2} \quad (a\in(0,1)) .$$

Now let us put

$$\kappa_{(x)} := \int\delta_{\delta_{x+y}+m\delta_\infty}(.)(p\otimes\pi_{f(x)})(d[y,m]) \quad \text{for } x\in\mathbf{R},$$

$$\kappa_{(\infty)} := \delta_o .$$

The intensity kernel J then is given by

$$J(\infty,(.)) = o \ ; \ J(a,\{\infty\}) = f(a) \quad (a\in\mathbf{R});$$

$$J(a,X) = p(X-a) \qquad (a\in\mathbf{R}, X\in\mathfrak{A}\cap\mathbf{R}) .$$

Denote Lebesgue measure on B by μ. Obviously the measure $\sigma := \mu+\delta_\infty$ belongs to N_κ^{st}, and $\Pi^\kappa{}_\sigma$ equals $(\Pi_\mu)_\kappa$.

For all natural numbers n and all $a \in A$ there holds

$$J^{[n]}(a,\{\infty\}) \geq 2^{-(n-1)}f(a)$$

and hence

$$\int(J^{[n]}(a,\{\infty\}))^2\,\sigma(da) \;\geq\; 2^{-2(n-1)}\int(f(a))^2\mu(da) = +\infty \quad .$$

On the other hand one has for any bounded subset $X \in B$

$$\int(\kappa^{[n]}_{(a)}(\chi(X)>0))^2\,\sigma(da) = \int(p^n(X-a))^2\mu(da) \xrightarrow[n\to\infty]{} 0$$

and also

$$\int(\kappa^{[n]}_{(a)}(\chi(\{\infty\})>0))^2\,\sigma(da) \leq \int(\kappa^{[n-1]}_{(a)}(\chi(\{0,1\})>0))^2\sigma(da) \xrightarrow[n\to\infty]{} 0 \quad ,$$

which together shows, in virtue of Theorem 4.7.19., that σ is regular. ■

Next we give an interesting example of a non-regular stable measure:

4.7.22. Example. Put $A := \mathbf{Z}$, $\rho_A(i,j) := |i-j|$ for $i, j \in \mathbf{Z}$, and $\kappa_{(i)} := \delta_{\delta_{i+1}}$ for all $i \in \mathbf{Z}$. Then

$$\mu := \sum_{i \in \mathbf{Z}} \delta_i$$

belongs to N_κ^{st}, and there holds

$$P := \Pi^\kappa_\mu = \Pi_\mu \quad ; \quad \mathcal{W}_{\Pi^\kappa_\nu} = \Pi_\mu \quad .$$

This shows, in view of 4.7.17., that μ is not regular.

The system $(M^{\mathbf{Z}}, \mathcal{F}, \underline{P}, \theta)$ obivously has the same ergodic properties as the system $(M, \mathcal{M}, P, \varphi)$,where φ is the automorphism on $[M, \mathcal{M}]$ induced by shift $i \to i+1$, $i \in \mathbf{Z}$; hence the distribution $\underline{P}$ is "mixing of all orders" (cf. [Fleischmann (1981), Prop. 2.3.]). ■

An immediate consequence of Proposition 4.1.14. is

4.7.23. Theorem. If the clustering field κ obeys the condition

$$\kappa_{(a)}(\chi \text{ is finite}) = 1 \qquad (a \in A)$$

then any Poisson type equilibrium distribution with regular intensity measure is ${}_\infty\mathcal{M}$-trivial.

The following example shows that the condition imposed on κ in the preceding theorem cannot be cancelled:

4.7.24. Example. Put $A := \mathbf{Z}$, $A' := \mathbf{Z}\times\{0,1\}$, and let ρ_A and $\rho_{A'}$ be the usual (Euclidean) metrics on A and A′, respectively. Let κ be a clustering field on $[A, \rho_A]$. Writing

$$\mu := \sum_{i \in \mathbf{Z}} \delta_{[i,1]} \quad ,$$

we define a clustering field κ' on $[A', \rho_{A'}]$ by

$$\kappa'_{([j,0])} := \kappa_{(j)}(\chi\otimes\delta_0 \in (.)) \qquad (j \neq 0)$$

$$\kappa'_{([0,0])} := \kappa_{(0)}((\chi\otimes\delta_0 + \mu) \in (.))$$

$$\kappa'_{([j,1])} := \delta_o \qquad (j \in \mathbf{Z}) \quad .$$

Consider, for a measure $\nu \in N_\kappa^{st}$, the measure $\nu' := \nu \otimes \delta_0 + \nu(\{0\})\mu$. Then ν' belongs to $N_{\kappa'}^{st}$, and $\Pi^{\kappa'}_{\nu'} = (\Pi^\kappa_\nu)_\omega$, where ω is the clustering field on $[A, \rho_A]$ with phase space $[A', \rho_{A'}]$ defined by $\omega_{(j)} := \kappa'_{([j,0])}$, $j \in \mathbf{Z}$. The distribution $\Pi^{\kappa'}_{\nu'}$ is, however, not ${}_\infty\mathcal{M}$-trivial unless $\nu(\{0\}) = 0$. Indeed, consider the ${}_\infty\mathcal{M}$-measurable set

$$Y := \{\Phi : \Phi \in M', \Phi(\mathbf{Z} \times \{1\}) \text{ is infinite}\} \quad ,$$

for which there holds

$$\Pi^{\kappa'}_{\nu'}(Y) = \Pi^\kappa_\nu(\Psi(\{0\}) > 0) \quad .$$

This latter quantity is strictly between zero and one iff $\nu(\{0\}) > 0$. On the other hand, for any $\nu \in N_\kappa^{reg}$, the measure ν' obeys condition b) of Theorem 4.7.19. and hence is regular. ■

An immediate consequence of Theorem 4.5.8. is the following "local structural result":

4.7.25. Theorem. For all $\nu \in N_\kappa^{reg}$, the mapping $Q \rightarrow V_{(Q)}$ provides a 1-1 correspondence between the sets $\mathbf{V}^{inv,\nu}$ and $\mathbf{P}^{inv,\nu}$. Its inverse mapping is given by $P \rightarrow \mathcal{W}_P$.

As a counterpart to Theorem 4.5.9. there holds

4.7.26. Theorem. For all $\nu \in N_\kappa^{reg}$, the following assertions are equivalent:

a) ν is regular.

b) Any J-invariant distribution Q with intensity measure ν arises as characteristic J-invariant distribution of some equilibrium distribution $P \in \mathbf{P}^{inv,\nu}$.

c) For all $c \in \mathbf{R}_+$ there exists an equilibrium distribution $P_c \in \mathbf{P}^{inv,c\cdot\nu}$, whose corresponding κ-process $\underline{P}_c$ is extremal.

Proof. Assertions a), b) and c) are "stationary versions" of assertions a), b), c), respectively, in Theorem 4.5.9. Correspondingly, the proofs of the implications a) ⇒ b), b) ⇒ c) and c) ⇒ a) are "stationary versions" of steps 1., 3. and 2., respectively, in the proof of Theorem 4.5.9. Let us just note that the extremal κ-process $\underline{P}_c$ which is constructed in step 2. of this proof, and whose characteristic J-process $\underline{\mathcal{W}}^{\underline{P}_c}$ now equals $\delta_{(\nu)}$, is in fact a stationary κ-process due to the inversion formula 4.2.7. ■

4.7.27. Example. Let $[A,\rho_A]$ be the direct product of $\{-1,1\}$ and the real axis $\mathbf{R}$. Further, let μ denote Lebesgue measure on B and γ denote uniform distribution on [0,1]. Then

$$\kappa_{([i,x])} := \int \delta_{[-i,x+y]}(.)\, \gamma(dy) \qquad (x \in \mathbf{R};\ i \in \{-1,1\})$$

defines a stochastic clustering field κ. The measure

$$\nu := 2^{-1}(\delta_{-1} \otimes \mu + \delta_1 \otimes \mu)$$

belongs to N_κ^{reg}; the equilibrium distribution of Poisson type with intensity measure ν is just Π_ν. The distribution

$$P := 2^{-1}(\Pi_{\delta_{-1}\otimes\mu} + \Pi_{\delta_1\otimes\mu})$$

is another equilibrium distribution with intensity measure ν and characteristic J-invariant distribution

$$\mathcal{W}_P = 2^{-1}(\delta_{\delta_{-1}\otimes\mu} + \delta_{\delta_1\otimes\mu}) \quad . ■$$

The preceding example shows that the characteristic J-invariant distribution of an equilibrium distribution P need not be concentrated on N_J, even if the intensity measure of P is regular. Due to Proposition 3.8.2., however, the J-invariant distribution of an equilibrium distribution P *is* concentrated on N_J under the additional assumptions that Λ_P is J-aperiodic. Together with Theorem 4.7.25. this implies

4.7.28. Theorem. For all $\nu \in N_\kappa^{reg} \cap N_J^{ap}$, the mapping

$$Q \rightarrow \int \Pi^\kappa_\gamma(.)\, Q(d\gamma)$$

provides a 1-1 correspondence between the sets $\{Q: Q \in V, \Lambda_Q = \nu, Q(N_J) = 1\}$ and $P^{inv,\nu}$.

If one assumes in addition that ν is extremal, then one arrives, in virtue of 3.7.14. and Theorem 4.7.28., at the following counterpart to Theorem 3.8.9.

4.7.29. Theorem. For all $\nu \in N_\kappa^{reg} \cap N_J^{ap} \cap N_J^{ex}$ different from o, the mapping

$$\sigma \rightarrow \int \Pi^\kappa_{c\cdot\nu}(.)\, \sigma(dc)$$

provides a 1-1 correspondence between the set of all distributions on the nonnegative real axis with unit mean and the set $P^{inv,\nu}$. For $P := \int \Pi^\kappa_{c\cdot\nu}(.)\sigma(dc)$ and $X \in \mathcal{B}$ such that $\nu(X) > 0$ there holds

$$\sigma = \mathcal{W}_P((\nu(X)^{-1} \cdot \gamma(X) \in (.)).$$

Any measure $\nu \in N_J^{reg}$ has the **weak deconcentration property**

$$\int (\kappa^{[n]}_{(a)}(\chi(\{\infty\}) > 0))^2 \nu(da) \longrightarrow 0 \qquad (X \in \mathcal{B}).$$

(cf. the remark after Proposition 4.4.3.). Hence Theorem 4.6.2. yields immediately

4.7.30. Theorem. Let L be a distribution on $\mathcal{M}$ with N_J^{reg}-bounded intensity measure, and assume that $\mathcal{W}_L$ exists. Then there holds

$$L_{\kappa^{[n]}} \Rightarrow V_{(\mathcal{W}_L)} \,.$$

We conclude this section with a "global structural result", which is a direct consequence of Theorem 4.5.5.:

4.7.31. Theorem. If κ is globally deconcentrating, then the mapping $Q \rightarrow V_{(Q)}$ provides a 1-1 correspondence between the set of all stable J-invariant distributions Q and the set of all equilibrium distributions P for which $\mathcal{W}_P$ exists. Its inverse mapping is given by $P \rightarrow \mathcal{W}_P$.

5. ERGODIC INTENSITY DISTRIBUTIONS

This chapter shows essential parallels to the preceding one. For example, the role of the σ-algebra $\mathcal{F}_{-\infty}$ of infinite past is taken over by the σ-algebra $\mathcal{J}$ of shift invariant sets in $\mathcal{M}^{\otimes \mathbf{Z}}$, and instead of $\mathcal{F}_{-\infty}$-triviality of arbitrary κ–processes one now considers ergodicity of stationary κ-processes $\underline{P}$, $P \in P^{inv}$.

If two distributions L, Q on $\mathcal{N}$ obey the convergence relation

$$L((\gamma * n^{-1} \sum_{0 \leq i < n} J^{[n]}) \in (.)) \underset{n \to \infty}{\Rightarrow} Q \quad ,$$

then we put $\mathcal{V}_L := Q$. If, in addition, Λ_L has a majorant in N_J, then $\mathcal{V}_L$ is concentrated on the set of J-invariant measures (Proposition 5.2.11.). According to 5.2.1., $\mathcal{V}_P$ exists for all equilibrium distributions P of first oder, and (Theorem 5.3.1.) in this case one has

$$\mathcal{V}_P = \mathcal{V}_{\mathcal{W}_P} \quad .$$

Due to Proposition 5.2.8., an equilibrium distribution P admits the disitribution $\mathcal{V}_P$ iff the conditional distribution $\underline{P}((.)|\mathcal{J})$ is of first order $\underline{P}$-almost surely, and in this case $\mathcal{V}_P$ coincides with the distribution of $\Lambda_{\underline{P}(\Phi_0 \in (.)|\mathcal{J})}$ under $\underline{P}$. The distribution $\mathcal{V}_P$ is then called the ergodic intensity distribution of P.

If, for an equilibrium distribution P of first order, $\underline{P}$ is ergodic, then one has by 5.2.4.

$$\mathcal{V}_P = \delta_{\Lambda_P} \quad ,$$

whereby the converse holds not in general, but under the additional assumption of infinite divisibility of P (Prop. 5.2.6.).

As a counterpart to the deconcentration property of J-invariant measures considered in the preceding chapter we now introduce the condition

$$\int (n^{-1} \sum_{0 \leq i < n} J^{[i]}(a,X)) \min(1, n^{-1} \sum_{0 \leq i < n} J^{[n]}(a,X)) \, \nu(da) \underset{n \to \infty}{\longrightarrow} 0 \qquad (X \in \mathcal{B}) \quad .$$

This is (Prop. 5.7.4.) equivalent to

$$\mathcal{V}_{\Pi_\nu} = \delta_\nu$$

and can be interpreted as a kind of "mobility condition" on the individuals of a random population with intensity measure ν.

The borders of parallelism to the preceding chapter arise especially in the fact that (cf. Example 5.9.3.) for a stable measure ν obeying $\mathcal{V}_{\Pi_\nu} = \delta_\nu$ there need not hold $\mathcal{V}_{\Pi^\kappa_\nu} = \delta_\nu$ (which is equivalent to ergodicity of the corresponding Poisson type κ-process $\underline{\Pi}^\kappa{}_\nu$).

This implication, however, does hold under the additional assumption that the intensity kernel J is substochastic (Theorem 5.9.2.).

5.1. Extremal Equilibrium Distributions

Let us recall that the mapping $P \rightarrow \underline{P}$, which assigns to each equilibrium distribution P its corresponding stationary κ-process $\underline{P}$, provides a 1-1 correspondence between the set P^{inv} and the set of all stationary κ-processes. Moreover, the mapping $P \rightarrow \underline{P}$ preserves convex combinations as well as absolute continuity.

On the set $M^{\mathbf{Z}}$, the **σ-algebra $\mathcal{J}$ of shift invariant events** is defined by

$$\mathcal{J} := \{Y : Y \in \mathcal{F}, \theta(Y) = Y\} .$$

A stationary distribution $\underline{H}$ on $\mathcal{F}$ is called **ergodic** if $\underline{H}$ takes only values zero and one on $\mathcal{J}$.

The following three lemmata are well known from the general theory of stationary distributions and Markov chains, respectively. We give their proofs only for the sake of completeness.

5.1.1. For any stationary distribution $\underline{H}$ on $\mathcal{F}$ there holds

$$\mathcal{J} \subseteq \mathcal{F}_{-\infty} \qquad \underline{H}\text{-almost surely}$$

(i. e. for any $Y \in \mathcal{J}$ there exists some $Y' \in \mathcal{F}_{-\infty}$ such that $\underline{H}((Y \setminus Y') \cup (Y' \setminus Y)) = 0$) .

Proof. 1. Let C be a cylinder set in $\mathcal{F}$, i.e. C is of the form

$$C = \{\Phi_{-m} \in B_{-m}, \ldots, \Phi_m \in B_m\}, \quad m \in \mathbf{Z}_+ \; ; \quad B_{-m}, \ldots B_m \in \mathcal{M} .$$

Birkhoff's ergodic theorem implies that

$$n^{-1} \sum_{1 \leq j \leq n} 1_C \circ \theta^j \xrightarrow[n \to \infty]{} E_{\underline{H}}(1_C | \mathcal{J}) \quad \underline{H}\text{-a.s.}$$

But for any $i \in \mathbf{Z}_+$ such that $j \geq m+i$, the mapping $1_C \circ \theta^j$ is $\mathcal{F}_{-i}$ –measurable. Hence follows

(*) $\qquad E_{\underline{H}}(1_C | \mathcal{J})$ is $\mathcal{F}_{-\infty}$ -measurable $\underline{H}$–a.s.

2. The set of all $C \in \mathcal{F}$ with property (*) is a monotone class which contains the algebra of cylinder sets; hence (*) holds for all $C \in \mathcal{F}$.

3. Now take $Y \in \mathcal{J}$, and let f denote an $\mathcal{F}_{-\infty}$-measurable version of $E_{\underline{H}}(Y | \mathcal{J})$ (such a version exists due to 2.). From

$$1_Y = E_{\underline{H}}(1_Y | \mathcal{J}) = f \qquad \underline{H}\text{–a.s.}$$

one concludes that $\underline{H}((\{f>0\} \setminus Y) \cup (Y \setminus \{f>0\})) = 0$. ■

5.1.2. Let $\underline{P}$ and $\underline{H}$ be stationary distributions on $\mathcal{F}$; assume that $\underline{P} \ll \underline{H}$ and that $\underline{H}$ is a κ-process. Then $\underline{P}$ is a κ-process as well.

Proof. By stationarity of $\underline{P}$ and $\underline{H}$, the Radon-Nikodym derivative $f := d\underline{P}/d\underline{H}$ is $\mathcal{J}$-measurable $\underline{H}$-a.s., and hence, by 5.1.1., also $\mathcal{F}_{-\infty}$-measurable $\underline{H}$-a.s. This implies, for all $n \in \mathbf{Z}$, $Y \in \mathcal{F}_n$, $B \in \mathcal{M}$, the equality

$$\begin{aligned} \underline{P}(\Phi_{n+1} \in B \, ; Y) &= E_{\underline{H}}(f \cdot 1_Y \, ; \Phi_{n+1} \in B) \\ &= E_{\underline{H}}(f \cdot 1_Y \cdot \kappa_{(\Phi_n)}(B)) = E_{\underline{P}}(1_Y \cdot \kappa_{(\Phi_n)}(B)), \end{aligned}$$

which shows that $\underline{P}$ is a κ–process. ■

The set of all those equilibrium distributions which are **extremal** in the convex set P^{inv} will be denoted by P^{ex} .

5.1.3. An equilibrium distribution **P** belongs to P^{ex} iff $\underline{P}$ is ergodic.

Proof 1. Assume that $P \notin P^{ex}$, i. e.

$$P = cP_1+(1-c)P_2$$

for two equilibrium distributions P_1, P_2 different from P, and some $c \in (0,1)$. Then there holds

$$\underline{P} = c\underline{P}_1+(1-c)\underline{P}_2 ,$$

and the Radon-Nikodym derivative $d\underline{P}_1/d\underline{P}$ is $\mathcal{J}$-measurable but not constant ($\underline{P}$-a.s.), which shows that $\underline{P}$ is not ergodic.

2. Conversely assume that $\underline{P}$ is not ergodic. Then there exists a set $Y \in \mathcal{J}$ with the property

$$0 < \underline{P}(Y) < 1 ,$$

and $\underline{P}$ is a convex combination of the two stationary distributions $\underline{P}_1 := \underline{P}(.|Y)$, $\underline{P}_2 := \underline{P}(.|M^Z \setminus Y)$. According to 5.1.2., $\underline{P}_1$ and $\underline{P}_2$ are also κ-processes. Being a convex combination of the corresponding equilibrium distributions, P does not belong to P^{ex}. ■

In view of 5.1.3. the following proposition says that a stationary κ-process is ergodic iff its associated family process has this property.

5.1.4. Proposition. An equilibrium distribution P is extremal iff $\mathcal{G}_P$ is extremal in the convex set $\{\mathcal{G}: \mathcal{G} \in \mathbb{P}, \mathcal{G}_\kappa = \mathcal{G}\}$.

Proof. 1. Assume that P is extremal and that $\mathcal{G}_P$ is a convex combination

$$\mathcal{G}_P = c\ \mathcal{G}_1 + (1-c)\ \mathcal{G}_2$$

where $0<c<1$ and $\mathcal{G}_1$, $\mathcal{G}_2$ are distributions on $\mathcal{M}$ with the property $(\mathcal{G}_i)_\kappa = \mathcal{G}_i$, i=1,2. Theorem 2.7.38. now allows to conclude that there exist $P_1, P_2 \in P^{inv}$ such that

$$\mathcal{G}_i = \mathcal{G}_{P_i} \quad (i=1,2)$$

and therefore also

$$P = c\ P_1+(1-c)\ P_2 \ .$$

Extremality of P guarantees that $P_1 = P_2$ and hence $\mathcal{G}_1 = \mathcal{G}_2$, which shows that $\mathcal{G}_P$ is extremal.

2. Conversely, $\mathcal{G}_P$ cannot be extremal unless P has this property, because any convex combination

$$P = c\ P_1+(1-c)\ P_2 \qquad (0<c<1;\ P_1, P_2 \in P^{inv})$$

carries over to a convex combination

$$\mathcal{G}_P = c\ \mathcal{G}_{P_1}+(1-c)\ \mathcal{G}_{P_2} \ . \ ■$$

Before we turn to a characterization of the infinitely divisible extremal equilibrium distributions, let us recall that a stationary distribution $\underline{H}$ on $\mathcal{F}$ is called **weakly mixing** (with respect to time shift) if for all $Y_1, Y_2 \in \mathcal{F}$ there holds

$$\lim_{n\to\infty} n^{-1} \sum_{1\le j\le n} |\underline{H}(Y_1 \cap \theta^j(Y_2)) - \underline{H}(Y_1)\underline{H}(Y_2)| = 0 \ .$$

Clearly, this is in general a stronger property than ergodicity.

5.1.5. Example. Let $[A, \rho_A]$, κ and P be as in Example 4.7.27. Then $\underline{P}$ is ergodic but not weakly mixing.

5.1.6. Proposition. For all $P \in \mathbf{P}^{inv} \cap T$ the following assertions are equivalent:

a) $P \in \mathbf{P}^{ex}$.

b) $\underline{P}$ is weakly mixing.

c) There exists no factorization $P = P_1 * \mathcal{E}_{c \cdot P_2}$, where $P_1, P_2 \in \mathbf{P}^{inv}$; $P_2(\{o\}) = 0$; $c>0$.

Proof. 1. Let $[\underline{A}, \rho_{\underline{A}}]$ denote the direct product of $[A, \rho_A]$ and $[\mathbf{Z}, \rho_{\mathbf{Z}}]$, where $\rho_{\mathbf{Z}}(i,j) := |i-j|$. The mapping

$$(\Phi_i)_{i \in \mathbf{Z}} \rightarrow \chi := \sum_{i \in \mathbf{Z}} \Phi_i \otimes \delta_i$$

establishes a 1-1 correspondence, which is measurable in both directions, between the space $[M^{\mathbf{Z}}, \mathcal{F}]$ and the space $[\underline{M}, \underline{\mathcal{M}}]$ of counting measures on the phase space $[\underline{A}, \rho_{\underline{A}}]$. Hereby the time shift θ on $M^{\mathbf{Z}}$ is carried into the automorphism

$$T(\sum_{i \in \mathbf{Z}} \Phi_i \otimes \delta_i) := \sum_{i \in \mathbf{Z}} \Phi_{i+1} \otimes \delta_i$$

on $[\underline{M}, \underline{\mathcal{M}}]$. For any $P \in \mathbf{P}^{inv}$, the image of $\underline{P}$ under the mapping $(\Phi_i) \rightarrow \chi$ will be denoted by $\mathcal{P}$; in this way there arises a 1-1 mapping $P \rightarrow \mathcal{P}$ from $\mathbf{P}^{inv}$ into the set of T-invariant distributions on $\underline{\mathcal{M}}$. This mapping preserves convolution due to 3.1.9.; together with 2.1.3. this yields that, for all $P \in \mathbf{P}^{inv} \cap T$, the distribution $\mathcal{P}$ is infinitely divisible.

2. In view of 1., Theorem 1.4. in [Fleischmann (1981)] allows to conclude that b) is equivalent to T-ergodicity of $\mathcal{P}$, i. e. to a).

3. In virtue of Proposition 1.6. in [Fleischmann (1981)], $\mathcal{P}$ is T-ergodic iff there exists no representation $\mathcal{P} = H_1 * \mathcal{E}_{rH_2}$ where H_1, H_2 are T-invariant distributions on $\underline{\mathcal{M}}$ obeying $H_2(\{o\}) = 0$, and r is some positive real number.

4. Assume that

$$P = P_1 * \mathcal{E}_{cP_2} ,$$

where $P_1, P_2 \in \mathbf{P}^{inv}$; $P_2\{o\}) = 0$; $c>0$.
According to 1. there follows

$$\mathcal{P} = \mathcal{P}_1 * \mathcal{E}_{c\mathcal{P}_2} ,$$

and 3. yields that $P \notin \mathbf{P}^{ex}$. Hence c) is a consequence of a) .

5. Conversely, let us assume that $\mathcal{P}$ is not T-ergodic, and let

$$\mathcal{P} = H_1 * \mathcal{E}_{rH_2}$$

be a factorisation of $\mathcal{P}$ with the properties stated in 3.

This factorisation may be rewritten as

$$\mathcal{P} = e^{-r} H_1 + (1 - e^{-r}) H_3 ,$$

where

$$H_3 := (1 - e^{-r})^{-1} e^{-r} \sum_{n>0} (n!)^{-1} r^n H_1 * (H_2)^n .$$

The absolute continuity $H_1 << \mathcal{P}$ now implies, by 5.1.2., the existence of a distribution

$P_1 \in \mathbf{P}^{inv}$ such that $H_1 = \underline{P}_1$. The distribution $L := H_2(\Phi_0 \in (.))$ is different from δ_o, because H_2 is T-invariant and has no mass in $\{o\}$. Thus we can define

$$P_2 := L((.)|\Phi \neq o)$$

Further putting

$$c := rL(\Phi \neq o)$$

one obtains

$$P = P_1 * \mathcal{E}_{cP_2} .$$

Together with P and P_1 also $\mathcal{E}_{cP_2}$ is an equilibrium distribution in virtue of 2.1.1. Due to 2.1.2., also P_2 belongs to $\mathbf{P}^{inv}$. Hence c) implies a). ■

5.1.7. Example. In the setting of Example 4.7.22. , it was stated that the κ-process $\underline{P}$ corresponding to the equilibrium distribution $P := \Pi_\mu$ is "mixing of all orders". Hence P belongs to $\mathbf{P}^{ex}$ by Proposition 5.1.6. Recall, however, that $\underline{P}$ is not extremal in the convex set of κ-processes. ■

From the general theory of ergodic decompositions (see e.g. Dynkin (1978), Kerstan/Wakolbinger (1981)) there results:

5.1.8. There exists a kernel Γ from $[M, \mathcal{M}]$ to $[M, \mathcal{M}]$ with the following properties:

a) $\{\Gamma(\Phi,(.))\}_{\Phi \in M} = \mathbf{P}^{ex}$.

b) For all $P \in \mathbf{P}^{ex}$ there holds
$$\Gamma(\Phi,(.)) = P \qquad \text{P-almost all } \Phi \quad .$$

c) For all $P \in \mathbf{P}^{inv}$ there holds
$$P(.) = \int \Gamma(\Phi,(.))\, P(d\Phi) \quad .$$

In the following we fix a kernel Γ with the just stated properties. For all $\Phi \in M$, we write $\underline{\Gamma}(\Phi,(.))$ for the stationary κ-process corresponding to the equilibrium distribution $\Gamma(\Phi,(.))$.

5.1.9. For all $P \in \mathbf{P}^{inv}$ and $m \in \mathbf{Z}$, the mapping

$$(\Phi_i) \rightarrow \underline{\Gamma}(\Phi_m, (.))$$

represents a version of $\underline{P}(.|\mathcal{J})$.

Proof. 1. Lemma 5.1.8. b) implies, for all $P \in \mathbf{P}^{ex}$ and $m \in \mathbf{Z}$

$$\Gamma(\Phi_m, (.)) = P = \Gamma(\Phi_0, (.)) \qquad \underline{P}\text{-almost all } (\Phi_i) \quad .$$

Due to 5.1.8. a) and c) there follows, for all $P \in \mathbf{P}^{inv}$ and $m \in \mathbf{Z}$

$$\underline{\Gamma}(\Phi_m, (.)) = \Gamma(\Phi_0, (.)) \qquad \underline{P}\text{-almost all } (\Phi_i) \quad .$$

Hence, for all $P \in \mathbf{P}^{inv}$ and $m \in \mathbf{Z}$, the mappings

$$(\Phi_i) \rightarrow \Gamma(\Phi_m, B) \qquad (B \in \mathcal{M})$$

are $\mathcal{J}$-measurable $\underline{P}$-almost surely, and so are the mappings

$$(\Phi_i) \rightarrow \underline{\Gamma}(\Phi_m, Y) \qquad (Y \in \mathcal{F}) \quad .$$

2. For all $P \in \mathbf{P}^{inv}$, $m \in \mathbf{Z}$, $Y_1 \in \mathcal{F}$, $Y_2 \in \mathcal{J}$ there holds in virtue of 5.1.3. and 5.1.8.

$$\int_{Y_2} \Gamma(\Phi_m, Y_1) \underline{P}(d(\Phi_i)) = \int\int_{Y_2} \Gamma(\Phi_m, Y_1) \Gamma(\Psi, d(\Phi_i)) P(d\Psi)$$

$$= \int \Gamma(\Psi, Y_1 \cap Y_2) P(d\Psi) = \underline{P}(Y_1 \cap Y_2) \quad . \blacksquare$$

For abbrevation, we put

$$I(\Phi,(.)) := \Lambda_{\Gamma(\Phi,(.))} \quad (\Phi \in M) \quad ,$$

thus obtaining a kernel I from $[M, \mathcal{M}]$ to $[A, \mathcal{A}]$. Immediate consequences of 5.1.8. and 5.1.9., respectively, are

5.1.10. $$P*I = \Lambda_P \qquad (P \in \mathbf{P}^{inv}) \quad .$$

5.1.11. For all $P \in \mathbf{P}^{inv}$, $m \in \mathbf{Z}$ and $X \in \mathcal{A}$ there holds

$$I(\Phi_m, X) = E_{\underline{P}}(\Phi_0(X) \mid \mathcal{J}) \qquad \underline{P} \text{ a.s.}$$

The individual ergodic theorem now implies

5.1.12. For all $P \in \mathbf{P}^{inv}$, $m \in \mathbf{Z}$ and $X \in \mathcal{A}$ one has

$$n^{-1} \sum_{0 \le i < n} \Phi_i(X) \xrightarrow[n \to \infty]{} I(\Phi_0, X) \qquad \underline{P}\text{-a.all } (\Phi_i) \quad .$$

Another basic property of the kernel I is given by

5.1.13. For all $P \in \mathbf{P}^{inv}$ there holds

$$\kappa_{(\Phi)} * I = I(\Phi,(.)) \qquad P \text{ a.all } \Phi \quad .$$

Proof. For all $X \in \mathcal{B}$ there holds in virtue of 5.1.11. and 5.1.1. for $\underline{P}$-almost all (Φ_i)

$$(\kappa_{(\Phi_0)} * I)(X) = \int I(\Psi, X) \kappa_{(\Phi_0)}(d\Psi)$$

$$= E_{\underline{P}}(I(\Phi_1, X) \mid \mathcal{F}_0) = E_{\underline{P}}\left(E_{\underline{P}}(\Phi_0(X) \mid \mathcal{J}) \mid \mathcal{F}_0\right) = I(\Phi_0, X) \quad . \blacksquare$$

Let us introduce the shorthand notation

$$G_n := n^{-1} \sum_{0 \le i < n} J^{[i]} \qquad (n=1,2,...) \quad .$$

5.1.14. For all $P \in \mathbf{P}^{inv}$ and $f \in F_{\Lambda_P}$ there holds

$$\int | \langle \Phi * G_n, f \rangle - \langle I(\Phi,(.)), f \rangle | P(d\Phi) \xrightarrow[n \to \infty]{} 0 \quad .$$

In particular, one has for any $X \in \mathcal{A}$ with the property $\Lambda_P(X) < +\infty$:

$$\int | (\Phi * G_n)(X) - I(\Phi, X) | P(d\Phi) \longrightarrow 0 \quad .$$

Proof. For all natural numbers n we obtain by means of 5.1.11. and 5.1.1.

$$\int | \langle \Phi * G_n, f \rangle - \langle I(\Phi,(.)), f \rangle | P(d\Phi)$$

$$= E_{\underline{P}} | E_{\underline{P}}(n^{-1} \sum_{0 \le i < n} f(\Phi_i) \mid \mathcal{F}_0) - E_{\underline{P}}(f(\Phi_0) \mid \mathcal{J}) |$$

$$\le E_{\underline{P}} | n^{-1} \sum_{0 \le i < n} f(\Phi_i) - E_{\underline{P}}(f(\Phi_0) \mid \mathcal{J}) | \quad .$$

The latter expression, however, tends towards zero by means of the statistical ergodic theorem. ■

By using not the L_1-, but the L_2-version of the statistical ergodic theorem, one derives in the same way

5.1.15. For all $P \in \mathbf{P}^{inv}$ and $X \in \mathcal{A}$ which obey $\Lambda^{(2)}_P(X \times X) < +\infty$, there holds

$$\int (I(\Phi,X))^2 P(d\Phi) < +\infty$$

and

$$\int ((\Phi * G_n)(X) - I(\Phi,X))^2 P(d\Phi) \longrightarrow 0 \quad .$$

5.2. Equilibrium Distributions with First Order Representation and Their Ergodic Intensity Distributions

Let us recall that 5.1.8. provides, for any equilibrium distribution P, in form of

$$P = \int \Gamma(\Phi,(.)) P(d\Phi)$$

an integral representation of P over the set $\mathbf{P}^{ex}$. P is said to have a **first order representation** if $\Gamma(\Phi(.))$ is of first order for P-almost all Φ. In this case we put

$$\mathcal{V}_P(Y) := P(I(\Phi,(.)) \in Y) \qquad (Y \in \mathcal{N}) \quad ,$$

thus obtaining a distribution $\mathcal{V}_P$ on $\mathcal{N}$, which we call **ergodic intensity distribution** of P. The following remarks are immediate:

5.2.1. Any equilibrium distribution P which is of first order, has a first order representation.

5.2.2. For any $\mathbf{P}^{inv}$ having a first order representation there holds

$$\Lambda_{\mathcal{V}_P} = \Lambda_P \quad ; \quad \mathcal{V}_P(N_\kappa^{st}) = 1 \quad .$$

5.2.3. The mapping $P \rightarrow \mathcal{V}_P$ preserves convex combination as well as convolution. If $(P_s)_{s \in S}$ is an at most countable family of equilibrium distributions such that $\sum_{s \in S} \Lambda_{P_s} \in N$, then there holds

$$\mathcal{V}_{\underset{s \in S}{*} P_s} = \underset{s \in S}{*} \mathcal{V}_{P_s} \quad .$$

Proof. 1. For all $P_1, P_2 \in \mathbf{P}^{inv}$ and $c \in (0,1)$ there holds

$$\mathcal{V}_{cP_1+(1-c)P_2} = (cP_1+(1-c)P_2)\,(I(\Phi,(.)) \in (.))$$

$$= cP_1(I(\Phi,(.)) \in (.)) + (1-c)P_2(I(\Phi,(.)) \in (.))$$

$$= c\mathcal{V}_{P_1} + (1-c)\mathcal{V}_{P_2} \quad .$$

2. Putting

$$P^S := \bigotimes_{s \in S} P_s \quad ,$$

we obtain for all $X \in \mathcal{B}$

$$\int |((\sum_{s\in S}\Phi_s) * G_n)(X) - \sum_{s\in S} I(\Phi_s,X)|\, P^S(d(\Phi_s))$$

$$\leq \sum_{s\in S} \int |(\Phi * G_n)(X) - I(\Phi,X)|\, P_s(d\Phi) \ .$$

Due to 5.1.14. all summands on the r.h.s. tend, for $n \to \infty$, towards zero. Moreover, there holds for all $n \in \mathbf{Z}_+$ and $s \in S$

$$\int |(\Phi * G_n)(X) - I(\Phi,X)|\, P_s(d\Phi) \leq 2\Lambda_{P_s}(X) \quad ,$$

which allows to conclude that the sequence $((\sum_{s\in S}\Phi_s) * G_n)(X)$ converges in $L_1(P^S)$ towards $\sum_{s\in S} I(\Phi_s, X)$. On the other hand, 5.1.14. yields $L_1(P^S)$ - convergence of the same sequence towards $I(\sum_{s\in S}\Phi_s, X)$, which reveals that

$$I(\sum_{s\in S}\Phi_s, (.)) = \sum_{s\in S} I(\Phi_s, (.)) \quad P^S\text{-a.s.}$$

and thus proves also the second part of our assertion . ■

5.1.8. b) yields immediately

5.2.4. Any extremal equilibrium distribution P, for which $\mathcal{V}_P$ exists, obeys

$$\mathcal{V}_P = \delta_{\Lambda_P} \quad .$$

As the following example shows, the converse of 5.2.4. is not true without further assumptions. In addition, this example reveals that the mapping $P \to \mathcal{V}_P$ is in general not 1-1.

5.2.5. Example. Let us return to Example 4.7.27. Both P and $\Pi^\kappa{}_\nu$ are extremal equilibrium distributions with intensity measure ν, i. e. there holds

$$\mathcal{V}_P = \mathcal{V}_{\Pi^\kappa{}_\nu} = \delta_\nu \quad .$$

Therefore

$$H := 2^{-1}(\Pi^\kappa{}_\nu + P)$$

is a non-extremal equilibrium distribution with the property

$$\mathcal{V}_H = \delta_\nu \quad . ■$$

The converse of 5.2.4. *is* true, however, under the assumption of infinite divisibility:

5.2.6. Proposition. A distribution $P \in P^{inv} \cap T$, for which $\mathcal{V}_P$ exists, belongs to P^{ex} iff

$$\mathcal{V}_P = \delta_{\Lambda_P} \quad .$$

Proof. In view of 5.2.4. it suffices to show that any non-extremal infinitely divisible equilibrium distribution P of first order obeys

$$\mathcal{V}_P \neq \delta_{\Lambda_P} \quad .$$

According to Proposition 5.1.6., a distribution P of this kind has a representation

$$P = P_1*(e^{-c}\sum_{m\geq 0}(m!)^{-1}c^m(P_2)^m) \quad ,$$

where $P_1, P_2 \in P^{inv}$; $P_2(\{o\}) = 0$; $c>0$.

Obviously also P_1 and P_2 are of first order, and there follows

$$\mathcal{V}_P = e^{-c}\sum_{m\geq 0}(m!)^{-1}c^m\mathcal{V}_{P_1}*(\mathcal{V}_{P_2})^m$$

Having intensity measure Λ_{P_2} , the distribution $\mathcal{V}_{P_2}$ is different from δ_o . Hence $\mathcal{V}_P$ cannot be equal to δ_{Λ_P} . ■

5.2.7. Example. With the assumptions and notations of Example 2.8.4. there holds

$$\mathcal{V}_{\Pi^\kappa_v} = e^{-1}\sum_{m\geq 0}(m!)^{-1}\delta_{m\cdot v} \quad (\text{whereas } \mathcal{V}_{\Pi^\kappa_v} = \Pi^\kappa_v) \quad .$$

Proof. For all natural numbers n we have

$$\widetilde{\Pi^\kappa_v}((\Phi*G_n)\in(.)) = p(\,(n^{-1}\sum_{0\leq j<n}\Phi_{z^{(j)}})\in(.)\,) \quad ,$$

where p denotes the distribution of the random sequence $z=(z_i)_{i\geq 0}$ introduced in Example 2.8.4. (There, also the "deterministic time evolution" $z^{(j)}$, $j\in \mathbf{Z}_+$, of z has been defined). If z_i equals zero for infinitely many i (which happens almost surely), then, for any $s\in \mathbf{Z}_+$, the "age" of the individual belonging to $\Phi_{z^{(j)}}$ and situated at position $-s$, runs, with increasing j, through the values $1,\dots, 2^s$ in a cyclic way. Thus there holds p-almost surely

$$n^{-1}\sum_{0\leq j<n}\Phi_{z^{(j)}} \xrightarrow[n\to\infty]{N} v \quad ,$$

which means that $\mathcal{V}_{\widetilde{\Pi^\kappa_v}} = \delta_v$, and hence

$$\mathcal{V}_{\Pi^\kappa_v} = e^{-1}\sum_{m\geq 0}(m!)^{-1}(\mathcal{V}_{\widetilde{\Pi^\kappa_v}})^m = e^{-1}\sum_{m\geq 0}(m!)^{-1}\delta_{m\cdot v} \quad . ■$$

Next we mention an alternative approach to the ergodic intensity distribution.

5.2.8. Proposition. An equilibrium distribution P has a first order representation iff it has the following two properties

a) $P((\Phi*G_n)\in N) = 1$ for $n=1,2,\dots$

b) $P((\Phi*G_n)\in(.))$ tends, for $n\to\infty$, weakly towards a distribution $Q\in V$.

In this case there holds $\mathcal{V}_P = Q$.

Proof. 1. For any $P\in P^{inv}$ of first order we obtain by means of 5.1.14. for all finite sequences $X_1,\dots,X_m$ of sets in $\mathcal{B}$

$$P([(\Phi*G_n)(X_1),\dots, (\Phi*G_n)(X_m)]\in(.))$$

$$\underset{n\to\infty}{\Rightarrow} P([I(\Phi, X_1),\dots, I(\Phi, X_m)]\in(.)) = (\mathcal{V}_P)_{X_1,\dots,X_m}$$

and thus

$$P((\Phi*G_n)\in(.)) \Rightarrow \mathcal{V}_P \quad .$$

2. If, more generally, P has a first order representation, we obtain from 1.

$$P((\Phi*G_n)\in(.)) = \int\Gamma(\Psi, (\Phi*G_n)\in(.))P(d\Psi) \Rightarrow \int\mathcal{V}_{\Gamma(\Psi,(.))}P(d\Psi) = \mathcal{V}_P \quad .$$

3. Let us now assume, conversely, that P has no first order representation. In this case there exists an $X \in \mathcal{B}_P$ such that

(*) $$P(I(\Psi, X) = +\infty) > 0 \quad .$$

If now L is some extremal equilibrium distribution with the property $\Lambda_L(X) = +\infty$, then we obtain by 5.1.12.

$$L(n^{-1} \sum_{0 \le i < n} \Phi_i(X) \xrightarrow[n\to\infty]{} +\infty) = 1$$

and therefore

$$L((\Phi * G_n)(X) \xrightarrow[n\to\infty]{} +\infty) = L(E_L(n^{-1} \sum_{0 \le i < n} \Phi_i(X) | \mathcal{F}_0) \xrightarrow[n\to\infty]{} +\infty) = 1 \quad .$$

Using (*), we conclude

$$P((\Phi * G_n)(X) \to +\infty) = \int \Gamma(\Psi, ((\Phi * G_n) \to +\infty)\, P(d\Psi) \ge P(I(\Psi,X) = +\infty) > 0 \quad .$$

Hence condition b) is violated. ■

In view of Proposition 5.2.8., the original definition of $\mathcal{V}_P$ may be extended as follows:

If a distribution L on $\mathcal{N}$ satisfies the conditions

$$L((\gamma * G_n) \in N) = 1 \quad \text{for } n=1,2,\dots; \quad L((\gamma * G_n) \in (.)) \Rightarrow Q \quad ,$$

then we put $\mathcal{V}_L := Q$.

A quick consequence of this definition is

5.2.9. Let L and H be distributions in V for which $\mathcal{V}_L$ and $\mathcal{V}_H$ exist. Then there holds

$$\mathcal{V}_{L*H} = \mathcal{V}_L * \mathcal{V}_H \quad ,$$

$$\mathcal{V}_{cL+(1-c)H} = c\, \mathcal{V}_L + (1-c)\, \mathcal{V}_H \qquad (0 \le c \le 1) \quad .$$

In a similar way as in Proposition 4.3.2. there follows

5.2.10. Let L be a distribution on $\mathcal{N}$ for which $\mathcal{V}_L$ exists and assume that there holds $\Lambda_L \le \lambda$ for some $\lambda \in N_J$. Then one has $\Lambda_{\mathcal{V}_L} \le \lambda$ and

$$\mathcal{L}_{L((\gamma * G_n) \in (.))}(f) = \mathcal{L}_L(G_n * f) \longrightarrow \mathcal{L}_{\mathcal{V}_L}(f) \qquad (f \in F_\lambda) \quad .$$

We are now in the position to prove

5.2.11. Proposition. Under the assumptions of 5.2.10. there holds

$$\mathcal{V}_L(N_J) = 1 \quad .$$

Proof. For all $f \in F_c$ and all natural numbers m one has

$$\langle \lambda, G_m * f \rangle = \langle \lambda * G_m, f \rangle = \langle \lambda, f \rangle < +\infty \quad .$$

Using 5.2.10. one derives

$$|\mathcal{L}_{\mathcal{V}_L}(G_m*f) - \mathcal{L}_{\mathcal{V}_L}(f)|$$

$$= \lim_{n\to\infty} |\mathcal{L}_L(G_n* G_m*f) - \mathcal{L}_L(G_n*f)|$$

$$\leq \limsup_{n\to\infty} \int |\exp(-\langle\gamma, G_n* G_m*f\rangle) - \exp(-\langle\gamma, G_n*f\rangle)| L(d\gamma)$$

$$\leq \limsup_{n\to\infty} \int |\langle\gamma, G_n* G_m*f\rangle - \langle\gamma, G_n*f\rangle| L(d\gamma)$$

$$\leq \limsup_{n\to\infty} \langle \Lambda_L , |G_n* G_m*f - G_n*f|\rangle$$

$$\leq \limsup_{n\to\infty} n^{-1}\langle\lambda, \sum_{0\leq i<m} (J^{[i]}*f+ J^{[i+n]}*f)\rangle$$

$$= 2m \limsup_{n\to\infty} n^{-1}\langle\lambda, f\rangle = 0 .$$

Thus there holds

$$\mathcal{V}_L((\gamma*G_m)\in(.)) = \mathcal{V}_L \qquad (m=1,2,...) ,$$

which for all $X\in\mathcal{B}$ and all natural numbers m leads to

$$\int |\gamma(X) - (\gamma*J)(X)| \mathcal{V}_L(d\gamma)$$

$$= \int |(\gamma*G_m)(X) - (\gamma*G_m*J)(X)| \mathcal{V}_L(d\gamma) \leq \langle\lambda, |G_m(a,X) - (G_m*J)(a, X)| \rangle$$

$$\leq m^{-1}\langle\lambda, J^{[0]}(a,X)+J^{[m]}(a,X)\rangle = 2m^{-1}\lambda(X) \xrightarrow[m\to\infty]{} 0 .$$

This, however, shows that $\mathcal{V}_L$ is concentrated on N_J. ■

All $Q\in V$ which are concentrated on N_J clearly obey $\mathcal{V}_Q=Q$. In order to show that $\mathcal{V}_Q$ exists for a larger class of J-invariant distributions Q, let us have a look at their "ergodic decomposition". To this purpose, we denote by V^{ex} the set of all J-invariant distributions, which are extremal in the convex set V^{inv}, and obtain exactly as in 5.1.8.:

5.2.12. There exists a kernel Δ from $[N, \mathcal{N}]$ to $[N, \mathcal{N}]$ with the following properties:

a) $\{\Delta(\gamma, (.))\}_{\gamma\in N} = V^{ex}$.

b) For all $Q\in V^{ex}$ there holds

$$\Delta(\gamma, (.)) = Q \qquad Q\text{-almost all } \gamma .$$

c) For all $Q\in V^{inv}$ there holds

$$Q(.) = \int\Delta(\gamma, (.))Q(d\gamma) .$$

Fixing a kernel Δ with these properties and paralleling the arguments leading to 5.1.12, one arrives at

5.2.13. For all $Q\in V^{inv}$ and $X\in\mathcal{A}$ one has

$$(\gamma*G_n)(X) \xrightarrow[n\to\infty]{} \Lambda_{\Delta(\gamma,(.))}(X) \qquad Q\text{-a. all } \gamma .$$

This together with Proposition 5.2.11. implies readily

5.2.14. For any $Q \in V^{inv}$, the distribution $\mathcal{V}_Q$ exists iff $\Delta(\gamma,(.))$ is of first order for Q-almost all γ. In this case there hold

$$\mathcal{V}_Q = Q(\Lambda_{\Delta(\gamma,(.))} \in (.)) \quad ; \quad \Lambda_{\mathcal{V}_Q} = \Lambda_Q \quad ; \quad \mathcal{V}_Q(N_J) = 1 \quad .$$

In the situation of 5.2.14. we say that the J-invariant distribution Q has a **first order representation**, and call $\mathcal{V}_Q$ its **ergodic intensity distribution.**

5.3. Relations between Ergodic Intensity Distribution and Characteristic J-invariant Distribution

We are now in the position to clarify the relation between the characteristic J-invariant distribution and the ergodic intensity distribution of an equilibrium distribution $P \in \mathbf{P}^{inv}$:

5.3.1. Theorem. For $P \in \mathbf{P}^{inv}$, the distribution $\mathcal{V}_P$ exists iff both $\mathcal{W}_P$ and $\mathcal{V}_{\mathcal{W}_P}$ exist. In this case there holds

$$\mathcal{V}_P = \mathcal{V}_{\mathcal{W}_P} \quad .$$

(Verbally, this reads as follows: An equilibrium distribution P has a first order representation iff both its corresponding κ-process $\underline{P}$ and its characteristic J-invariant distribution $\mathcal{W}_P$ have a first order representation. In this case, the ergodic intensity distribution of P equals that of its characteristic J-invariant distribution.)

Proof. 1. In view of the relations

$$\underline{P} = \int \underline{\Gamma}(\Psi, (.)) P(d\Psi)$$

and

$$\underline{\Gamma}(\Psi, (.)) = \int \underline{K}((\Phi_i, (.)) \, \underline{\Gamma}(\Psi, d(\Phi_i)) \quad ,$$

$\underline{P}$ has certainly a first order representation (i. e. $\underline{K}((\Phi_i), (.))$ is of first order for $\underline{P}$-a.all (Φ_i)) if P has a first order representation (i. e. $\underline{\Gamma}(\Phi, (.))$ is of first order for P-almost all Φ).

2. Due to 1., the existence of $\mathcal{V}_P$ implies that of $\mathcal{W}_P$. Henceforth we assume that $\mathcal{W}_P$ exists.

3. Proposition 4.7.2. tells that

$$\mathcal{W}_P = \underline{P}(S_0^{(\Phi_i)} \in (.)) \, .$$

On the other hand, one has for all natural numbers n and all $X \in \mathcal{B}$ in view of 5.1.12. and 5.1.1. for $\underline{P}$ almost all (Φ_i) (note that interchange of limit and conditional expectation is justified by a similar argument as in the first step of the proof of 4.1.6.)

$$\begin{aligned} (S_0^{(\Phi_i)} * G_n)(X) &= E_{\underline{P}}((\Phi_0 * G_n)(X) \mid \mathcal{F}_{-\infty}) \\ &= E_{\underline{P}}(E_{\underline{P}}(n^{-1} \sum_{0 \le i < n} \Phi_i(X) \mid \mathcal{F}_0) \mid \mathcal{F}_{-\infty}) \\ &\longrightarrow E_{\underline{P}}(\Phi_0(X) \mid \mathcal{J}) = I(\Phi_0, X) \quad . \end{aligned}$$

Hence, for $X \in \mathcal{B}$, $((\gamma * G_n)(X))$ converges $\mathcal{W}_P$-almost surely in $\mathbf{R}$ iff $I(\Phi_0, X) < +\infty$.

In virtue of 5.2.13. and 5.2.14. it is now clear that $\mathcal{V}_{\mathcal{W}_P}$ exists iff $\mathcal{V}_P$ exists and that both distributions coincide in case of there existence. ■

The next statement arises as a corollary of Theorems 5.3.1. and 4.7.15.

5.3.2. For all $v \in N_\kappa^{st}$ there holds

$$\mathcal{V}_{\Pi^\kappa_v} = \mathcal{V}_{\mathcal{W}_{\Pi_v}} \quad .$$

This says, in particular, that the ergodic intensity distribution of a Poisson type equilibrium distribution depends only on its intensity measure and on the intensity kernel J of the branching dynamics.

The characteristic J-invariant distribution $\mathcal{W}_P$ of an equilibrium distribution P whose intensity measure is J-aperiodic, in fact is concentrated on N_J (cf. Proposition 3.8.2.), hence coincides with $\mathcal{V}_{\mathcal{W}_P}$ and - due to Theorem 5.3.1. - also with $\mathcal{V}_P$. If we assume in addition that P is extremal, then we obtain by 5.2.4. that $\mathcal{W}_P = \mathcal{V}_P = \delta_{\Lambda_P}$. Combined with Theorem 4.7.16., 5.1.1. and 5.1.3. this yields:

5.3.3. Proposition. Assume that $P \in \mathbf{P}^{inv}$ obeys $\Lambda_P \in N_J^{ap}$. Then $\underline{P}$ is an extremal κ-process iff P is an extremal equilibrium distribution (or,in other words, $\underline{P}$ is $\mathcal{F}_{-\infty}$ -trivial iff $\underline{P}$ is ergodic).

As we have seen in Example 4.7.27., the characteristic J-invariant distribution $\mathcal{W}_P$ of an equilibrium distribution P is not necessarily concentrated on N_J. The following proposition characterizes the situation in which $\mathcal{W}_P(N_J) = 1$.

5.3.4. Proposition. Let P be an equilibrium distribution for which $\mathcal{W}_P$ exists. Then the following assertions are equivalent:

a) $\mathcal{W}_P(N_J) = 1$.

b) $\underline{K}((\Phi_i),(.))$ is stationary for $\underline{P}$-almost all (Φ_i).

c) $I(\Phi_0,(.)) = S_0^{(\Phi_i)}(.)$ $\underline{P}$-almost all (Φ_i).

d) $\mathcal{W}_P = \mathcal{V}_P$.

Proof 1. Assume that $\mathcal{W}_P(N_J) = 1$. In this case $\underline{\mathcal{W}}^{\underline{P}}$ is concentrated on constant J-sequences, and hence, for $\underline{P}$-almost all (Φ_i) the characteristic J-process of $\underline{K}((\Phi_i),(.))$ is of the form $\delta_{(v)}$, $v \in N_\kappa^{st}$. Thus the inversion formula 4.2.7. guarantees stationarity of $\underline{K}(\Phi_i),(.))$ for $\underline{P}$-a. all (Φ_i), and b) turns out to be a consequence of a).

2. Next we assume the validity of b). Together with 4.1.4., this readily implies that the mapping $(\Phi_i) \to \underline{K}((\Phi_i),(.))$ is $\mathcal{J}$-measurable $\underline{P}$-almost surely. In view of 5.1.1. there follows

$$\underline{K}((\Phi_i),(.)) = \underline{P}(.\mid \mathcal{J}) \quad \underline{P}\text{-a.s.,}$$

which together with 5.1.11. yields c).

3. The implication "c) $\Rightarrow$ d)" is immediate from the definitions, and so is, in view, of 5.2.2., the implication "d) $\Rightarrow$ a)", which closes the circle. ■

In case of a first order equilibrium distribution P, the preceding proposition may be extended as follows:

5.3.5. Proposition. For any $P \in \mathbf{P}^{inv}$ of first order, the following assertions are equivalent:

a) $\mathcal{W}_P(N_J) = 1$.

b) $\int |(\Phi * J^{[n]})(X) - I(\Phi, X)| \, P(d\Phi) \longrightarrow 0 \qquad (X \in \mathcal{B})$.

c) $\int |(\Phi * J^{[n]})(X) - (\Phi * J^{[n+1]})(X)| \, P(d\Phi) \longrightarrow 0 \qquad (X \in \mathcal{B})$.

Proof. 1. Due to the martingale convergence theorem, the following chain of equalities holds true for all $X \in \mathcal{B}$:

$$\begin{aligned} &\limsup_{n\to\infty} \int |(\Phi * J^{[n]})(X) - (\Phi * J^{[n+1]})(X)| \, P(d\Phi) \\ &= \limsup_{n\to\infty} \int |(\Phi_{-n} * J^{[n]})(X) - (\Phi_{-n} * J^{[n+1]})(X)| \, \underline{P}(d(\Phi_i)) \\ &= \limsup_{n\to\infty} \int | \, E_{\underline{P}}((\Phi_0(X) | \, \mathcal{F}_{-n})((\Phi_i)) - E_{\underline{P}}((\Phi_1(X) | \, \mathcal{F}_{-n})((\Phi_i)) | \, \underline{P}(d(\Phi_i)) \\ &= \int |(S_0^{(\Phi_i)}(X) - S_1^{(\Phi_i)}(X)| \, P(d(\Phi_i)) \\ &= \int |\gamma(X) - (\gamma * J)(X)| \, \mathcal{W}_P(d\gamma) \ . \end{aligned}$$

This reveals that a) and c) are equivalent.

2. Now assume the validity of a), and choose some $X \in \mathcal{B}$. The martingale convergence theorem assures that $((\Phi_{-n} * J^{[n]}(X))$ converges, for $n \to \infty$, in $L_1(\underline{P})$ towards $(S_0^{(\Phi_i)})(X)$, which by Proposition 5.3.4. equals $I(\Phi_0, X)$ for $\underline{P}$-almost all (Φ_i). In summary, we obtain

$$\begin{aligned} &\int |(\Phi * J^{[n]})(X) - I(\Phi, X)| \, P(d\Phi) \\ &= \int |(\Phi_{-n} * J^{[n]})(X) - I(\Phi_0, X)| \, \underline{P}(d(\Phi_i)) \longrightarrow 0 \qquad (X \in \mathcal{B}) \ , \end{aligned}$$

which is condition b)

3. The implication "b) $\Rightarrow$ c)" , finally, is a quick consequence of the triangle inequality. ■

5.4. Extremal Equilibrium Distributions of Poisson Type

A stable measure ν is called **of ergodic type** if the Poisson type equilibrium distribution with intensity measure ν belongs to P^{ex}. The set of all stable measures of ergodic type will be denoted by N_κ^{erg}.

5.4.1. Theorem. For all $\nu \in N_\kappa^{st}$ the following assertions are equivalent

a) ν is of ergodic type.

b) Π^κ_ν is weakly mixing with respect to time shift.

c) There exists no $\sigma \in N_\kappa^{fin}$ with the properties $\sigma \leq \nu$, $\sigma \neq o$.

d) $\mathcal{V}_{\Pi^\kappa_\nu} \doteq \delta_\nu$.

Proof. 1. Proposition 5.1.6. ensures the equivalence of a) and b), whereas Proposition 5.2.6. yields that of a) and d).

2. Now let us prove the equivalence of a) and c). In virtue of Proposition 5.1.6. a stable measure ν is not of ergodic type iff there exists a factori ation

$$\Pi^\kappa_\nu = P_1 * \mathcal{E}_{cP_2} ,$$

where $P_1, P_2 \in P^{inv}$; $c>0$; $P_2(\{o\}) = 0$. By 2.1.2. this is equivalent to the existence of a factorization

$$\Pi^\kappa_\nu = P_1 * P_3$$

with the properties $P_1, P_3 \in P^{inv}$; $P_3 \in T$ and $\widetilde{P_3}(M) \in (0, +\infty)$.
Due to the generalized Raikov theorem 2.7.27. this amounts to

$$\Pi^\kappa_\nu = \Pi^\kappa_\rho * \Pi^\kappa_\sigma$$

for some $\sigma, \rho \in N_\kappa^{st}$, where σ is of finite type and different from o. ■

According to 5.1.1., any regular stable measure is of ergodic type. In virtue of Proposition 5.3.3.,for J-aperiodic measures also the converse is true:

5.4.2. $N_\kappa^{erg} \cap N_J^{ap} = N_\kappa^{reg} \cap N_J^{ap}$.

Proposition 5.1.4. yields immediately the following pendant to Proposition 4.4.13.

5.4.3. A stable measure ν is of ergodic type iff $\widetilde{\Pi^\kappa_\nu}$ belongs to N_κ^{erg}.

As an analogue to Proposition 4.4.1. and 4.4.7. one gets

5.4.4. Proposition. N_κ^{erg} is a descending subset of N_κ^{st}. If $(\nu_i)_{i \in I}$ is an at most countable family of measures in N_κ^{erg} whose sum ν belongs to N, then ν is also of ergodic type. Together with ρ, also its multiples $c \cdot \rho$, $c \in \mathbf{R}_+$, belong to N_κ^{erg}.

Proof. 1. For all $\sigma, \gamma \in N_\kappa^{st}$ with the property $\sigma \le \gamma$, 5.2.3. yields

$$\mathcal{V}_{\Pi^\kappa_\sigma} * \mathcal{V}_{\Pi^\kappa_{\gamma-\sigma}} = \mathcal{V}_{\Pi_\gamma} ,$$

which in the case $\mathcal{V}_{\Pi^\kappa_\gamma} = \delta_\gamma$ implies that $\mathcal{V}_{\Pi^\kappa_\sigma} = \delta_\sigma$, and hence $\sigma \in N_\kappa^{erg}$ by Theorem 5.4.1.

2. If the sequence (ν_i) has the above stated properties, then its sum ν is stable by 2.3.8. and obeys due to 5.2.3.

$$\mathcal{V}_{\Pi^\kappa_\nu} = \mathop{*}_{i\in I} \delta_{\nu_i} = \delta_\nu .$$

Hence ν is of ergodic type due to Theorem 5.3.1.

3. For $\rho \in N_\kappa^{erg}$ and $c \in \mathbf{R}_+$, the inequality $c\cdot\nu \le ([c]+1)\cdot\nu$ together with 1. implies that $c\cdot\nu$ is of ergodic type. ■

We conclude this section with a criterion of ergodicity which follows directly from Theorem 5.3.1. and Proposition 2.8.7.

5.4.5. Theorem. If the intensity kernel J is substochastic, then a Poisson type equilibrium distribution P is extremal iff its intensity measure has, apart from o, no finite minorant in $[N_J, \le]$.

5.5. Purely Non-Extremal Equilibrium Distributions of Poisson Type

A Poisson type equilibrium distribution is called **purely non extremal** if P has, apart from δ_o, no convolution divisor which belongs to $\mathbf{P}^{ex}$. The main result of this section is

5.5.1. Theorem. For any Poisson type equilibrium distribution P, the following assertions are equivalent:

a) P is purely non-extremal.

b) Λ_P has, apart from o, no minorant in $[N_\kappa^{st}, \le]$ which is of ergodic type.

c) Λ_P is a (finite or countable) sum of stable measures of finite type.

Theorem 5.5.1. motivates to say that a stable measure ν is **of purely non-ergodic type** if ν has a representation

$$\nu := \sum_{i\in I} \nu_i ,$$

where (ν_i) is an at most countable set of measures in N_κ^{fin}. In view of 2.8.2., this is the case iff there exists a sequence (σ_n) of measures in N_κ^{fin} which obeys $\sigma_n \uparrow \nu$.

The set of all stable measures which are of purely non-ergodic type will be denoted by N_κ^{nerg}.

For the proof of Theorem 5.5.1., and also lateron, we need a series of propositions:

5.5.2. Proposition. N_κ^{nerg} is a descending subset of N_κ^{st}. If $(\nu_i)_{i\in I}$ is an at most countable family of measures in N_κ^{nerg} whose sum ν belongs to N, then ν even belongs to N_κ^{nerg}. Furthermore, any multiple $c\cdot\gamma$, $c\in \mathbf{R}_+$, of a measure $\gamma\in N_\kappa^{nerg}$ is again of purely non-ergodic type.

Proof. 1. Assume that $\sigma, \rho\in N_\kappa^{st}$, $\sigma\le\rho$ and $\rho_n\uparrow\rho$, whereby $\rho_n\in N_\kappa^{fin}$ for n=1,2,... In virtue of 2.2.6. and 2.8.2. we obtain

$$\sigma\cap\rho_n\uparrow\sigma\cap\rho=\sigma \;;\; \sigma\cap\rho_n\in N_\kappa^{fin} \quad \text{for } n=1,2,...$$

and hence $\sigma\in N_\kappa^{nerg}$.

2. If a measure $\nu\in N$ can be represented as an infinite sum of measures in N_κ^{nerg}, then it can as well be represented as an infinite sum of measures in N_κ^{fin}, which means that ν is of purely non ergodic type.

3. Any $\gamma\in N_\kappa^{nerg}$ has a representation

$$\gamma=\sum_{n>0}\gamma_n \;,\; \gamma_n\in N_\kappa^{fin} \qquad (n=1,2,...)\;.$$

For c>0, there holds

$$c\cdot\gamma=\sum_{n>0}c\cdot\gamma_n \;.$$

As all summands of this latter series are of finite type by 2.8.2., the measure $c\cdot\gamma$ turns out to be of purely non-ergodic type. ■

Using the preceding result, we now prove

5.5.3. Proposition. For all $\nu\in N_\kappa^{st}$ there exists

$$\nu_{nerg}:=\max\{\rho:\rho\in N_\kappa^{nerg}, \rho\le\nu\}\;.$$

Proof. Let us put

$$S:=\{\rho:\rho\in N_\kappa^{nerg}, \rho\le\nu\}\;.$$

Obviously, zero measure o belongs to S. For $\rho_1, \rho_2\in S$ there holds $\rho_1\cup\rho_2\le\nu$. On the other hand one has $\rho_1\cup\rho_2\le\rho_1+\rho_2\in N_\kappa^{nerg}$, which by Proposition 5.5.2. yields $\rho_1\cup\rho_2\in N_\kappa^{nerg}$ and, in summary, $\rho_1\cup\rho_2\in S$. Hence S is a non-void, directed, bounded subset of N. Denoting its least upper bound in N by γ, we obtain

$$\gamma(X)=\sup\{\rho(X)\}_{\rho\in S} \qquad (X\in\mathcal{A})\;.$$

Let $\mathcal{B}'$ be some at most countable generating subring of $\mathcal{B}$. Then there exist an at most countable non-void subset $\{\rho_1,\rho_2,...\}$ of S such that

$$\rho(X)=\sup_{n=1,2,...}\rho_n(X) \qquad (X\in\mathcal{B}')$$

and hence

$$\rho_1\cup...\cup\rho_n\uparrow\rho.$$

This shows, in virtue of Proposition 5.5.2., that ρ is of purely non-ergodic type, i. e. there holds $\rho=\max S$. ■

The following remark is immediate:

5.5.4. For all $\nu\in N_\kappa^{st}$ and $c\in\mathbf{R}_+$ there holds

$$(c\cdot\nu)_{nerg}=c\cdot\nu_{nerg}\;.$$

5.5.5. Proposition. Let $(\nu_i)_{i\in I}$ be an at most countable family of measures in N_κ^{st}, whose sum ν belongs to N. Then there holds

$$\nu_{nerg} = \sum_{i\in I} (\nu_i)_{nerg} \ .$$

Proof. Without loss of generality we assume $I=\{1,2,...\}$. By 2.3.8., ν is stable. Due to Proposition 2.8.5.,

$$\gamma := \sum_{n\geq 1} (\nu_n)_{nerg}$$

belongs to N_κ^{nerg}. To prove our assertion, it is thus sufficient to show that $\gamma \geq \nu_{nerg}$.

We now put

$$\rho_n := \sum_{1\leq i\leq n} \nu_i \qquad (n=1,2,...)$$

From

$$\nu_{nerg} \cap \rho_n \leq \sum_{1\leq i\leq n} \nu_{nerg} \cap \nu_i \qquad (n= 1,2...)$$

we get by Proposition 5.5.2.

$$\nu_{nerg} \cap \rho_n \leq \sum_{1\leq i\leq n} (\nu_i)_{nerg} \leq \gamma \qquad (n= 1,2...) \ .$$

Hence we obtain by means of 2.2.6.

$$\nu_{nerg} \cap \rho_n \uparrow \nu_{nerg} \cap \nu = \nu_{nerg} \leq \gamma \ . \blacksquare$$

In view of the equivalence of a) and c) in Theorem 5.4.1. one now gets immediately:

5.5.6. A stable measure ρ is of ergodic type iff $\rho_{nerg} = o$.

Now we turn to the

Proof of Theorem 5.5.1. 1. The equivalence of a) and b) follows from Theorem 2.7.27.

2. Now assume the validity of c), i. e. Λ_P is of purely non-ergodic type. Consider some minorant $\rho\neq o$ of Λ_P in $[N_\kappa^{st}, \leq]$. Due to Proposition 5.5.2., ρ belongs to N_κ^{nerg} as well, hence $\rho_{nerg} = \rho \neq o$, revealing that ρ cannot be of ergodic type.

3. Conversely assume that c) fails to hold. This amounts to $(\Lambda_P)_{nerg} \neq \Lambda_P$, i.e. $\rho := \Lambda_P - (\Lambda_P)_{nerg} \neq o$. Proposition 5.5.5. entails the equality $\rho_{nerg} = o$. By 5.5.6., however, this is equivalent to $\rho \in N_\kappa^{erg}$, which altogether violates condition b). $\blacksquare$

Let us still mention that for any $\nu \in N_\kappa^{st}$ the measure ν_{nerg} may be expressed, due to Theorem 2.8.6., via the intensity kernel J. A fortiori, the question whether a stable measure ν is of purely non-ergodic type, and in view of 5.4.1. also the question whether a stable measure ν is of ergodic type, may be decided " by inspection of ν and J ".

In the case of a substochastic intensity kernel, all purely non extremal Poisson type equilibrium distributions are in fact Poissonian:

5.5.7. Theorem. If J is substochastic, then there holds

$$\Pi^{\kappa}_{\nu} = \Pi_{\nu} \qquad (\nu \in N_{\kappa}^{nerg}) \quad .$$

Proof. Due to 2.3.8. it sufficies to consider measures ν of finite type. Then there holds by Proposition 2.8.7.

$$\nu(A) = \int \Psi(A)\, \widetilde{\Pi^{\kappa}_{\nu}}(d\Psi) = \widetilde{\Pi^{\kappa}_{\nu}}(M) \quad .$$

As $\widetilde{\Pi^{\kappa}_{\nu}}$-almost all counting measures Ψ obey

$$\Psi(A) \geq 1 \,,$$

there follows

$$\widetilde{\Pi^{\kappa}_{\nu}}(\Psi(A) = 1) = 1 \quad ,$$

i. e.

$$\widetilde{\Pi^{\kappa}_{\nu}} = \mathcal{Q}_{\nu} \quad . \blacksquare$$

5.6. A Natural Factorization of Poisson Type Equilibrium Distributions

We are going to show in this section that any Poisson type equilibrium distribution P has a unique representation $P = P_1 * P_2$, where P_1 and P_2 are an extremal and a purely non-extremal Poisson type equilibrium distribution, respectively. It will turn out that on the level of ergodic intensity distributions this factorization coincides with the canonical representation of the infinitely divisible distribution $\mathcal{V}_P$.

Let us start the investigation "on the intensity level":

5.6.1. For all $\gamma \in N_{\kappa}^{erg}$ and $\rho \in N_{\kappa}^{nerg}$ there holds

$$\gamma \cap \rho = o \quad , \quad \gamma \cup \rho = \gamma + \rho \quad .$$

Proof: Both N_{κ}^{erg} and N_{κ}^{nerg} are descending subsets of $[N_J, \leq]$, hence follows $\gamma \cap \rho \in N_{\kappa}^{erg} \cap N_{\kappa}^{nerg}$. This implies $\gamma \cap \rho = o$ in virtue of 5.5.6. The second asserted relation now follows from Proposition 2.2.5. ■

For all $\nu \in N_{\kappa}^{st}$ we put

$$\nu_{erg} := \nu - \nu_{nerg} \quad .$$

5.6.2. Proposition. For all $\nu \in N_{\kappa}^{st}$ there holds

a) $\nu_{erg} = \max \{\rho : \rho \in N_{\kappa}^{erg}, \rho \leq \nu\}$

b) $\nu = \nu_{erg} + \nu_{nerg}$ is the unique representation of ν as a sum of two measures in N_{κ}^{erg} and N_{κ}^{nerg}, respectively.

Proof: 1. By Proposition 5.5.5. there follows

$$\nu_{nerg} = (\nu_{erg})_{nerg} + \nu_{nerg} ,$$

i. e. $(\nu_{erg})_{nerg} = o$, which by 5.5.6. amounts to $\nu_{erg} \in N_\kappa^{erg}$.

If now ρ is measure in N_κ^{erg} with the property $\rho \le \nu$, then one obtains

$$(\nu - \rho)_{nerg} = \nu_{nerg} - \rho_{nerg} = \nu_{erg} = \nu - \nu_{erg}$$

and hence

$$\nu - \rho \ge \nu - \nu_{erg} \quad ,$$

i. e. $\rho \le \nu_{erg}$. Thus we have proved a).

2. Concerning b), it remains to check uniqueness.

For $\nu = \nu_1 + \nu_2 = \nu_{erg} + \nu_{nerg}$, $\nu_1 \in N_\kappa^{erg}$, $\nu_2 \in N_\kappa^{erg}$, there follows by a) and by definition of ν_{nerg}

$$\nu_1 \le \nu_{erg} \quad ; \quad \nu_2 \le \nu_{nerg} \quad ,$$

which altogether implies equality in both cases. ■

The preceding result immediately translates into

5.6.3. Theorem. Any equilibrium distribution P of Poisson type has a unique representation $P = P_1 * P_2$, where P_1 and P_2 are an extremal and a purely non-extremal Poisson type equilibrium distribution, respectively.

Let us recall that due to 5.2.3. the ergodic intensity distribution $\mathcal{V}_P$ of any Poisson type equilibrium distribution of first order is an infinitely divisible distribution on $\mathcal{N}$.

5.6.4. Theorem. Let P be an equilibrium distribution of Poisson type, and

$$\mathcal{V}_P = \delta_\gamma * \mathcal{E}_{\widetilde{\mathcal{V}}_P}$$

be the canonical representation of its ergodic intensity distribution.
Then there holds

$$(\Lambda_P)_{erg} = \gamma \quad ; \quad (\Lambda_P)_{nerg} = \Lambda_{\widetilde{\mathcal{V}}_P} \quad .$$

Proof. Putting $\nu := \Lambda_P$, one obtains from

$$P = \Pi^\kappa_\nu = \Pi^\kappa_{\nu_{erg}} * \Pi^\kappa_{\nu_{nerg}}$$

and 5.2.3. the factorization

$$\mathcal{V}_P = \delta_{\nu_{erg}} * \mathcal{V}_{\Pi^\kappa_{\nu_{nerg}}} \quad .$$

ν_{nerg} is of the form

$$\nu_{nerg} = \sum_{n>0} c_n \sigma_n \quad ,$$

where $c_n \in \mathbf{R}_+$ and $\widetilde{\Pi^\kappa_{\sigma_n}}(M) = 1$ for n= 1,2,... .

One more application of 5.2.3. leads to

$$\mathcal{V}_P = \delta_{\nu_{erg}} * \mathop{\ast}_{n>0} \mathcal{V}_{e^{-cn} \sum_{i\ge 0} (i!)^{-1} (c_n)^i (\widetilde{\Pi^\kappa_{\sigma_n}})^i}$$

$$= \delta_{\nu_{erg}} * \mathop{\ast}_{n>0} \left(e^{-c_n} \sum_{i\geq 0} (i!)^{-1} (c_n)^i (\mathcal{V}_{\widetilde{\Pi^\kappa_{\sigma_n}}})^i\right)$$

$$= \delta_{\nu_{erg}} * \mathop{\ast}_{n>0} \mathcal{E}_{c_n \mathcal{V}_{\widetilde{\Pi^\kappa_{\sigma_n}}}}$$

$$= \delta_{\nu_{erg}} * \mathcal{E}_{\sum_{n>0} c_n \mathcal{V}_{\widetilde{\Pi^\kappa_{\sigma_n}}}} .$$

Uniqueness of the canonical representation of $\mathcal{V}_P$ now yields $\gamma = \nu_{erg}$. On the other hand there holds

$$\nu_{erg} + \nu_{nerg} = \nu = \gamma + \Lambda_{\widetilde{\mathcal{E}_{\mathcal{V}_P}}} = \gamma + \Lambda_{\widetilde{\mathcal{V}_P}} . \blacksquare$$

The mapping $V \to \gamma$, which assigns to an infinitely divisible distribution on $\mathcal{N}$ the measure γ figuring in its canonical representation $V = \delta_\gamma * \mathcal{E}_{\tilde{V}}$, is V, $\mathcal{N}$-measurable. Bearing in mind the measurability of the mappings $\nu \to \Pi^\kappa_\nu$, $\nu \in N_\kappa^{st}$, and $P \to \mathcal{V}_P$, $P \in \mathbf{P}^{inv}$ of first order, we obtain

5.6.5. The mappings $\nu \to \nu_{erg}$ and $\nu \to \nu_{nerg}$ from N_κ^{st} into itself are $\mathcal{N}$-measurable.

As a counterpart to 2.6.14. one gets immediately:

5.6.6. $N_\kappa^{st} \cap N_J^{ex} \subseteq N_\kappa^{erg} \cup N_\kappa^{nerg}$.

The following two remarks are direct consequences of 5.5.5. and Proposition 5.5.6., respectively.

5.6.7. For all $\nu \in N_\kappa^{st}$ and $c \in \mathbf{R}_+$ there holds

$$(c\cdot\nu)_{erg} = c\cdot\nu_{erg} .$$

5.5.8. Let $(\nu_i)_{i\in I}$ be an at most countable family of measures whose sum ν belongs to N. Then there holds

$$\nu_{erg} = \sum_{i\in I} (\nu_i)_{erg} .$$

The mappings $\nu \to \nu_{erg}$ and $\nu \to \nu_{nerg}$ even "commute with mixtures":

5.6.9. Proposition. Let Q be a distribution on $\mathcal{N}$ which is of first order and satisfies $Q(N_\kappa^{st}) = 1$. Then there holds

$$(\Lambda_Q)_{erg} = \int \nu_{erg}(.) Q(d\gamma) .$$

Proof. 1. The measure $\nu := \Lambda_Q$ is stable (cf. Thm 3.7.9.) and there holds (see 3.4.7.)

$$\widetilde{\Pi^\kappa_\nu} = \int \widetilde{\Pi^\kappa_\gamma}(.) Q(d\gamma) .$$

For any $f \in F_c$ and any $\sigma \in N_\kappa^{st}$ one has by 1.3.10. and Proposition 5.2.8.

$$\int (1 - \exp(-\langle \Psi * G_n, f\rangle)) \widetilde{\Pi^\kappa_\sigma}(d\Psi)$$

$$= -\ln \mathcal{L}_{\widetilde{\Pi^\kappa_\sigma}} (G_n * f) \xrightarrow[n\to\infty]{} -\ln \mathcal{L}_{\mathcal{V}_{\widetilde{\Pi^\kappa_\sigma}}} (f) .$$

In summary, one recognizes the validity of

$$\ln \mathcal{L}_{\mathcal{V}_{\Pi^\kappa_v}}(f) = \int \ln \mathcal{L}_{\mathcal{V}_{\Pi^\kappa_\gamma}}(f) Q(d\gamma) \quad .$$

2. Theorem 5.6.4. yields, for all $\sigma \in N_\kappa^{st}$, the relation

(o) $$\mathcal{V}_{\Pi^\kappa_\sigma} = \delta_{\sigma_{erg}} * \mathcal{E}_{\widetilde{\mathcal{V}_{\Pi^\kappa_\sigma}}} \quad .$$

This implies, for all $f \in F_c$,

$$-\ln \mathcal{L}_{\mathcal{V}_{\Pi^\kappa_\sigma}}(f) = \langle \sigma_{erg}, f\rangle + \int (1-\exp(-\langle \rho, f\rangle)) \widetilde{\mathcal{V}_{\Pi^\kappa_\sigma}}(d\rho) \quad .$$

In virtue of 1. we thus obtain

$$-\ln \mathcal{L}_{\mathcal{V}_{\Pi^\kappa_v}}(f) = \int \langle \gamma_{erg}, f\rangle Q(d\gamma) + \int\int (1-\exp(-\langle \rho, f\rangle)) \widetilde{\mathcal{V}_{\Pi^\kappa_\gamma}}(d\rho) Q(d\gamma) \quad .$$

3. Putting

$$\sigma := \int \gamma_{erg}(.) Q(d\gamma) \quad ; \quad W := \int \widetilde{\mathcal{V}_{\Pi^\kappa_\gamma}}(.) Q(d\gamma) \quad ,$$

we may reformulate the relation at the end of step 2. as

$$-\ln \mathcal{L}_{\mathcal{V}_{\Pi^\kappa_v}}(f) = \langle \sigma, f\rangle + \int (1-\exp(-\langle \rho, f\rangle)) W(d\rho) \quad .$$

In view of the relations

$$\sigma \leq \int \gamma(.) Q(d\gamma) = v \quad ; \quad W(\{o\}) = 0 \quad ;$$
$$\int (1-\exp(-\langle \rho, f\rangle)) W(d\rho) \leq -\ln \mathcal{L}_{\mathcal{V}_{\Pi^\kappa_v}}(f) < +\infty \qquad (f \in F_c) \quad ,$$

one obtains in form of

$$\mathcal{V}_{\Pi^\kappa_v} = \delta_\sigma * \mathcal{E}_W$$

the canonical representation of the infinitely divisible distribution $\mathcal{V}_{\Pi^\kappa_v}$. A comparison with (o) in step 2. now reveals that $\sigma = v_{erg}$. ■

In virtue of the preceding proposition, any distribution Q on $\mathcal{N}$ with the properties $Q(N_J) = 1$, $\Lambda_Q \in N_\kappa^{erg}$, is even concentrated on N_κ^{erg}. This implies readily

5.6.10. Proposition. Let Q be a disttibution on $\mathcal{N}$ with the properties $Q(N_J) = 1$, $\Lambda_Q \in N_\kappa^{erg}$. Then there holds

$$\mathcal{V}_{V_{(Q)}} = Q$$

(i. e. in this case the ergodic intensity distribution and the intensity distribution of $V_{(Q)}$ coincide.)

If P is an equilibrium distribution whose intensity measure is of ergodic type, then $Q := \mathcal{V}_P$ has all the properties required in the preceding proposition ; therefore one ends up with

$$\mathcal{V}_{V_{(\mathcal{V}_P)}} = \mathcal{V}_P \quad .$$

5.7. Cesaro Clustering and J-ergodic Measures

For all natural numbers n we put

$$\omega_n := n^{-1} \sum_{0\le i<n} \kappa^{[i]} \quad ,$$

thus obtaining a sequence $\omega_1, \omega_2, \ldots$ of clustering fields on $[A, \rho_A]$. Obviously there holds

5.7.1. $$G_n = J_{\omega_n} \qquad (n=1,2,\ldots) \quad .$$

For all $\nu \in N_\kappa^{st}$ one has for n=1,2,...

$$\widetilde{(\Pi_\nu)_{\omega_n}} = (\mathcal{Q}_\nu)_{\omega_n}((.)\setminus\{o\}) = \int (\omega_n)_{(a)}((.)\setminus\{o\})\nu(da)$$

$$= n^{-1} \sum_{0\le i<n} (\mathcal{Q}_\nu)_{\kappa^{[i]}}((.)\setminus\{o\})$$

$$= n^{-1} \sum_{0\le i<n} \widetilde{(\Pi_\nu)_{\kappa^{[i]}}} \quad .$$

As the convergence of $\widetilde{(\Pi_\nu)_{\kappa^{[n]}}}$ towards $\widetilde{\Pi^\kappa}_\nu$ in **W** carries over to the convergence of the corresponding Cesaro-means, one gets

5.7.2. For all $\nu \in N_\kappa^{st}$ there holds

$$(\Pi_\nu)_{\omega_n} \Rightarrow \Pi^\kappa_\nu \quad .$$

As an analogue to Theorem 4.7.1. we obtain

5.7.3.Theorem. Let L be a distribution in V with N_J -bounded intensity measure and such that $\mathcal{V}_L$ exists. Then there holds

$$(\mathfrak{I}(L))_{\omega_n} \Rightarrow V_{(\mathcal{V}_L)} \quad .$$

Proof. 1. Let $\lambda \in N_J$ be some majorant of Λ_L . Further let f be some function in F_b. For arbitrary, fixed natural numbers m, n we put

$$\tau := \omega_n \quad , \quad \sigma := ((\omega_m)_{(.)})_{\kappa^{[n]}} = \kappa^{[n]} \circ \omega_m$$

and obtain

$$\int |\mathcal{L}_{(\Pi_\gamma)_{\omega_{m+n}}}(f) - \mathcal{L}_{((\Pi_\gamma)_{\omega_m})_{\kappa^{[n]}}}(f)|\, L(d\gamma)$$

$$= \int |\exp(-\langle\gamma, 1-\mathcal{L}_{(\omega_{m+n})_{(.)}}(f)\rangle) - \exp(-\langle\gamma, 1-\mathcal{L}_{\sigma_{(.)}}(f)\rangle|\, L(d\gamma)$$

$$\le \int \langle\gamma, |\mathcal{L}_{(\omega_{m+n})_{(.)}}(f) - \mathcal{L}_{\sigma_{(.)}}(f)|\rangle L(d\gamma)$$

$$\le \langle\lambda, |\mathcal{L}_{(m+n)^{-1}(n\tau_{(.)}+m\sigma_{(.)})}(f) - \mathcal{L}_{\sigma_{(.)}}(f)|\rangle$$

$$= \langle \lambda, |(m+n)^{-1} n\, \mathcal{L}_{\tau_{(.)}}(f) + (m+n)^{-1} m\, \mathcal{L}_{\sigma_{(.)}}(f) - \mathcal{L}_{\sigma_{(.)}}(f)| \rangle$$

$$= (m+n)^{-1} n \langle \lambda, |\mathcal{L}_{\tau_{(.)}}(f) - \mathcal{L}_{\sigma_{(.)}}(f)| \rangle$$

$$\leq (m+n)^{-1} n \langle \lambda, (1-\mathcal{L}_{\sigma_{(.)}}(f)) + (1-\mathcal{L}_{\tau_{(.)}}(f)) \rangle$$

$$\leq (m+n)^{-1} n \langle \lambda, (G_m * J^{[n]})((.), \text{supp } f) + G_n((.), \text{supp } f) \rangle$$

$$= (m+n)^{-1} 2n\, \lambda(\text{supp } f) \quad .$$

Hence results for all natural numbers n:

$$\int |\mathcal{L}_{(\Pi_\gamma)_{\omega_{m+n}}}(f) - \mathcal{L}_{((\Pi_\gamma)_{\omega_m})_{\kappa^{[n]}}}(f)| L(d\gamma) \underset{m\to\infty}{\longrightarrow} 0 \quad .$$

2. For all m, $n \in \mathbf{Z}_+$ we obtain by paralleling the arguments in the first step of the proof of Theorem 4.6.1. (where $J^{[m]}$, $\kappa^{[m]}$ and λ_i, $i \in \mathbf{Z}$, have to be replaced by G_m, ω_m and λ, respectively):

$$\int |\mathcal{L}_{((\Pi_\gamma)_{\omega_m})_{\kappa^{[n]}}}(f) - \mathcal{L}_{(\Pi_{\gamma * G_m})_{\kappa^{[n]}}}(f)| L(d\gamma)$$

$$\leq \int (1-\mathcal{L}_{\kappa^{[n]}_{(b)}}(f))(1-\mathcal{L}_{\kappa^{[n]}_{(\chi)}}(f)) C^!_{\widetilde{\Pi^{\kappa}_{\lambda_{st}}}}(d[b,\chi]) + \int (1-\mathcal{L}_{\kappa^{[n]}_{(b)}}(f)) \lambda_{nst}(db)$$

$$\underset{n\to\infty}{\longrightarrow} 0 \quad .$$

3. The same reasoning as in the third step of the proof of Theorem 4.6.1. tells that

$$\int \mathcal{L}_{(\Pi_{\gamma * G_m})_{\kappa^{[n]}}}(f)\, L(d\gamma) \underset{m\to\infty}{\longrightarrow} \mathcal{L}_{(\mathfrak{I}(\mathcal{V}_L))_{\kappa^{[n]}}}(f) \quad .$$

4. Combining 1., 2., 3., one arrives at

$$\limsup_{j\to\infty} |\mathcal{L}_{(\mathfrak{I}(L))_{\omega_j}}(f) - \mathcal{L}_{V_{(\mathcal{V}_L)}}(f)|$$

$$\leq \limsup_{n\to\infty} \limsup_{m\to\infty} \int |\mathcal{L}_{(\Pi_\gamma)_{\omega_{m+n}}}(f) - \mathcal{L}_{((\Pi_\gamma)_{\omega_m})_{\kappa^{[n]}}}(f)| L(d\gamma)$$

$$+ \limsup_{n\to\infty} \sup_{m=1,2,\ldots} \int |\mathcal{L}_{((\Pi_\gamma)_{\omega_m})_{\kappa^{[n]}}}(f) - \mathcal{L}_{(\Pi_{\gamma * G_m})_{\kappa^{[n]}}}(f)| L(d\gamma)$$

$$+ \limsup_{n\to\infty} \limsup_{m\to\infty} |\int \mathcal{L}_{(\Pi_{\gamma * G_m})_{\kappa^{[n]}}}(f) L(d\gamma) - \mathcal{L}_{(\mathfrak{I}(L))_{\kappa^{[n]}}}(f)|$$

$$+ \limsup_{n\to\infty} |\mathcal{L}_{(\mathfrak{I}(\mathcal{V}_L))_{\kappa^{[n]}}}(f) - \mathcal{L}_{V_{(\mathcal{V}_L)}}(f)|$$

$$= 0+0+0+0 \quad . \blacksquare$$

To get a similar convergence result as in Theorem 4.7.30., we first have to care for an "ergodic" pendant to the set N_J^{reg}.

5.7.4. Proposition. For any measure $\nu \in N_J$, the following assertions are equivalent:

a) $\int G_n(a,X)\min(1,G_n(a,X))\nu(da) \xrightarrow[n\to\infty]{} 0 \qquad (X\in\mathcal{B})$.

b) $\mathcal{V}_{\Pi_\nu} = \delta_\nu$.

Proof: follows along the same lines as that of Proposition 4.4.3., where only ν_i, $i\in \mathbf{Z}_+$ and $J^{[n]}$, $n\in\mathbf{Z}_+$, have to be replaced by ν and G_n, respectively . ■

A measure $\nu\in N_J$ having the properties stated in Proposition 5.7.4. will be called J-ergodic; the set of all these measures will be denoted by N_J^{erg} .

5.7.5. $N_\kappa^{erg} \subseteq N_J^{erg}$.

Proof. For any $\nu\in N_\kappa^{erg}$ and $f\in F_c$ we get, using 4.4.10. and Jensen´s inequality

$$\mathcal{L}_{\Pi^\kappa_\nu}(G_n*f) \geq \mathcal{L}_{\Pi_\nu}(G_n*f) = \int\exp(-\langle\Phi, G_n*f\rangle)\Pi_\nu(d\Phi)$$

$$\geq \exp(-\int\langle\Phi, G_n*f\rangle\Pi_\nu(d\Phi)) = \exp(-\langle\nu,f\rangle) \ .$$

Hence the convergence

$$\mathcal{L}_{\Pi^\kappa_\nu}(G_n*f) \xrightarrow[n\to\infty]{} \mathcal{L}_{\delta_\nu}(f) \qquad (f\in F_c)$$

implies

$$\mathcal{L}_{\Pi_\nu}(G_n*f) \xrightarrow[n\to\infty]{} \mathcal{L}_{\delta_\nu}(f) \qquad (f\in F_c) \ . \ ■$$

5.7.6. Theorem . Let L be a distribution on $\mathcal{M}$ with N_J^{erg}-bounded intensity measure and such that $\mathcal{V}_L$ exists. Then there holds

$$L_{\omega_n} \Rightarrow V_{(\mathcal{V}_L)} \ .$$

Proof. 1. Let $\lambda\in N_J^{erg}$ be some majorant of Λ_L. Then λ obeys (cf. Proposition 5.7.4.)

$$\int((\omega_n)_{(a)}(\chi(X)>0))^2\lambda(da) \xrightarrow[n\to\infty]{} 0 \qquad (X\in\mathcal{B}) \ .$$

2. For all $f\in F_b$ and all $n\in\mathbf{Z}_+$ we obtain

$$\int|\mathcal{L}_{(\omega_n)_{(\Phi)}}(f) - \mathcal{L}_{(\Pi_\Phi)_{\omega_n}}(f)|\, L(d\Phi)$$

$$= \int\Big|\prod_{a\in\text{supp}\Phi}(\mathcal{L}_{(\omega_n)_{(a)}}(f))^{\Phi(\{a\})} - \prod_{a\in\text{supp}\Phi}\big(\exp(-(1-\mathcal{L}_{(\omega_n)_{(a)}}(f)))^{\Phi(\{a\})}\big)\Big|\, L(d\Phi)$$

$$\leq \iint|\exp(-(1-\mathcal{L}_{(\omega_n)_{(a)}}(f))) - (1-(1-\mathcal{L}_{(\omega_n)_{(a)}}(f)))|\,\Phi(da)\, L(d\Phi)$$

$$\leq \iint(1-\mathcal{L}_{(\omega_n)_{(a)}}(f))^2\Phi(da)\, L(d\Phi) \ \leq \ \langle\lambda, (1-\mathcal{L}_{(\omega_n)_{(.)}}(f))^2\rangle$$

$$\leq \int((\omega_n)_{(a)}(\chi(X)>0))^2\,\lambda(da) \ ,$$

which tends, for $n\to\infty$, towards zero due to 1.

3. Combining 2. with Theorem 5.7.3. we arrive immediately at our assertion. ■

Next, we characterize the set of all those Cox type equilibrium distributions, whose intensity measure is of ergodic type and whose intensity distribution is concentrated on J-invariant measures:

5.7.7. Theorem. For all $\nu \in N_\kappa^{erg}$ there holds

$$\{V_{(Q)}\}_{Q \in V, \Lambda_Q = \nu; Q(N_J)=1} = \{P : P \in \mathbf{P} \; ; \; \Lambda_P = \nu \; ; \; P_{\omega_n} \Rightarrow P\} \; .$$

Proof. 1. From 5.7.5. and Theorem 5.7.6. one gets immediately for all $\gamma \in N_\kappa^{erg}$

$$(\Pi^\kappa_\gamma)_{\omega_n} \Rightarrow V_{(\mathcal{V}_{\Pi^\kappa_\gamma})} = V_{(\gamma)} = \Pi^\kappa_\gamma \; .$$

2. Let Q be a distribution in $V^{inv,\nu}$ with the property $Q(N_J) = 1$. Then Q is even concentrated on N_κ^{erg} (cf. the remark after Proposition 5.6.9.), and 1. yields

$$(V_{(Q)})_{\omega_n} = \int (\Pi^\kappa_\gamma)_{\omega_n}(.)Q(d\gamma) \Rightarrow \int \Pi^\kappa_\gamma(.)Q(d\gamma) = V_{(Q)} \; .$$

Finally, Theorem 3.7.10. guarantees that $\Lambda_{V_{(Q)}} = \nu$.

3. Conversely, let us choose a distribution $P \in \mathbf{P}$ obeying $\Lambda_P = \nu$ and $P_{\omega_n} \Rightarrow P$. Because of

$$\Lambda_{P_{\omega_n}} = \nu * G_n = \nu \qquad (n=1,2,\ldots) \; ,$$

1.5.11. yields the convergence

$$(P_{\omega_n})_\kappa \Rightarrow P_\kappa \; .$$

On the other hand one gets by means of [MKM; 4.2.7.] for all $X \in \mathcal{B}$ and $n \in Z_+$

$$
\begin{aligned}
& {}_X\|(P_{\omega_n})_\kappa - P_{\omega_n}\| \\
& \leq \int {}_X\|((\omega_n)_{(a)})_\kappa - (\omega_n)_{(a)}\| \, \nu(da) \\
& = 2n^{-1}\langle \nu, {}_X\|\kappa^{[n]}_{(.)} - \kappa^{[0]}_{(.)}\| \rangle \\
& = 2n^{-1}\langle \nu, \kappa^{[n]}_{(.)}(\chi(X)>0) + \kappa^{[0]}_{(.)}(\chi(X)>0)\rangle \\
& \leq 2n^{-1}\langle \nu, J^{[n]}((.),X) + J^{[0]}((.),X)\rangle \\
& = 4n^{-1}\nu(X) \xrightarrow[n\to\infty]{} 0 \; .
\end{aligned}
$$

Hence follows $(P_{\omega_n})_\kappa \underset{n\to\infty}{\Rightarrow} P$ and, in summary, $P = P_\kappa$. A fortiori, the distribution $\mathcal{V}_P$ exists, and Theorem 5.7.6. can be applied to yield

$$(P_{\omega_n})_\kappa \Rightarrow V_{(\mathcal{V}_P)} \; .$$

This leads to $P = V_{(\mathcal{V}_P)}$; finally 5.2.2. tells that

$$\mathcal{V}_P \in V^{inv,\nu} \quad ; \quad \mathcal{V}_P(N_J) = 1 \; . \blacksquare$$

We conclude the section with

5.7.8. Proposition. A stable measure ν is of ergodic type iff $P := \Pi^{\kappa}_{\nu}$ satisfies

(o) $$\int | \mathcal{L}_{(\omega_n)_{(\Phi)}}(f) - \mathcal{L}_P(f) | P(d\Phi) \longrightarrow 0 \qquad (f \in F_c) \quad .$$

Proof: 1. Assume the validity of (o). First there follows, for any fixed $f \in F_c$, the convergence

$$\mathcal{L}_{(\omega_{m_i})_{(\Phi)}}(f) \underset{i\to\infty}{\longrightarrow} \mathcal{L}_P(f) \qquad \text{P-a. all } \Phi$$

for a strictly increasing sequence (m_i) depending on f. Choosing an appropriate countable subset of F_c and using the method of diagonal sequences, one arrives at

$$(\omega_{n_j})_{(\Phi)} \underset{j\to\infty}{\Rightarrow} P \quad \text{P-a. all } \Phi$$

for some suitable strictly increasing sequence (n_j) of natural numbers.
For any $X \in \mathcal{B}_\nu$ and $k \in \mathbb{Z}_+$ there results

(*) $$\lim_{j\to\infty} (\omega_{n_j})_{(\Phi)}(\Psi(X) = k) \;=\; P(\Psi(X) = k) \quad .$$

Hence follows for all m>0:

$$\limsup_{j\to\infty} \int |(\Phi * G_{n_j})(X) - \nu(X)| \, P(d\Phi)$$

$$= \limsup_{j\to\infty} \int | \sum_{k\geq 0} k \left((\omega_{n_j})_{(\Phi)}(\Psi(X) = k) - P(\Psi(X) = k)\right) | P(d\Phi)$$

$$\leq \limsup_{j\to\infty} \sum_{k\leq m} k \int | (\omega_{n_j})_{(\Phi)}(\Psi(X) = k) - P(\Psi(X) = k) | P(d\Phi)$$

$$+ \limsup_{j\to\infty} \sum_{k>m} k \int (\omega_{n_j})_{(\Phi)}(\Psi(X) = k)\, P(d\Phi) \;+\; \sum_{k>m} k P(\Psi(X) = k)$$

The first summand on the r.h.s. vanishes due to (*), the second one equals zero due to uniform integrability of $((\Phi * J^{[n]})(X))$ - and hence also uniform integrability of $((\Phi * G_n)(X))$ - with respect to P, and the third summand becomes arbitrarily small for large m.

In virtue of 5.1.14. there follows

$$I(\Phi, (.)) = \nu \quad \text{P-a. all } \Phi \quad ,$$

which due to Proposition 5.2.6., means that ν is of ergodic type.

2. Conversely, let us assume that the measure ν is of ergodic type. For all $f \in F_b$ we obtain

$$\limsup_{i\to\infty} \int |\mathcal{L}_{(\omega_i)_{(\Phi)}}(f) - \mathcal{L}_P(f)| \, P(d\Phi)$$

$$\leq \limsup_{i\to\infty} \int |\mathcal{L}_{(\omega_i)_{(\Phi)}}(f) - \mathcal{L}_{(\Pi_\Phi)_{\omega_i}}(f)| \, P(d\Phi)$$

$$+ \limsup_{n\to\infty} \limsup_{m\to\infty} \int | \mathcal{L}_{(\Pi_\Phi)_{\omega_{m+n}}}(f) - \mathcal{L}_{((\Pi_\Phi)_{\omega_m})_{\kappa^{[n]}}}(f) | P(d\Phi)$$

$$+ \limsup_{n\to\infty} \sup_{m=1,2,\ldots} \int | \mathcal{L}_{((\Pi_\Phi)_{\omega_m})_{\kappa^{[n]}}}(f) - \mathcal{L}_{(\Pi_{\Phi * G_m})_{\kappa^{[n]}}}(f) | P(d\Phi)$$

$$+ \limsup_{n\to\infty} \limsup_{m\to\infty} \int | \mathcal{L}_{(\Pi_{\Phi * G_m})_{\kappa^{[n]}}}(f) - \mathcal{L}_P(f) | P(d\Phi)$$

The first term on the right hand side of the preceding estimate vanishes according to 5.7.5. and the second step in the proof of Theorem 5.7.6.

Steps 1. and 2. in the proof of Theorem 5.7.3. reveal that also the second and the third term on the right hand side equal zero.

It remains to show that also the last term vanishes.

For short, we put

$$u_n := 1 - \mathcal{L}_{\kappa^{[n]}_{(.)}}(f) \qquad (n = 1,2,...)$$

and obtain

$$\langle \nu, u_n \rangle \leq \iint \langle \Phi, f \rangle \, \kappa^{[n]}_{(a)}(d\Phi) \, \nu(da)$$

$$= \iint f(b) \, J^{[n]}(a,db) \, \nu(da) = \langle \nu, f \rangle < +\infty ,$$

which shows that $u_n \in F_\nu$ for n= 1,2,... By means of 5.1.14 one gets for all natural numbers n

$$\limsup_{m\to\infty} \int | \mathcal{L}_{(\Pi_{\Phi * G_m})_{\kappa^{[n]}}}(f) - \mathcal{L}_P(f) | P(d\Phi)$$

$$= \limsup_{m\to\infty} \int | \mathcal{L}_{(\Pi_{\Phi * G_m})_{\kappa^{[n]}}}(f) - \mathcal{L}_{(\Pi_{I(\Phi,(.))})_{\kappa^{[n]}}}(f) | P(d\Phi)$$

$$= \limsup_{m\to\infty} \int | \exp(-\int (1 - \mathcal{L}_{\kappa^{[n]}_{(a)}}(f))(\Phi * G_m)(da)) - \exp(-\int (1 - \mathcal{L}_{(\kappa^{[n]}_{(a)}}(f)) I(\Phi,da)) | P(d\Phi)$$

$$\leq \limsup_{m\to\infty} \int | \langle G_m * \Phi, u_n \rangle - \langle I(\Phi,(.)), u_n \rangle | P(d\Phi) = 0 \quad . \blacksquare$$

5.8. Ergodic Intensity Distributions of Poisson Type Equilibria

The analogies of $\mathcal{V}_{\Pi^\kappa_\nu}$ to $\mathcal{W}_{\Pi^\kappa_\nu}$ and of $\mathcal{V}_{\Pi_\nu}$ to $\mathcal{W}_{\Pi_\nu}$ seduce to conjecture that for any stable measure ν the distribution $\mathcal{V}_{\Pi_\nu}$ exists and coincides with $\mathcal{V}_{\Pi^\kappa_\nu}$. This conjecture turns out to be false, however, as the following example shows:

5.8.1. Example. With the assumptions and notations of Example 2.8.4 , the distribution $\mathcal{V}_{\Pi_\nu}$ does not exist.

Proof. 1. Let us recall the time evolution starting with a single individual at site $[i,j] \in A$, that is , at "position" $-i$ and "age" j.

In case $j > 2^{i-1}$, this individual inhabits the position $-i$ up to a certain time and then dies, without ever having produced an offspring at some other positions.

In case $1 \le j \le 2^{i-1}$, this individual produces, at time $2^{i-1}-j+1$, an offspring at positions $\{0, -1, ..., -i+1\}$, which regenerates at least the (single) inhabitant of position 0 exactly until the original individual dies, i. e. up to time 2^i-j.

Thus we obtain, for all $[-i, j] \in A$ and $n \in \mathbf{Z}_+$

$$J^{[n]}([-i,j], \{a_0\}) = \begin{cases} 1 & \text{if } j \le 2^{i-1} \text{ and } 2^{i-1} < j+n \le 2^i \\ 0 & \text{otherwise.} \end{cases}$$

where here and below a_0 denotes the site $[0,1] \in A$ (corresponding to position zero and age one).

2. For short, we put

$$p_n := \Pi_v((\Phi * G_n)(\{a_0\}) \in (.)) \qquad (n=1,2,...)$$

and denote, for a fixed real number $c \in [1/2, 1)$ and any $m \in \mathbf{Z}_+$, the integer part of $c \cdot 2^m$ by $n(m)$.

We are going to show in step 3. that

$$p_{n(m)} \underset{m \to +\infty}{\Rightarrow} q_c ,$$

where $q_c = \mathcal{E}_{\beta_c}$ is an infinitely divisible distribution on $\boldsymbol{B}_+$ with canonical measure

$$\beta_c := c\big(\mu((.) \cap [0,1]) + \mu((.) \cap [1-1/(2c), 1/(2c)])\big) + (c-1/2)\delta_{1/(2c)} + (1/2) \sum_{k \ge 1} \delta_{1/(2^{k+1}c)}$$

and μ denotes Lebesgue measure on the real axis.

Because of

$$\beta_c(\{0\}) = 0 \quad ; \quad \int x \, \beta_c(dx) = 1 \; ,$$

β_c indeed satisfies the conditions which are characteristic for canonical measures of infinitely divisible distributions on $\boldsymbol{B}_+$, hence q_c is well-defined. Obviously there holds $\beta_c \ne \beta_{c'}$ for $1/2 \le c < c' < 1$, which reveals that not all subsequences of $\big(\Pi_v(\Phi * G_n \in (.))\big)_{\{a_0\}}$ converge towards one and the same limit and therefore shows the non-existence of $\mathcal{V}_{\Pi_v}$.

3. The distribution $\Pi_v((\Phi * G_n \in (.))$ is infinitely divisible, and so is the distribution p_n, obeying

$$p_n = \mathcal{E}_{\alpha_n} \; , \quad \alpha_n(X) = v(g_n(a) \in X) \; , \quad X \in \boldsymbol{B}_+ \; ,$$

where we use the shorthand notation

$$g_n(a) := G_n(a, \{a_0\}) \qquad (a \in A) \; .$$

In order to verify the convergence asserted in 2. it thus suffices to check the following conditions

a) For all continuous nonnegative real functions f on $(0, +\infty)$ with compact support there holds

$$\lim_{m \to +\infty} \int f(x) \alpha_{n(m)}(dx) = \int f(x) \beta_c(dx) \; .$$

b)
$$\lim_{s \to +\infty} \limsup_{m \to +\infty} \alpha_{n(m)}([s, +\infty)) = 0 \; .$$

c)
$$\lim_{\varepsilon \to 0+} \limsup_{m \to +\infty} \int_{(0,\varepsilon)} x \, \alpha_{n(m)}(dx) = 0 \; .$$

Let us start in proving a), and first extend f continuously to $(0, +\infty)$ by putting $f(0) := 0$.

For all natural numbers m there holds

$$(*) \qquad \int f(x)\alpha_{n(m)}(dx) = \sum_{i\geq 0} 2^{-i} \sum_{1\leq j\leq 2^i} f(g_{n(m)}([-i,j])) := S_m$$

where

$$g_{n_m}([-i,j]) = n(m)^{-1} \sum_{0\leq n<n(m)} J^{[n]}([-i,j]),\{a_0\}) \ .$$

According to step 1., $J^{[n]}([-i,j]),\{a_0\})$ - and hence also $f(g_{n(m)}([-i,j]))$ - vanishes if $j> 2^{i-1}$. Hence (*) may be rewritten as

$$(**) \qquad S_m = \sum_{i\geq 0} 2^{-i} \sum_{1\leq j\leq 2^{i-1}} f((g_{n(m)}([-i,j])) =: \sum_{i\geq 0} s_{m,i} \ .$$

For all natural numbers j between 1 and 2^{i-1}, step 1. tells that $J^{[n]}([-i,j]),\{a_0\})$ equals one if $2^{i-1}-j<n\leq 2^i-j$, and equals zero otherwise. For $1\leq j\leq 2^{i-1}$, the above stated expression for $g_{n(m)}([-i,j])$ may therefore be rewritten as

$$g_{n(m)}([-i,j]) = n(m)^{-1} \max\{0, \min\{n(m)-1-2^{i-1}+j, 2^{i-1}\}\} \ .$$

In distinguishing the cases i=m, i>m and i<m, this boils down to the following:

for i=m : $\quad g_{n(m)}([-i,j]) = n(m)^{-1}\min\{n(m)-1-2^{i-1}+j, 2^{i-1}\}$,

for i>m : $\quad g_{n(m)}([-i,j]) = n(m)^{-1}\max\{0, n(m)-1-2^{i-1}+j\}$,

for i<m : $\quad g_{n(m)}([-i,j]) = n(m)^{-1} 2^{i-1}$.

Hence the summands $s_{m,i}$ occuring in (**) may be rewritten as follows

for i=m:

$$s_{m,m} = 2^{-m}\Big(\Big(\sum_{1\leq j\leq 2^m-n(m)} f(n(m)^{-1}(n(m)-1-2^{m-1}+j))\Big)+(n(m)-2^{m-1})f(n(m)^{-1}2^{m-1})\Big) ,$$

for i=m+k, k>0:

$$s_{m,m+k} = 2^{-m-k}\Big(\sum_{2^{m+k-1}-n(m)<j\leq 2^{m+k-1}} f(n(m)^{-1}(n(m)-1-2^{m+k-1}+j))\Big) ,$$

for i=m−k, 0<k≤m:

$$s_{m,m-k} = 2^{-m+k}\cdot 2^{m-k-1} f(n(m)^{-1} 2^{m-k-1}) \ .$$

The sum S_m can now be split as follows:

$$S_m = s_{m,m} + \sum_{k>0} s_{m,m+k} + \sum_{0<k\leq m} s_{m,m-k} \ .$$

In view of the assumed continuity of f, the following convergence relations are fairly obvious:

$$s_{m,m} \xrightarrow[m\to\infty]{} c \int_{[1-1/(2c),1/(2c)]} f(x)\mu(dx) + (c-1/2)f(1/(2c))$$

$$s_{m,m+k} \xrightarrow[m\to\infty]{} 2^{-k}\cdot c \int_{[0,1]} f(x)\mu(dx) \qquad (k>0)$$

$$s_{m,m-k} \xrightarrow[m\to\infty]{} 2^{-1}f(1/(2^{k+1}c)) \qquad (k>0) \ .$$

For k>0 there further holds

$$s_{m,m+k} \leq 2^{-k} \max_{x\in\mathbf{R}_+} f(x) \quad ,$$

whereas for $0<k\leq m$ one has the estimate

$$s_{m,m-k} \leq 2^{-1} \max_{x\in\mathbf{R}_+} f(x)\, 1_{\operatorname{supp} f}(2^{-k}) \ .$$

Therefore, Lebesgue's dominated convergence theorem guarantees that summation and passage to the limit may be interchanged in the computation of $\lim_{m\to\infty} S_m$, which leads immediately to a).

Condition b) is clearly fulfilled, because $g_{n(m)}$ is bounded from above by the constant one.

Finally let us turn to c). Inserting $x\cdot 1_{(0,\varepsilon)}(x)$ instead of f(x) into formula (*) and exploiting the above computations, one gets

$$\int_{(0,\varepsilon)} x\alpha_{n_m}(dx) \ \leq \ 2\varepsilon + 2^{-1} \sum_{k:k>0,2^{-(k+1)}<\varepsilon} 2^{-k} \ < \ 4\varepsilon \ ,$$

which shows the validity of c) and thus completes the proof. ■

Henceforth, N_κ^{reas} will denote the set of all those ("reasonable") measures ν in N_κ^{st} for which there holds

$$\mathcal{V}_{\Pi_\nu} = \mathcal{V}_{\Pi^\kappa_\nu} \quad .$$

This set contains N_κ^{erg}, because in virtue of Theorem 5.4.1. and 5.7.5. there holds

5.8.2. Proposition. $N_\kappa^{erg} = N_J^{erg} \cap N_\kappa^{reas}$.

"Reasonableness" of a stable measure ν can be decided from the intensity kernel alone, because in view of 5.3.2. one has

5.8.3. A measure ν in N_κ^{st} belongs to N_κ^{reas} iff there holds

$$\mathcal{V}_{\Pi_\nu} = \mathcal{V}_{\mathcal{W}_{\Pi_\nu}} \quad .$$

By a simple trick we deduce

5.8.4. Proposition. If the kernel J is substochastic, then N_κ^{st} equals N_κ^{reas} .

Proof. Putting

$$\kappa' := \tau(J) \quad ,$$

we obtain a substochastic translation κ'. Obviously there holds

$$N_\kappa^{st} \subseteq N_J = N_{\kappa'}^{st} \quad .$$

For all $\nu\in N_\kappa^{st}$ one has $\Pi^{\kappa'}_\nu = \Pi_\nu$ and hence clearly $\nu\in N_{\kappa'}^{reas}$. Since the intensity kernels of κ and κ' coincide, this implies our assertion in view of the remark in front of 5.8.3. ■

The following theorem characterizes "reasonableness" of stable measures:

5.8.5. Theorem. For any measure $\nu \in N_\kappa^{st}$ the folowing assertions are equivalent:

a) $\nu \in N_\kappa^{reas}$.

b) $(\Pi^\kappa{}_\nu)_{\omega_n} \Rightarrow \Pi^\kappa{}_\nu$.

c) $\int \left(\left(\widetilde{\Pi^\kappa{}_\nu}\right)^!{}_a \right)_{\omega_n} (\Phi(X)>0)\, G_n(a,X)\, \nu(da) \longrightarrow 0 \qquad (X \in \mathcal{B})$.

Proof. 1. To show the equivalence of b) and c), it suffices to mimic the proof of Theorem 3.3.8. in replacing there P_n, $\kappa^{[n]}$ and $J^{[n]}$, $n \in \mathbf{Z}_+$, by $\Pi^\kappa{}_\nu$, ω_n and G_n, respectively.

2. For all $X \in \mathcal{B}$, together with $((\Phi * J^{[n]})(X))$ also the sequence $((\Phi * G_n)(X))$ is uniformly integrable under $\Pi^\kappa{}_\nu$, i. e. there holds

$$(*) \qquad \sup_{n=1,2,\ldots} \int_{\{\Phi:(\Phi*G_n)(X)>d\}} (\Phi*G_n)(X)\, (\Pi^\kappa{}_\nu)(d\Phi) \xrightarrow[d\to\infty]{} 0 \qquad (X \in \mathcal{B}) .$$

On the other hand we get for n=1,2,...

$$\int_{\{\Phi:(\Phi*G_n)(X)>d\}} (\Phi*G_n)(X)\, \Pi^\kappa{}_\nu(d\Phi)$$

$$= \int_{\{\Phi:(\Phi*G_n)(X)>d\}} \int G_n(a,X)\, \Phi(da)\, (\Pi^\kappa{}_\nu)(d\Phi)$$

$$= \int \int_{\{\Phi:(\Phi*G_n)(X)>d\}} G_n(a,X)\, (\Pi^\kappa{}_\nu)_a(d\Phi)\, \nu(da)$$

$$= \int G_n(a,X) \left(\left(\widetilde{\Pi^\kappa{}_\nu}\right)_a * \Pi^\kappa{}_\nu \right) ((\Phi*G_n)(X)>d)\, \nu(da)$$

$$\geq \int G(a,X) \left(\widetilde{\Pi^\kappa{}_\nu}\right)_a ((\Phi*G_n)(X)>d)\, \nu(da)$$

$$= \int_{\{\Psi:(\Psi*G_n)(X)>d\}} (\Psi*G_n)(X)\, \widetilde{\Pi^\kappa{}_\nu}(d\Psi) .$$

Hence (*) implies the convergence relation

$$(**) \qquad \sup_{n=1,2,\ldots} \int_{\{\Psi:(\Psi*G_n)(X)>d\}} (\Psi*G_n)(X)\, \widetilde{\Pi^\kappa{}_\nu}(d\Psi) \xrightarrow[d\to\infty]{} 0 \qquad (X \in \mathcal{B}) .$$

3. Let us now show that a) is a consequence of b). We choose some $f \in F_c$, and put $\|f\| := \sup_{a \in A} f(a)$; $X := \operatorname{supp} f$. Further we write $P := \Pi^\kappa{}_\nu$ for abbreviation. Replacing ν_i, $i \in \mathbf{Z}_+$, by ν, $\kappa^{[n]}$ by ω_n and $J^{[n]}$ by G_n, $n \in \mathbf{Z}_+$, in the proof of Theorem 4.3.5. and paralleling all arguments line by line, we get for all c, d>0

$$| \mathcal{L}_P(G_n*f) - \mathcal{L}_{\Pi_\nu}(G_n*f) |$$

$$\leq \|f\|^2 \cdot c \int G_n(a,X) (\widetilde{P}^!{}_a)_{\omega_n} (\Phi(X)>0)\, \nu(da)$$

$$+ (1+\|f\|)^2 \int_{\{\Psi:(\Psi * G_n)(X)>d\}} (\Psi * G_n)(X)\,\tilde{P}(d\Psi)$$

$$+ (1+\|f\|)^2 \cdot d \int_{\{\Phi:(\Phi(X)>c\}} \Phi(X)\,(\tilde{P})_{\omega_n}(d\Phi) \quad .$$

Due to 1. the first term on the right hand side tends, for all $c>0$, towards zero for $n\to\infty$. According to (**) in 2., the second term can be made, by a suitable choice of d, arbitrarily small uniformly in n.

We now show that the third term can, for all $d>0$, be made arbitrary small uniformly in n by a suitable choice of c. Indeed, the relations

$$\Lambda_{\widetilde{P_{\omega_n}}} = \nu = \Lambda_{\tilde{P}} \quad \text{for } n=1,2,\dots \;;\quad \widetilde{P_{\omega_n}} \overset{w}{\Rightarrow} \tilde{P}$$

imply

$$\sup_{n=1,2,\dots} \int_{\{\Phi:(\Phi(X)>c\}} \Phi(X)(\tilde{P})_{\omega_n}(d\Phi)$$

$$= \sup_{n=1,2,\dots} \int_{\{\Phi:(\Phi(X)>c\}} \Phi(X)\,\widetilde{P_{\omega_n}}(d\Phi) \underset{c\to\infty}{\longrightarrow} 0 \quad .$$

In summary we obtain

$$|\mathcal{L}_{\Pi^\kappa_\nu}(G_n * f) - \mathcal{L}_{\Pi_\nu}(G_n * f)| \longrightarrow 0 \qquad (f\in F_c) \quad ,$$

i. e. the validity of a).

4. Finally we prove that a) implies c). Let g be the mapping defined in 4.3.6., and X be some set in $\mathcal{B}$. For all natural numbers n there results

$$|\ln \mathcal{L}_{\Pi_\nu}(G_n * 1_X) - \ln \mathcal{L}_{\Pi^\kappa_\nu}(G_n * 1_X)|$$

$$= |\ln \mathcal{L}_{\Pi_\nu}(G_n((.),X)) - \ln \mathcal{L}_P(G_n((.),X))|$$

$$= \Big|\int (1-\exp(-G_n(a,X)))\nu(da) - \int (1-\exp(-\langle\Psi, G_n((.),X)\rangle))\tilde{P}(d\Psi)\Big|$$

$$= \iint \big(g(G_n(a,X)) - g(G_n(a,X) + (\Psi * G_n)(X))\big) G_n(a,X)\,(\tilde{P}^!_a)(d\Psi)\,\nu(da)$$

$$\geq \iint \big(g(G_n(a,X)) - g(G_n(a,X) + (\omega_n)_{(\Psi)}(\chi(X)>0))\big)\, G_n(a,X)\,(\tilde{P}^!_a)(d\Psi)\,\nu(da) \quad ,$$

where we used the monotonicity of the mapping g in the last estimate. In virtue of inequality (δ) in 4.3.6. we may proceed, for all $c>0$, by

$$\geq \iint (\omega_n)_{(\Psi)}(\chi(X)>0)\, |g'(G_n(a,X) + (\omega_n)_{(\Psi)}(\chi(X)>0))|\, G_n(a,X)\,(\tilde{P}^!_a)(d\Psi)\,\nu(da)$$

$$\geq \int (\tilde{P}^!_a)_{\omega_n}(\chi(X)>0)\, |g'(G_n(a,X)+1)|\, G_n(a,X)\,\nu(da)$$

$$\geq |g'(c+1)| \int_{\{a:G_n(a,X)\leq c\}} (\tilde{P}^!_a)_{\omega_n}(\chi(X)>0)\, G_n(a,X)\,\nu(da) \quad .$$

Bearing in mind that $|g'(c+1)|$ is strictly positive, we infer by means of 5.2.10. from a) the validity of

$$(+) \qquad \int_{\{a:G_n(a,X)\leq c\}} (\tilde{P}^!_a)_{\omega_n}(\chi(X)>0)\, G_n(a,X)\,\nu(da) \underset{c\to\infty}{\longrightarrow} 0 \quad .$$

On the other hand, the convergence relation (*) in step 2. yields

$$\limsup_{c\to\infty} \sup_{n=1,2,\dots} \int_{\{a:G_n(a,X)>c\}} G_n(a,X)\,\nu(da)$$

$$= \limsup_{c\to\infty} \sup_{n=1,2,\dots} \int_{\{a:G_n(a,X)>c\}} G_n(a,X)\,\Phi(da)\,P(d\Phi)$$

$$\leq \limsup_{c\to\infty} \sup_{n=1,2,\dots} \int_{\{\Phi:(\Phi * G_n)(X)>c\}} (\Phi * G_n)(X)\,P(d\Phi) = 0 \quad .$$

Together with (+), this shows the validity of c). ■

5.8.6.Proposition. N_κ^{reas} is a descending subset of $[\,N_\kappa^{st}, \leq]$. If $(\nu_i)_{i\in I}$ is an at most countable family of measures in N_κ^{reas} whose sum ν belongs to N, then ν even belongs to N_κ^{reas}. Together with γ, any multiple $c\cdot\gamma$, $c\in\mathbf{R}_+$, belongs to N_κ^{reas}.

Proof. 1. For all $\nu\in N_\kappa^{st}$, all natural numbers n and $X\in\mathcal{B}$ there holds

$$\int ((\widetilde{\Pi^\kappa_\nu})^!_a)_{\omega_n} (\Phi(X)>0)\, G_n(a,X)\,\nu(da)$$

$$= \iint (\omega_n)_{(\Psi)}(\Phi(X)>0) G_n(a,X) \mathcal{C}^! _{\widetilde{\Pi^\kappa_\nu}}(d[a,\Psi]) \quad .$$

The latter expression obviously is additive in ν; together with Theorem 5.8.5. this shows that N_κ^{reas} is a descending subset of $[N_\kappa^{st}, \leq]$.

2. The sum of finitely many measures in N_κ^{reas} belongs to N_κ^{reas} due to Theorem 5.8.5.

3. Let $(\nu_i)_{i\geq 1}$ be a sequence of measures in N_κ^{reas} whose sum ν belongs to N and hence by 2.3.8. to N_κ^{st}. For abbreviation, we write

$$\lambda_m := \sum_{1\leq i\leq m} \nu_i \qquad (m\geq 1) \quad .$$

For fixed $f\in F_c$ we choose, for any given $\varepsilon>0$, a natural number m such that $\langle \nu-\lambda_m, f\rangle < \varepsilon$.

The following estimate, which holds true for all $P\in\mathbf{P}$, is a simple consequence of Jensen's inequality:

$$1 \geq \mathcal{L}_P(f) \geq e^{-\langle \Lambda_P, f\rangle} \geq 1-\langle \Lambda_P, f\rangle \quad .$$

By means of the preceding estimate we arrive at

$$\limsup_{n\to\infty} | \mathcal{L}_{(\Pi^\kappa_\nu)_{\omega_n}}(f) - \mathcal{L}_{\Pi^\kappa_\nu}(f) |$$

$$= \limsup_{n\to\infty} | \mathcal{L}_{(\Pi^\kappa_{\lambda_m})_{\omega_n}}(f)\, \mathcal{L}_{(\Pi^\kappa_{\nu-\lambda_m})_{\omega_n}}(f) - \mathcal{L}_{\Pi^\kappa_{\lambda_m}}(f)\, \mathcal{L}_{\Pi^\kappa_{\nu-\lambda_m}}(f) |$$

$$\leq 2\varepsilon + \limsup_{n\to\infty} | \mathcal{L}_{(\Pi^\kappa_{\lambda_m})_{\omega_n}}(f) - \mathcal{L}_{(\Pi^\kappa_{\lambda_m})}(f) | \quad .$$

Now 2. together with Theorem 5.8.5. yields that $\nu\in N_\kappa^{reas}$.

4. For any $\gamma \in N_\kappa^{reas}$ and $c \in \mathbf{R}_+$, $c \cdot \gamma$ is majorized by some integer multiple of γ and hence belongs to N_κ^{reas} due to 2. and 1. ■

Combining the above result with 5.8.2. and Proposition 5.6.2., one gets

5.8.7. Proposition. A stable measure ν belongs to N_κ^{reas} iff this holds true for all those minorants of ν which are of finite type.

Finally we show :

5.8.8. Proposition. $N_\kappa^{ff} \subseteq N_\kappa^{reas}$.

Proof: Let ν be a stable measure which is finitely fed. We put $P := \Pi^\kappa_\nu$, and choose some set $S \in \mathcal{A}_{(P)}^{stoch}$. Writing $\gamma := {}_{A\setminus S}\nu$, we obtain according to Theorem 2.10.1. the convergence $\gamma * J^{[n]} \downarrow o$, implying that $\gamma * G_n \downarrow o$.

For any function $f \in F_c$ and $n \in \mathbf{Z}_+$ there follows in virtue of Proposition 2.10.10.

$$
\begin{aligned}
& | \mathcal{L}_P(G_n * f) - \mathcal{L}_{\Pi_\nu}(G_n * f) | \\
\leq \; & | \mathcal{L}_P(G_n * f) - \mathcal{L}_{{}_S P}(G_n * f) | + | \mathcal{L}_{\Pi_\nu}(G_n * f) - \mathcal{L}_{\Pi_{{}_S\nu}}(G_n * f) | \\
\leq \; & \int \langle {}_{A\setminus S}\Phi, G_n * f \rangle P(d\Phi) + \int \langle {}_{A\setminus S}\Phi, G_n * f \rangle \Pi_\nu(d\Phi) \\
= \; & 2\langle \gamma, G_n * f \rangle = 2\langle \gamma * G_n, f \rangle \downarrow 0 \;.
\end{aligned}
$$

Hence results $\quad \mathcal{V}_{\Pi_\nu} = \mathcal{V}_P$. ■

Note that Proposition 5.8.7. and 5.8.8. together with Proposition 2.10.9. offer a "trickless" approach to Proposition 5.8.4.

5.9. An Example of a Non-extremal Poisson Type Equilibrium Distribution whose Intensity Measure is J-ergodic

5.9.1. Proposition. A stable measure ν belongs to N_J^{erg} iff it obeys

$$(+) \qquad \nu(\{a: a \in A, G_n(a,X) > \varepsilon\}) \xrightarrow[n\to\infty]{} 0 \qquad (X \in \mathcal{B}; \varepsilon > 0) \;.$$

Proof. Obviously, condition (+) is fulfilled for all $\nu \in N_J^{erg}$.

Now assume that ν belongs to N_κ^{st}. Similar to 2.2.9. there results

$$\lim_{c\to\infty} \sup_{n=0,1,2,\dots} \int_{\{a: G_n(a,X) > c\}} G_n(a,X)\, \nu(da) = 0 \qquad (X \in \mathcal{B}) \;.$$

If ν, in addition, satisfies condition (+), then we conclude like in the fourth step of the proof of Theorem 4.4.5. - only replacing (ν_i) by the constant sequence (ν) and $J^{[n]}$ by G_n - that ν belongs to N_J^{erg}. ■

Proposition 5.8.2. and 5.8.4. imply directly

5.9.2.Theorem. If the intensity kernel J is substochastic, then there holds

$$(*) \qquad N_\kappa^{erg} = N_J^{erg} \cap N_\kappa^{st} \quad .$$

It is a natural question whether the restrictive assumption in the preceding theorem is superfluous, and indeed the answer has been conjectured to be affirmative for quite a long period. A first indication that (*) might not be true in general is provided by Example 5.8.1., which rules out one tool of proving (*), namely the general validity of

$$\mathcal{V}_{\Pi_\nu} = \mathcal{V}_{\Pi^\kappa_\nu} \quad , \quad \nu \in N_\kappa^{st} \quad .$$

The following more sophisticated scenario provides a counterexample to (*):

5.9.3. Example. Let us put

$$A := \bigcup_{k\geq 1} Z\times\{k\}\times\{1,...,2^k\}$$

and choose some metric ρ_A such that just the finite subsets of A are ρ_A - bounded. For all $a = [i, k, j] \in A$ we interpret δ_a as description of an individual at position [i, k] and age j.

Now we introduce, step by step, a deterministic branching dynamics κ.

a) For $[i, k, j] \in A \;\; ; \;\; j \notin \{2^{k-1}, 2^k\}$ we put

$$\kappa_{(i,k,j)} := \delta_{\delta_{[i+1,k,j+1]}} \quad .$$

This describes transition from [i, k, j] to [i+1, k, j+1].

b) For $[i, k, j] \in A \;\; , \;\; j = 2^k$ we put

$$\kappa_{(i,k,j)} := \delta_o \quad ,$$

i. e. an individual at position [i, k] dies after reaching the age level 2^k.

An individual at position [i, k] with "medium age" 2^{k-1} survives according to the mechanism in a) and, in addition, generates an offspring of individuals of age 1. In order to describe this offspring in a transparent way, let us denote by $\angle$ the minimal semi-order in $Z\times\{1, 2,...\}$ with the properties

$$[i, k] \angle [i, k+1]$$

and

$$[i-2^{(k/2)-1}, k-1] \;\; , \;\; [i+2^{(k/2)-1}, k-1] \angle [i, k]$$

for all $i \in Z$ and all even natural numbers k.

For abbreviation we put for all $i \in Z$ and all natural numbers k

$$\chi[i,k] := \sum_{[x,y]\angle[i,k]} \delta_{[x,y,1]} \quad .$$

The definition of κ is now completed by

c) For $[i, k, j] \in A \;\; , \;\; j = 2^{k-1}$ we put

$$\kappa_{(i,k,j)} := \delta_{\delta_{[i+1,k,j+1]}} + \chi_{[i+1,k]} \quad .$$

For any $c \in \mathbf{R}_+$, its integer part will, as usual, be denoted by $[c]$.

With the notation

$$\mu := \sum_{i \in \mathbf{Z}} \delta_i \quad ; \quad \mu_k := 2^{-k} \sum_{1 \leq j \leq 2^k} \delta_j \quad \text{for } k = 1,2,...;$$

$$\nu := \sum_{k \geq 1} 2^{-([k/2]+1)} \mu \otimes \delta_k \otimes \mu_k$$

there holds $\nu \neq o$, $\nu \in N_J^{erg} \cap N_K^{fin}$ and hence according to Theorem 5.4.1. also $\nu \notin N_K^{erg}$.

Proof. 1. Let $\xi := (\xi_s)_{s \in \mathbf{Z}_+}$, $\eta := (\eta_s)_{s \in \mathbf{Z}_+}$ be two stochastically independent i. i. d.-sequences of random elements of $\{0,1\}$, where ξ_s as well as η_s has distribution $2^{-1}(\delta_0 + \delta_1)$. If we put, in the sense of binary addition "carrying to the right",

$$\xi' := \xi + (1,0,0,...) \quad ; \quad \eta' := \eta + (1,0,0,...) \quad ,$$

then $[\xi', \eta']$ has the same distribution as $[\xi, \eta]$.

2. For all natural numbers k we put

$$\sigma_k(\xi) := 1 + \sum_{0 \leq s < k} 2^s \xi_s \quad ,$$

$$\zeta_k(\eta) := 1 + \sum_{0 \leq s < [k/2]} 2^s \eta_s \quad ,$$

$$Z_k := \{r \cdot 2^{[k/2]+1} + 2^{[k/2]}\}_{r \in \mathbf{Z}} \quad ,$$

noting, in particular, that $\sigma_k(\xi)$ and $\zeta_k(\eta)$ have distributions μ_k and $\mu_{[k/2]+1}$, respectively.

Obviously, $\bigcup_{k \leq 1} Z_k$ is a descending set with respect to the semi-order $\angle$. Further, we put for k= 1, 2,...

$$\Psi_k[\eta] := \sum_{z \in Z_k} \delta_{z + \zeta_k(\eta)} \quad ,$$

thus getting random counting measures with phase space $\mathbf{Z}$. Note that $\Psi_k[\eta]$ is obtained by shifting $\sum_{z \in Z_k} \delta_z$ to the right with "phase" $\zeta_k(\eta)$.

Via

$$\Psi[\eta] := \sum_{k \geq 1} (\Psi_k[\eta]) \otimes \delta_k$$

one gets a random counting measure with phase space $\mathbf{Z} \times \{1, 2,...\}$.

Finally we put

$$\Phi[\xi,\eta] := \sum_{k \geq 1} (\Psi_k[\eta]) \otimes \delta_k \otimes \delta_{\sigma_k(\xi)} \quad ,$$

obtaining in this way a random counting measure with phase space $[A, \rho_A]$, whose distribution will be denoted by P.

Note that the construction of $\Phi[\xi, \eta]$ can be split into two steps, in first forming the "spatial" population $\Psi[\eta]$, which afterwards is "marked" by means of ξ.

For all $a = [i, k, j] \in A$ we obtain

$$\Lambda_P(\{a\}) = 2^{-([k/2]+1)} \mu_k(\{j\}) = \nu(\{a\}) \quad .$$

Hence there holds $\Lambda_P = \nu$.

3. For all natural numbers k, $\zeta_k(\eta')$ is the "cyclic successor" of $\zeta_k(\eta)$ in the set $\{1,...,2^{[k/2]+1}\}$, and hence

$$\Psi_k[\eta'] = \sum_{z \in Z_k} \delta_{z+\zeta_k(\eta)+1} \quad .$$

Therefore, $\Psi[\eta']$ arises from $\Psi[\eta]$ by the shift $[i, k] \rightarrow [i+1, k]$.

Bearing in mind that also $\sigma_k(\xi')$ is, for all natural numbers k, the cyclic successor of $\sigma_k(\xi)$ in the set $\{1,...,2^k\}$, one recognises:

$\Phi[\xi', \eta']$ arises from $\Phi[\xi, \eta]$ by shifting $[i, k, j]$ into $[i+1, k, j']$, where j' denotes the cyclic successor of j in the set $\{1,...,2^k\}$.

4. We are now going to show that, apart from the case $\xi = (1,1,...)$, there holds

(o) $$\kappa_{(\Phi[\xi,\eta])} = \delta_{\Phi[\xi',\eta']} \quad ,$$

which implies due to 1.

$$\nu \in N_\kappa^{st} \quad ; \quad P \in P^{inv,\nu} \quad .$$

If all entries of the sequence ξ equal zero, then the relation (o) is trivially satisfied because of 3.

Otherwise we put

$$k(\xi) := \min \{s : s \in \mathbf{Z}_+, \xi_s = 0\} \quad .$$

Obviously there holds

$$\sigma_s(\xi) = 2^s \qquad \text{if} \qquad s < k(\xi) \ ;$$
$$\sigma_s(\xi) = 2^{s-1} \qquad \text{if} \qquad s = k(\xi) \ ;$$
$$\sigma_s(\xi) \notin \{2^{s-1}, 2^s\} \qquad \text{if} \qquad s > k(\xi) \ .$$

Relying on step 3., we now determine the action of the deterministic clustering field κ on the three parts

$$\Phi_1 := \sum_{s>k(\xi)} (\Psi_s[\eta]) \otimes \delta_s \otimes \delta_{\sigma_s(\xi)} \quad ,$$

$$\Phi_2 := (\Psi_{k(\xi)}[\eta]) \otimes \delta_{k(\xi)} \otimes \delta_{2^{k(\xi)-1}}$$

$$\Phi_3 := \sum_{1 \le s < k(\xi)} (\Psi_s[\eta]) \otimes \delta_s \otimes \delta_{2^s} \quad .$$

of $\Phi[\xi, \eta]$.

a) Φ_1 is subject to the shift $[i, k, j] \rightarrow [i+1, k, j+1]$ and hence is turned into

$$\sum_{s>k(\xi)} (\Psi_s[\eta']) \otimes \delta_s \otimes \delta_{\sigma_s(\xi')} \quad .$$

b) Φ_3 dies as a whole, i. e. is turned into o .

c) Φ_2 is transformed into the sum

$$(\Psi_{k(\xi)}[\eta']) \otimes \delta_{k(\xi)} \otimes \delta_{\sigma_{k(\xi)}(\xi)+1}$$
$$+ \sum_{1 \le s < k(\xi)} \Big(\sum_{[i,k] \angle \operatorname{supp} \Psi_{k(\xi)}[\eta]} \delta_{[i+1,k]} \Big) \otimes \delta_1 \quad ,$$

i. e. into

$$\sum_{1 \le s \le k(\xi)} (\Psi_s[\eta']) \otimes \delta_s \otimes \delta_{\sigma_s(\xi')} \quad .$$

Hence, (o) is valid, provided $\xi \neq (1,1,...)$.

5. By means of Theorem 2.8.3. we are now going to show that P equals $\widetilde{\Pi^{\kappa}_{\nu}}$ and hence ν is of finite type.

To begin with, it is clear that $P(\{o\}) = 0$. Like at the end of the discussion of Example 2.8.4. one sees that also

$$\operatorname{sing}(\Phi, n, X) \xrightarrow[n\to\infty]{P} 1 \qquad (X \in \mathcal{B}) \quad .$$

6. Since κ is deterministic, there holds

$$\kappa^{[n]}_{(a)} = \delta_{J^{[n]}(a,(.))} \qquad (a\in A \quad ; \quad n\in \mathbf{Z}_+) \quad .$$

The definition of κ implies readily for all $[i, j, k]\in A$ and all natural numbers n:

$$J^{[n]}([i, k, j], (.)) = \begin{cases} \delta_{[i+n,\, k,\, j+n]} & \text{if } j+n \le 2^{k-1} \text{ or } 2^{k-1} < j \le 2^k - n \\ \sum\limits_{[x,y]\angle[i+n,k]} \delta_{[x,\, y,\, ((j+n-1) \bmod 2^y)+1]} & \text{if } j \le 2^{k-1} < j+n \le 2^k \\ o & \text{otherwise} \quad . \end{cases}$$

This has the following consequences:

(u) There holds

$$J^{[n]}(a, \{a'\}) \in \{0, 1\} \qquad (a,a'\in A \quad ; \quad n\in \mathbf{Z}_+)$$

and hence

$$G_m(a, \{a'\}) \le 1 \qquad (a,a'\in A \quad ; \quad m = 1, 2,\ldots)$$

(v) For any $[i, k, j]$, $[x, y, z]\in A$ and $n\in \mathbf{Z}_+$ with the property

$$J^{[n]}([i, k, j] \ , \ \{[x, y, z]\}) > 0$$

there holds $[x, y] \angle [i+n, k]$.

In the sequel we still need the elementary estimate

(w) Assume that $1\le y < k$ and $x\in \mathbf{Z}$. Then there holds

$$\#\{i: i\in \mathbf{Z} \ , \ [x, y] \angle [i, k]\} \ \le \ 2^{[k/2]} \quad .$$

7. Now let $[x,y,z]$ be an arbitrary element of A and put $X := \{[x, y, z]\}$. In virtue of 6. we obtain for all $[i, k, j]\in A$ and all natural numbers n

$$G_n([i, k, j], X) \ \le \ n^{-1}\, 2^{[k/2]} \quad .$$

For any natural number m there results

$$\int (G_n([i, k, j], X))^2\, \nu(d[i, k, j])$$

$$\le \int_{\{[i,k,j]:2^{k-m}<n\}} n^{-1/2}\, 2^{m/2}\, G_n([i, k, j], X)\, \nu(d[i, k, j])$$

$$+ \int_{\{[i,k,j]:2^{k-m}\ge n\}} G_n([i, k, j], X)\, \nu(d[i, k, j])$$

$$\le n^{-1/2}\, 2^{m/2}\, \nu(X) + n^{-1} \sum_{0\le r<n} \ \int_{\{[i,k,j]:2^{k-m}>r\}} J^{[r]}([i, k, j], X)\, \nu(d[i, k, j])$$

$$\le n^{-1/2}\, 2^{m/2}\, \nu(X)$$

$$+ n^{-1} \sum_{1\le r<n} \nu(\{[i, k, j] : [x, y] \angle [i+r, k] \ ; \ 2^{k-m} > r \ ; \ 2^{k-1} - r < j \le 2^{k-1}\})$$

$$+ n^{-1} \sum_{0\le r<\min(2^{y-m},z)} \nu(\{[x-r, y, z-r]\}) \quad .$$

Remembering the concrete form of ν we may proceed in our estimate by

$$\le n^{-1/2}\, 2^{m/2}\, \nu(X)$$

$$+\ n^{-1/2} \sum_{1\le r<n} \ \sum_{k>m+[\log_2 r]} r 2^{[k/2]}\, 2^{-k}\, 2^{-([k/2]+1)} + n^{-1}\, 2^{y-m}\, \nu(X)$$

$$\le\ n^{-1/2}\, \nu(X)\, (2^{m/2} + n^{-1/2}\, 2^{y-m}) \ +\ (2n)^{-1} \sum_{1\le r<n} r\, 2^{-m}\, 2^{-[\log_2 r]}$$

$$\le n^{-1/2}\, \nu(X)\, (2^{m/2} + n^{1/2}\, 2^{y-m}) + 2^{-m} \quad .$$

Since m was arbitrary, there results

$$\int (G_n(a, X))^2\, \nu(da) \longrightarrow 0$$

for all singletons X and hence, in our case, for all $X\in \mathcal{B}$. ■

Finally we turn to a property which is weaker than J-ergodicity and, in case of a substochastic intensity kernel, may be used to replace J-ergodicity in Theorem 5.9.2. Moreover, this property has a simple intuitive interpretation as a kind of *transience* of Π_ν^{κ}.

Let N_J^{trans} denote the set of all measures $\nu\in N_J$ having the property

$$\int_X G_n(a,X)\, \nu(da) \longrightarrow 0 \quad (X\in \mathcal{B}) \quad .$$

N_J^{erg} is a subset of N_J^{trans}, since there holds

5.4.9. A measure $\nu\in N_J$ belongs to N_J^{trans} iff it obeys

$$\int_X G_n(a,X) \min(1, G_n(a,X))\, \nu(da) \longrightarrow 0 \qquad (X\in \mathcal{B}) \quad .$$

Proof. Obviously any $\nu\in N_J^{trans}$ obeys the just stated condition.

Conversely, assume that a J-invariant measure ν satisfies this condition. Then we obtain for all $X\in \mathcal{B}$ and all $\varepsilon\in (0,1)$

$$\limsup_{n\to\infty} \int_X G_n(a,X)\, \nu(da)$$

$$\le \limsup_{n\to\infty} \int_{\{a:\, a\in X, G_n(a,X)\le\varepsilon\}} G_n(a,X)\, \nu(da) + \limsup_{n\to\infty} \int_{\{a \ldots\, a\in X, G_n(a,X)>\varepsilon\}} G_n(a,X)\, \nu(da)$$

$$\le\ \varepsilon\nu(X) + \varepsilon^{-1} \limsup_{n\to\infty} \int_X G_n(a,X) \min(1, G_n(a,X))\, \nu(da)$$

$$=\ \varepsilon\nu(X) \quad .$$

Hence follows $\nu\in N_J^{trans}$. ■

Similar to Proposition 5.9.1. there follows

5.9.5. Proposition A measure $\nu\in N_\kappa^{st}$ belongs to $\nu\in N_J^{trans}$ iff there holds

$$\nu(\{a : a\in X,\, G_n(a, X) > \varepsilon\}) \underset{n\to\infty}{\longrightarrow} 0 \qquad (\varepsilon>0,\, X\in \mathcal{B}) \quad .$$

The remark in front of Example 2.8.4. states, a fortiori, that the stable measure ν constructed there belongs to N_J^{trans}, whereas Example 5.8.1. reveals that ν does not belong to N_J^{erg}.

According to Theorem 5.4.1. a stable measure ν is of ergodic type iff it has, apart from o, no minorant belonging to N_κ^{fin}. It turns out that all stable measures in N_J^{trans} have the following weaker property:

5.9.6. Assume that ν belongs to $N_J^{trans} \cap N_\kappa^{st}$. Then there exists no $\sigma \in N_\kappa^{ff}$ with the properties $\sigma \neq o$, $\sigma \leq \nu$.

Proof. Assume that $\sigma \leq \nu$ and $\sigma \in N_\kappa^{ff}$, and choose some $S \in \mathcal{A}_{(\Pi^\kappa{}_\sigma)}^{stoch}$. Due to Proposition 2.10.10., σ(S) is finite, and we obtain for all $X \in S \cap \mathcal{B}$ and all $m \in \mathbf{Z}_+$

$$\sigma(S) - \int_X J^{[m]}(a, X)\,\sigma(da)$$

$$= \int_S J^{[m]}(a, S)\,\sigma(da) - \int_X J^{[m]}(a, X)\,\sigma(da)$$

$$= \int_S J^{[m]}(a, S\setminus X)\,\sigma(da) - \int_{S\setminus X} J^{[m]}(a, X)\,\sigma(da)$$

$$\leq 2\sigma(S\setminus X) \quad .$$

Hence results for all natural numbers n

$$\sigma(S) \leq \int_X G_n(a,X)\,\nu(da) + 2\sigma(S\setminus X) \quad .$$

From the assumptions there follows

$$\sigma(S) \leq 2\sigma(S\setminus X) \qquad (X \in S \cap \mathcal{B}) \quad .$$

and thus σ(S) = 0, which by Proposition 2.10.10. yields $\sigma = o$. ■

Combining Proposition 2.10.9. and 5.9.6. one arrives at

5.9.7. Theorem. In case of a substochastic intensity kernel J there follows

$$N_\kappa^{erg} = N_J^{trans} \cap N_\kappa^{st} \quad .$$

Comments and Remarks

2.1. If A is finite, say $A = \{1,...,m\}$, it suggests itself to interpret the elements of A as possible *types* of individuals (rather than their locations). In this case, a population Φ is also described by the m-tuple $[\Phi(\{1\}),...,\Phi(\{m\})] \in \mathbf{Z}_+^m$, where the j-th coordinate is interpreted as number of individuals of type j, and our model reduces to the classical scheme of branching models with a finite set of types and in discrete time.

The model can be refined in using as a phase space the direct product of a "position space" and a *bounded* (not necessarily discrete) "mark space". We have renounced a refinement of this kind, because the gain of results would not match the need of notation. This latter point of view, however, does not come true if the "position space" has symmetry properties which enable substantial simplifications, like in the case of spatially homogeneous branching processes with different types of individuals.

Let us stress once more that the aim of this monograph is not to establish a general model of branching processes (models of this kind exist for quite some time, see e. g. the last chapter of Sewastjanov (1974) or chapter 6 of Jagers (1975)) but to study the equilibrium distributions.

Repeatedly, instead of $\mathbf{P}^{inv}$, the subsets $\mathbf{P}^{inv,\nu}$, $\nu \in N_K^{st}$, will be investigated separately; results of this kind are refered to as *local* ones. As it turns out, one can fix, instead of ν, $\nu \in N_K^{st}$, the whole set $\{r\nu\}_{r \in \mathbf{R}_+}$, since the $\mathbf{P}^{inv,r\cdot\nu}$, $\nu \in N_K^{st}$, are "equally behaved " for different $r \in (0, +\infty)$. In the spatially homogeneous case, clearly the set of translation invariant measures in N is a distinguished one.

2.4. Essentially, this section is a generalization of concepts and results developed in Kallenberg (1977) for the spatially homogeneous case.

2.5. This section, as well, contains the generalization of results in the spatially homogeneous case (cf. [MKM, Section 12.6.]) There, however, still more efficient results are available (cf. Kallenberg (1977) Section 4), rendering, under certain assumptions on the branching dynamics necessary *and* sufficient criteria for stability.

2.6. Both assertions 2.6.6. and 2.6.14. generalize results (cf. Kallenberg (1977)) from the spatially homogeneous case .

The stability assertion in Example 2.6.5. which here is derived directly in a simple way for the sake of illustration, is an easy consequence of Theorem 12.6.4. in [MKM] and the well-known characterization of transience of random walks in locally compact Abelian groups.

2.7. The transformation $P \rightarrow \mathcal{G}_P$ allows, at least in principle, to reduce assertions about equilibrium distributions for branching dynamics κ to assertions about equilibrium distributions for the substochastic family dynamics $\boldsymbol{\kappa}$. Thus, an idea formulated by Kerstan in the late sixties is realized.

2.8. In the representative special case E(M)=1, condition b) in Theorem 2.8.3. is equivalent to

b)′ $\qquad E \in \mathbf{P}^{inv} \quad ; \quad \mathcal{G}_E(\Phi(A) = 1) = 1 \quad .$

Further notice that in virtue of 2.7.17. all $E \in \mathbf{P}^{inv}$ with the property

$$\mathcal{G}_E(\Phi(A) = 1) = 1$$

are prime elements on $\mathbf{P}^{inv}$ in the following sense:

$$E = P_1 * P_2 \quad ; \quad P_1 \neq \delta_o \quad ; \quad P_1, P_2 \in \mathbf{P}^{inv}$$

imply

$$P_1 = E \quad ; \quad P_2 = \delta_o \quad .$$

Example 2.8.4., which is repeatedly called upon in the sequel, and which realizes properties of equilibrium distributions considered as impossible for years, is due to Siegmund-Schultze. Notice that the distribution $P \in \mathbf{P}^{inv,v}$ which is constructed in 2.8.4. is also an equilibrium distribution with respect to the following simplified branching dynamics κ':

$$\kappa'_{(-i,j)} := \begin{cases} \delta_{\delta_{[-i,j+1]}} & \text{for } j < 2^i \\ \delta_{\delta_{[-i,1]}} & \text{for } j = 2^i \end{cases}$$

The same holds true for $\mathcal{E}_P = \Pi^{\kappa}_{v}$.

2.9. Theorem 2.9.17. generalizes results from the theory of branching processes with a finite set of types. Together with 2.9.14. it verifies the assertion stated in the preface, according to which the branching effect reduces to a "transient secondary effect" for any equilibrium distribution P with P(Φ finite) = 1, whereas, restricted to an essentially unique set in $\mathcal{A}_{(P)}{}^{stoch}$, the dynamics is a stochastic translation.

Notice that, in contrast to $\mathcal{A}_{(P)}{}^{fin}$, $\mathcal{A}_{(P)}$ need not have an essential minimum: for the equilibrium distribution P constructed in Example 2.1.6. all sets $\{\ldots, k-1, k\}$, $k \in \mathbf{Z}$, belong to $\mathcal{A}_{(P)}$.

3.2. This section generalizes setting and results from Liggett (1978). In this paper the role of entrance laws for the investigation of equilibrium distribution has been clarified (in the special case of stochastic translations).

3.8. The topic of this section is based on ideas and results of Kerstan (cf. Kerstan/Debes (1969), Debes/Kerstan/Liemant/Matthes (1971) and Fichtner (1974)).

Proposition 3.8.4. allows to interpret the relation $v \in N_J{}^{ap}$ as "($J^{[n]}$) being weakly asymptotically backward-equidistributed with respect to the semigroup $\{J^{[0]}, J^{[1]}, \ldots\}$and its invariant measure v " (cf. section 3 in Debes/Kerstan/Liemant/Matthes (1971)).

Theorem 3.8.6. generalizes Lemma 3 in Kerstan/Debes (1969). Corollary 3.7. in Liggett (1978) says that under the assumptions of Proposition 3.8.5. the assertion $Q(N_J) = 1$ is true for *all* $Q \in V^{inv}$. Proposition 3.8.5. itself is an easy consequence of assertion 3.1. in Kerstan/Debes (1969) and, in addition, is contained implicitely in Theorem 4.1.2. in Liggett (1978). Theorem 3.8.9. may be conceived as a far-reaching generalization of known results in the case of stochastic translations (see, e. g., section 9 in Debes/Kerstan/Liemant/Matthes (1970)).

It is an open **problem** whether there holds

$$N_\kappa^{fin} \cap N_J^{ap} \subseteq N_\kappa^{ff} \quad .$$

4.1. Proposition 4.1.7. generalizes a result proved for stochastic translations in Shiga/Takahashi (1974).

Proposition 4.1.10. is a classical result; see Lanford/Ruelle (1969).

Example 4.1.12. is due to Siegmund-Schultze, and Proposition 4.1.13. is taken from Fleischmann/Hermann/Matthes (1982). Proposition 4.1.14. generalizes an assertion in Fleischmann/Liemant/Matthes (1982).

Apparently, it is an open **problem** in the general theory of random point fields to characterize $_\infty\mathcal{M}$-triviality of arbitrary infinitely divisible distributions in terms of their canonical measures.

4.5. It seems to be an open **problem** whether condition

d) Any κ-process with intensity sequence (v_i) is of Cox type

can be added as a further equivalent condition to those of Theorem 4.5.9.

Theorem 4.5.11. is a generalization of Theorem 4.6. in Liggett (1978).

4.7. Theorem 4.7.19. has been proved in Fleischmann/Liemant/Matthes (1982). For the special case of stochastic translations it had been derived earlier in Shiga/Takahashi (1974). Theorem 4.7.23., as well, stems from Fleischmann/Liemant/Matthes (1982). Proposition 4.7.12. is a refinement of assertion 6.2. in Fleischmann/Matthes (1981). Example 4.7.22. is taken from Shiga/Takahashi (1974).

The results about convergence and structure (Theorems 4.7.1., 4.7.30., and Theorems 4.7.25., 4.7.28. and 4.7.29., respectively) have two sources: on one hand the concepts and estimation techniques developed by Kerstan et al. in the special case of substochastic translations (cf. Kerstan/Debes (1969) and Kerstan/Liemant/Matthes (1971)) (these are "backward" approaches), and on the other hand the approximation method in the spatially homogeneous case developed in Debes/Kerstan/Liemant/Matthes (1970) and presented in [MKM] (this is a "forward" technique which requires already in the case of arbitrary locally compact Abelian groups the utilization of the structural properties of these groups and cannot be extended to the spatially inhomogeneous case). It should be remarked, however, that in the spatially homogeneous case not only stronger assertions are available (cf. Theorems 12.4.5. and 12.4.6. in [MKM]) but also the majorizability conditions on the initial distribution´s intensity measure can essentially be removed (cf. Theorem 12.4.2 in [MKM]). In the frame of substochastic translations, an elimination of this majorizability condition has been achieved in another way (cf. Fichtner/Kerstan (1980)).

Theorem 4.7.31. is a generalization of Theorem 4.6. in Liggett (1978).

Up to now, it is an unsolved **problem** to characterize, for arbitrary $\nu \in N_\kappa^{st}$, those $Q \in V^{inv,\nu}$ which arise as J-invariant distributions $Q=\mathcal{W}^P$ of some equilibrium distribution $P \in P^{inv,\nu}$.

5.1. In the case of substochastic branching dynamics, the following result has been shown in Kerstan/Debes (1969) and in Debes/Kerstan/Liemant/Matthes (1971) for discrete and general phase spaces, respectively:

For any $P \in P^{inv}$ of first order there exists a kernel I_P from $[M, \mathcal{M}]$ to $[A, \mathcal{A}]$ with the properties

a) $\int |(\Phi * G_n(X) - I_P(\Phi,X)| P(d\Phi) \longrightarrow 0 \qquad (X \in \mathcal{B})$

b) $I_P(\Phi,(.)) \in N_\kappa^{st} \qquad (\Phi \in M)$

c) $\kappa_{(\Phi)} * I_P = I_P(\Phi,(.))$ P a.all Φ.

Proposition 5.1.6. orginates from Fleischmann/Matthes (1981).

5.2. The distributions $\mathcal{V}_P$, $P \in \mathbf{P}^{inv}$ of first order, have been introduced in Kerstan/Debes (1969) and Debes/Kerstan/Liemant/Matthes (1971) for substochastic translations, and likewise also a more stringent version of $\mathcal{V}_L$ for more general distributions $L \in \mathbf{P}$.

Example 5.2.7. is due to Siegmund-Schultze.

5.2.4. and Proposition 5.2.6. stem from Fleischmann/Matthes (1981).

5.4. The validity of 5.4.3. was to be expected from Theorem 2.5. in Fleischmann (1981). For substochastic translations, Theorem 5.4.5. has been proved in Shiga/Takahashi (1974).

For an arbitrary substochastic kernel J from $[A, \mathcal{A}]$ to $[A, \mathcal{A}]$ one infers, putting $\kappa := \tau(J)$, from Theorem 4.7.19., 5.4.2. and Theorem 5.4.5. the validity of

5.4.6. If J is a substochastic kernel from $[A, \mathcal{A}]$ to $[A, \mathcal{A}]$ and ν is a measure in N_J^{ap}, then there holds

$$\int (J^{[n]}(a,X))^2 \, \nu(da) \longrightarrow 0 \qquad (X \in \mathcal{B})$$

iff there exists no finite minorant of ν in N_J which is different from o.

This assertion is a special case of Satz 1′in section 3 in Debes/Kerstan/Liemant/Matthes (1971).

5.8. Example 5.8.1. is due to Siegmund-Schultze.

It is an interesting open **problem** whether there holds

$$N_\kappa^{fin} \cap N_\kappa^{reas} = N_\kappa^{ff} \quad .$$

5.9. Example 5.9.3., as well, has been constructed by Siegmund-Schultze. It is an open **question** if the converse of asssertion 5.9.6. is valid. Theorem 5.9.7. has been proved for stochastic translations already in Shiga/Takahashi (1974).

References

Asmussen, S. and Hering, H. (1983): Branching Processes. Birkhäuser: Boston-Basel-Stuttgart.

Billingsley, P. (1968): Convergence of Probability Measures. Wiley: New York.

Chung, K.L. (1967): Markov Chains With Stationary Transition Probabilities, 2nd ed. Springer: Berlin-Heidelberg-New York.

Debes, H., Kerstan, J., Liemant, A. and Matthes, K. (1970): Verallgemeinerungen eines Satzes von Dobruschin, I. Math. Nachr. **47**, 183-244.

- (1971): Verallgemeinerungen eines Satzes von Dobruschin, III. Math. Nachr. **50**, 99-139.

Dynkin, E.B. (1978): Sufficient statistics and extreme points. Ann. Probab. **6**, 705-730.

Fichtner, K.H. (1974): Gleichverteilungseigenschaften substochastischer Kerne und zufällige Punktfolgen. Math. Nachr. **62**, 251-260.

- (1975): Charakterisierung Poissonscher zufälliger Punktfolgen und infinitesimaler Verdünnungsschemata. Math. Nachr. **68**, 93-104 .

Fichtner, K. H. and Kerstan, J. (1978): Approximation von Mischungen Poissonscher Punktprozesse durch verschobene Punktprozesse. Serdica **4**, 367-376.

- (1980): Invarianz von Punktprozessen bei zufälligen Bewegungen, I. Serdica **6**, 324-340.

Fleischmann, K. (1981): Ergodic properties of infinitely divisible stochastic point processes. Math. Nachr. **102**, 127-135.

Fleischmann, K., Hermann, K. and Matthes, K. (1982): Kritische Verzweigungsprozesse mit allgemeinem Phasenraum, VIII. Math. Nachr. **109**, 239-247.

Fleischmann, K., Liemant, A. and Matthes, K. (1982): Kritische Verzweigungsprozesse mit allgemeinem Phasenraum, VI. Math. Nachr. **105**, 307-316.

Fleischmann, K., and Matthes, K. (1981): Kritische Verzweigungsprozesse mit allgemeinem Phasenraum, III. Math. Nachr. **102**, 223-233.

Harris, T.E. (1963): The Theory of Branching Processes. Springer: Berlin - Göttingen - Heidelberg.

Hermann, K. (1981): Critical measure-valued branching processes in discrete time with arbitrary phase space, I. Math. Nachr. **103**, 63-107.

Heyer, H. (1977): Probability Measures on Locally Compact Groups.Springer: Berlin - Heidelberg - New York.

Ikeda, N., Nagasawa, M. and Watanabe, S. (1968): Branching Markov Processes, II. Jour. Math. Kyoto Univ. **8**, 365-410.

- (1969): Branching Markov Processes, III. Jour. Math. Kyoto Univ. **9**, 95-160.

Jagers, P. (1975): Branching Processes with Biological Applications. Wiley: London - New York.

Kallenberg, O. (1977): Stability of critical cluster fields. Math. Nachr. **77**, 7-43.

- (1983): Random Measures, 3rd ed. Akademie-Verlag: Berlin / Academic Press: New York.

Kerstan, J. and Debes, H. (1969): Zufällige Punktfolgen und Markoffsche Übergangsmatritzen ohne stationäre Verteilungsgesetze. Wiss. Z. Friedrich- Schiller-Universität Jena **18**, 349-359.

Kerstan, J. and Fichtner, K.H. (1979): Asymptotische Invarianz bei Halbgruppen stochastischer Kerne, I. Serdica **5**, 229-251.

- (1979): Asymptotische Invarianz bei Halbgruppen stochastischer Kerne, II. Serdica **5**, 310-315.

- (1981): Invarianz von Punktprozessen bei zufälligen Bewegungen, II. Serdica **7**, 15-33.

Kerstan, J., Matthes, K. and Mecke, J. (1982): Infinitely Divisible Point Processes. (Russian.) Nauka: Moscow.

Kerstan, J. and Wakolbinger, A. (1981): Ergodic decomposition of probability laws. Z. Wahrscheinlichkeitstheorie verw. Gebiete **56**, 399-414.

Lanford, O.E. and Ruelle, D. (1969): Observables at infinity and states with short range correlations in statistical mechanics. Commun. math. Phys. **13**, 194-215.

Liemant, A. (1980 a) Kritische Verzweigungsprozesse mit allgemeinem Phasenraum, I. Math. Nachr. **96**, 119-124.

- (1980 b): Kritische Verzweigungsprozesse mit allgemeinem Phasenraum, II. Math. Nachr. **99**, 115-120.
- (1981): Kritische Verzweigungsprozesse mit allgemeinem Phasenraum, IV. Math. Nachr. **102**, 225-254.
- (1982): Kritische Verzweigungsprozesse mit allgemeinem Phasenraum, VII. Math. Nachr. **107**, 273-281.
- (1983 a): Kritische Verzweigungsprozesse mit allgemeinem Phasenraum, IX. Math. Nachr. **113**, 209-223.
- (1983 b) Structure-and convergence theorems for critical branching processes with general phase space, I. Math. Nachr. **112**, 209-226.
- (1984): Structure-and convergence theorems for critical branching processes with general phase space, III. Math. Nachr. **119**, 195-207.

Liemant, A. and Matthes, K. (1982): Kritische Verzweigungsprozesse mit allgemeinem Phasenraum V. Math. Nachr. **106**, 235-259.

- (1983): Structure- and convergence theorems for critical branching processes with general phase space, II. Math. Nachr. **113**, 259-266.
- (1984): Ergodic properties of stationary spatially non-homogeneous branching processes with infinitely many particles. Mathematical structures - computational mathematics - mathematical modelling, **2**, 253-261, Sofia.

Liggett, T.M. (1978): Random invariant measures for Markov chains and independent particle systems. Z. Wahrscheinlichkeitstheorie verz. Gebiete **45**, 297-313.

Matthes, K. (1984): Kritische Verzweigungsprozesse mit allgemeinem Phasenraum, X. Math. Nachr. **119**, 209-218.

- (1986): Kritische Verzweigungsprozesse mit allgemeinem Phasenraum, XI. Math. Nachr. **128**, 141-149.

Matthes, K. Kerstan, J. and Mecke, J. (1978): Infinitely Divisible Point Processes. Wiley: Chichester-New York

Meyer, P.A. (1966): Probability and Potentials. Blaisdell: Waltham, Mass.

Nagasawa, M. (1977): Basic models of branching processes. Bull. Int. Stat. Inst. **XLVII**, 423-445.

Parthasarathy, K.R. (1967): Probability Measures on Metric Spaces. Academic Press: New York-London

Prehn, U. and Röder, B. (1977): Limit theorems for spatially homogeneous branching processes with a finite set of types, I. (Russian.) Math. Nachr. **80**, 37-86.

Sewastjanow, B.A. (1974): Verzweigungsprozesse. Akademie-Verlag: Berlin. R. Oldenbourg: München-Wien.

Shiga, T. and Takahashi, Y. (1974): Ergodic properties of the equilibrium process associated with infinitely many Markovian particles. Publ. RIMS, Kyoto Univ. **9**, 505-516.

Spitzer, F. (1976): Principles of Random Walk, 2nd ed. Springer: New York - Berlin - Heidelberg.

Stoyan, D. (1983): Comparison Methodes for Queues and Other Stochastic Models. Wiley: Chichester-New York.

Symbol Index

$\mathcal{G}_L$	69	
$\mathcal{G}_{L,n}$	69	
$\mathcal{G}_P$	70	family equilibrium distribution associated with P
$\underline{\mathcal{G}}^{\underline{P}}$	115	family process associated with $\underline{P}$
$\mathcal{G}_m^{\underline{P}}$	115	
${}_XG$	12	$:= G({}_X\nu \in (.))$
G_n	191	$:= n^{-1} \sum_{0 \le i < n} J^{[i]}$
Γ	190	"decomposing kernel" of P^{inv}
$\Gamma^{(k)}{}_H$	19	k-th factorial moment measure of H
H^	34	
H°	36	
H_a	37	Palm distribution of H
$H^!{}_a$	39	reduced Palm distribution of H
H_κ	23	$:= \int \kappa_{(\Phi)}(.)\, H(d\Phi)$ ("clustered measure")
I	191	$:= \Lambda_{\Gamma(\Phi,(.))}$
$\mathbf{I}$	14	infinitely divisible distribuntions on $\mathcal{N}$
$\mathcal{J}$	187	σ-algebra of shift-invariant sets
J	47	intensity kernel of the underlying clustering field κ
J_τ	25	intensity kernel of a clustering field τ
$J_\tau^{[n]}$	26	n-th convolution power of J_τ
$\mathfrak{Z}(G)$	19	Cox transform of G
K_κ	33	
$\underline{K}$	142	"decomposing kernel" of $\underline{P}^\kappa$
$\mathcal{K}_{(a_i)}$	57	
$\mathcal{K}_{(v_i),a}$	128	
κ	47	the underlying clustering field
$\kappa_{(a)}$	22	offspring distribution of an individual at site a
$\kappa_{(\Phi)}$	22	offspring distribution of a population Φ
$\kappa^{[n]}$	24	n-th clustering power of κ
$\boldsymbol{\kappa}$	33	family dynamics associated with κ
κ_m	110	
$\kappa_{[X]}$	87	restriction of κ to the set X
$\mathcal{L}_G(f)$	20	Laplace transform of G, evaluated at f
Λ_H	12	intensity measure of H
$\Lambda^{(k)}{}_H$	12	k-th moment measure of H
M	15	counting measures on $\mathcal{A}$ ("populations")
$\xrightarrow{M}$	16	vague convergence in M
$\mathcal{M}$	16	σ-algebra of Borel subsets of M
$\mathbb{M}$	29	counting measures on $\boldsymbol{\mathcal{A}}$ ("family systems")

${}_{\kappa}M$	22	set of populations whose offspring distribution exists
M^{κ}	75	set of populations whose n-th generation offspring distribution exists for all n = 1, 2, ...
${}_{X}\mathcal{M}$	147	σ-algebra of events "observable within the set X"
${}_{\infty}\mathcal{M}$	147	spatial tail σ-algebra
N	10	"locally finite" measures on $\mathcal{A}$
$\xrightarrow{N}$	10	vague convergence in N
$\mathcal{N}$	11	σ-algebra of Borel subsets of N
$\mathbb{N}$	29	"locally finite" measures on $\mathcal{A}$
N_J	50	J-invariant measures
N_J^{entr}	117	J-sequences
$\mathcal{N}_J^{entr}$	117	
$N_J^{entr,reg}$	162	regular J-sequences
N_J^{ap}	133	J-aperiodic measures
N_J^{erg}	210	J-ergodic measures
N_J^{ex}	50	extremal J-invariant measures
N_J^{trans}	225	
$N_{\kappa}^{entr,reg}$	162	regular stable sequences of measures
$N_{\kappa}^{entr,st}$	116	stable sequences of measures
N_{κ}^{erg}	200	stable measures of ergodic type
N_{κ}^{fin}	77	stable measures of finite type
N_{κ}^{ff}	101	stable measures which are finitely fed
N_{κ}^{nerg}	201	stable measures of purely non-ergodic type
N_{κ}^{nst}	62	purely unstable measures
N_{κ}^{reas}	216	
N_{κ}^{reg}	181	regular stable measures
N_{κ}^{st}	49	stable measures
$N_{\kappa}^{st(k)}$	60	measures which are stable of k-th order
$N_{(\nu)}^{fin}$	101	
ν_{erg}	204	
ν_{nerg}	202	
ν_{nst}	64	purely unstable component of a J-invariant measure ν
ν_{st}	64	stable component of a J-invariant measure ν
$\nu^{st}{}_{m}$	123	
${}_{X}\nu$	12	restriction of the measure ν to the set X
o	11	zero measure on $\mathcal{A}$ ("the void population")
$\boldsymbol{o}$	30	zero measure on $\mathcal{A}$ ("the void family system")
ω_n	208	$:= n^{-1} \sum_{0 \le i < n} \kappa^{[i]}$
$\mathcal{P}(\mathbf{Z})$		power set of $\mathbf{Z}$

Subject Index

Analytic Methods of Probability Theory

by Hans-Joachim Rossberg, Bernd Jesiak, Gerhard Siegel

1985. 311 pages, large octavo.
GDR M 48, – , abroad DM 48, –
Order No.: 763 420 3 (6868)

This book aims at acquainting the reader with some new results on the normal distribution Φ and the central limit theorem. In particular, an infinitely divisible distribution F coinciding with Φ on a half axis proves to be uniquely determined so that it is everywhere equal to Φ; moreover, if we assume the convergence of distribution functions of sums of independent random variables to Φ on a half line, then this implies convergence to Φ everywhere. These two observations are intimately connected and have stimulated further research on the continuation of distribution functions and on 'restricted convergence'. A considerable part of these investigations is fairly elementary. Therefore this is both a textbook and a monograph; it leads the reader from introductury chapters (devoted to distributions, concentration functions, characteristic functions) up to unsolved problems in various fields. Analytic function methods, in particular the theory of Phragmén and Lindelöf, are widely used.

Akademie-Verlag Berlin

Stochastic Geometry and Its Applications

by Dietrich Stoyan, Wilfrid S. Kendall, and Joseph Mecke
with a foreword by D. G. Kendall

1987. 345 pages, large octavo.
GDR M 50, - , abroad DM 50, -
Order No.: 763 535 3 (6920)
ISBN 3-05-500032-3

Stochastic geometry involves the statistical analysis of complicated geometrical patterns which occur in many areas of science and technology. This book aims to make the results and methods of stochastic geometry accessible to non-theoreticians, while at the same time serving as an introduction to the subject for mathematicians. The exposition is mathematically exact and takes into account the latest results, but in most cases proofs are omitted, and it is intended that applied scientists who may not wish to follow the mathematical arguments in detail will still be able to interpret and use the formulae.

Topics covered include the basic theories of point processes, random sets, fibre processes, tessellations, stereology and the statistical theory of shape.

THEORIA CUM PRAXI

Akademie-Verlag Berlin

Band 1

Inverse and Improperly Posed Problems in Differential Equations

Proceedings of the Conference on Mathematical and Numerical Methods, held in Halle/Saale (GDR) from May 29 to June 2, 1979

Edited by Gottfried Anger

296 Seiten, 1979 (vergriffen)

Band 2

Fundamentals of Computation Theory FCT '79

Proceedings of the Conference on Algebraic, Arithmetic and Categorial Methods in Computation Theory, held in Berlin/Wendisch-Rietz (GDR) September 17–21, 1979

Edited by Lothar Budach

576 Seiten, 1979

Band 3

Decidability and Generalized Quantifiers

By Andreas Baudisch, Detlef Seese, Hans-Peter Tuschik, and Martin Weese

XII, 235 Seiten, 1980

Band 4

Zur Elastizitäts- und Thermoelastizitätstheorie

Potentialtheoretische Methoden in Anwendung auf verschiedene Modelle

Von Lothar Jentsch und Johannes Maul

157 Seiten, 1980

Band 5

Systems Analysis and Simulation 1980

Proceedings of the International Symposium held at Berlin (GDR) September 1–5, 1980

Edited by Achim Sydow

496 Seiten, 1980 (vergriffen)

Band 6

Nichtlineare diskrete Optimierung

von Eberhard Girlich und Michail M. Kowaljow

218 Seiten, 1981

Band 7

Completeness Problems in the Structural Theory of Automata

By Jürgen Dassow

148 Seiten, 1981

Band 8

Darstellung von Lösungen linearer elliptischer Differentialgleichungen

Von Günther Wildenhain

92 Seiten, 1981

Band 9

Introduction to the Spectral Theory of Operators in Spaces with an Indefinite Metric

By I. S. Iohvidov, M. G. Krein, and H. Langer

120 Seiten, 1982 (vergriffen)

Band 10

Nonlinear Time-discrete Systems

A General Approach by Nonlinear Superposition

By Michael Gössel

115 Seiten, 1982 (vergriffen)

Band 11

Preprimal Algebras

By Klaus Denecke

162 Seiten, 1982

Band 12

Ergodic Theory and Related Topics

Proceedings of the Conference held in Vitte/Hiddensee (GDR) October 19–23, 1981

Edited by Horst Michel

232 Seiten, 1982

Band 13

Singularly Perturbed Differential Equations

By H. Goering, A. Felgenhauer, G. Lube, H.-G. Roos, and L. Tobiska

176 Seiten, 1983

Band 14

Leben wir in einer Volterra-Welt?

Ein ökologischer Zugang zur angewandten Systemanalyse

Von Manfred Peschel und Werner Mende

233 Seiten, 1983

Band 15

Markov Decision Problems with Countable State Spaces

Optimality Criteria – Algorithms – Clustering

By Hans Michael Dietz and Volker Nollau

174 Seiten, 1983

Band 16

Stochastische Verfahren zur Suche nach dem Minimum einer Funktion

Von Ryszard Zieliński und Peter Neumann

133 Seiten, 1983

Band 17

Invertibility and Asymptotics of Toeplitz Matrices

By Albrecht Böttcher and Bernd Silbermann

200 Seiten, 1983

Band 18

Structural Induction on Partial Algebras

Introduction to Theory and Application of Partial Algebras – Part II

By Horst Reichel

205 pages, 1984

Band 19

Algebraic Methods for Toeplitz-like Matrices and Operators

By Georg Heinig and Karla Rost

212 pages, 1984

Band 20

Frege Conference 1984

Proceedings of the International Conference held at Schwerin (GDR), September 10–14, 1984
Edited by Gerd Wechsung
408 pages, 1984

Band 21

Numerical Analysis of Selected Semilinear Differential Equations

Edited by Christian Großmann, Thomas Riedrich, and Hartmut Schönheinz
203 pages, 1984

Band 22

Singular Ordinary Differential Operators and Pseudodifferential Equations

By Johannes Elschner
200 pages, 1985

Band 23

Lotka-Volterra-Approach to Cooperation and Competition in Dynamic Systems

Proceedings of the 5th Meeting of UNESCO's Working Group on Systems Theory held on the Wartburg, Eisenach (GDR) at March 5–9, 1984
Edited by Werner Ebeling and Manfred Peschel
271 pages, 1985

Band 24

Convergence Structures 1984

Proceedings of the Conference on Convergence held in Bechyně (Czechoslovakia), September 24–28, 1984
Edited by Josef Novák, Werner Gähler, Horst Herrlich, and Jan Mikusiński
254 pages, 1985

Band 25

Parcella '84

Proceedings of the II. International Workshop on Parallel Processing by Cellular Automata and Arrays held in Berlin (GDR), September 25–27, 1984
Edited by Wolfgang Händler, Tamás Legendi, and Gottfried Wolf
239 pages, 1985 (vergriffen)

Band 26

Multiobjective and Stochastic Optimization Based on Parametric Optimization

By Jürgen Guddat, Francisco Guerra Vasquez, Klaus Tammer, and Klaus Wendler
175 pages, 1985

Band 27/28

Systems Analysis and Simulation 1985
I: Theory and Foundations
II: Applications

Proceedings of the International Symposium held in Berlin (GDR), August 26–31, 1985
Edited by Achim Sydow, Manfred Thoma, and Robert Vichnevetsky
per volume 495 pages, 1985

Band 29

Parcella '86

Proceedings of the III. International Workshop on Parallel Processing by Cellular Automata and Arrays held in Berlin (GDR), September 9–11, 1986
Edited by Tamás Legendi, Dennis Parkinson. Roland Vollmar, and Gottfried Wolf
279 pages, 1986

Band 30

Fuzzy Sets Applications, Methodological Approaches, and Results

Proceedings of the International Workshop on Fuzzy Sets Applications cosponsored by IIASA and convened in cooperation with IAEA held on the Wartburg, Eisenach (GDR), at March 3–8, 1985
Edited by Steffen Bocklisch, Sergei Orlovski, Manfred Peschel, and Yasushi Nishiwaki
233 pages, 1986

Band 31

Mathematical Methods of Specification and Synthesis of Software Systems '85

Proceedings of the International Spring School, held in Wendisch-Rietz (GDR), April 22–26, 1985
Edited by Wolfgang Bibel and Klaus P. Jantke
245 pages, 1986 (vergriffen)

Band 32

A Model Theoretic Oriented Approach to Partial Algebras

Introduction to Theory and Application of Partial Algebras – Part I
By Peter Burmeister
319 pages, 1986

Band 33

Finite Difference Method on Irregular Networks

A Generalized Approach to Elliptic Problems of Second Order
By Bernd Heinrich
206 pages, 1986

Band 34

Modern Algorithms for Large Sparse Eigenvalue Problems

By Arnd Meyer
125 pages, 1987

Band 35

Parametric Optimization and Related Topics

Edited by Jürgen Guddat, Hubertus Th. Jongen, Bernd Kummer, and František Nožička
407 pages, 1987

Band 36

Discretization in Differential Equations and Enclosures

Edited by Ernst Adams, Rainer Ansorge, Christian Großmann, and Hans-Görg Roos
250 pages, 1987

FSC
www.fsc.org
MIX
Papier | Fördert gute Waldnutzung
FSC® C083411

Kolibri 360 GmbH
Ferdinand-Jühlke-Straße 7
99095 Erfurt, Deutschland
produktsicherheit@kolibri360.de